Membrane Filtration

Focusing on the application of membranes in an engineering context, this hands-on computational guide makes previously challenging problems routine. It formulates problems as systems of equations solved with MATLAB, encouraging active learning through worked examples and end-of-chapter problems.

The detailed treatments of dead-end filtration include novel approaches to constant rate filtration and filtration with a centrifugal pump. The discussion of crossflow microfiltration includes the use of kinetic and force balance models. Comprehensive coverage of ultrafiltration and diafiltration processes employs both limiting flux and osmotic pressure models. The effect of fluid viscosity on the mass transfer coefficient is explored in detail; the effects of incomplete rejection on the design and analysis of ultrafiltration and diafiltration are analysed; and quantitative treatments of reverse osmosis and nanofiltration process analysis and design are explored. Includes a chapter dedicated to the modelling of membrane fouling.

Greg Foley is a Lecturer at Dublin City University, with over 25 years' experience of teaching all aspects of chemical and bioprocess engineering. He is a dedicated educator, who has been nominated for the DCU President's Award for Excellence in Teaching on numerous occasions.

'Membrane Filtration: A Problem Solving Approach with MATLAB' is truly a most comprehensive, interesting and effective approach to this topic. The writing style is engaging, difficult concepts are explained with a clarity that is enviable and the coverage of the topics is quite comprehensive. When reading the text you can clearly hear Greg Foley speaking in every line.

I think the book is needed. It pulls together a diverse range of topics some of which are given only a cursory coverage in other texts and presents the topics in a connected manner with each chapter building on what was covered before. It is clearly a text aimed at teaching and the style will, I think, entice students to "learn by doing". Some of the material, particularly the mathematical aspects of the subject will prove challenging for students who are weak in this area, however, Greg does offer a number of alternative solution methodologies which should overcome some of the complexities for those less interested in the mathematical "tricks" that often intrigue those of a more mathematical bent.

The book will also be very useful to research students and academic staff, like, myself who are involved in teaching modules with aspects of membrane separations. In fact this book proposal reminded me of those famous Lectures in Physics by Richard Feynman, in that while the original intention of that series of books was to introduce physics to undergraduates, that the main beneficiaries of the publications were Feynman's colleagues, who gained fresh insights into their subject from the clarity of Feynman's thought processes as revealed by his pedagogical style.

Dr Dermot M. Malone, University College Dublin

I enjoyed reading this book. Its style is rigorous yet easily understandable, with many clarifying examples and problems. Although it is primarily focused on students, it presents not only standard information but recent results in membrane engineering field as well.

Miroslav Fikar, Slovak University of Technology

Membrane Filtration

A Problem Solving Approach with MATLAB

GREG FOLEY

Dublin City University

CAMBRIDGE
UNIVERSITY PRESS

CAMBRIDGE
UNIVERSITY PRESS

University Printing House, Cambridge CB2 8BS, United Kingdom

One Liberty Plaza, 20th Floor, New York, NY 10006, USA

477 Williamstown Road, Port Melbourne, VIC 3207, Australia

314-321, 3rd Floor, Plot 3, Splendor Forum, Jasola District Centre, New Delhi - 110025, India

79 Anson Road, #06-04/06, Singapore 079906

Cambridge University Press is part of the University of Cambridge.

It furthers the University's mission by disseminating knowledge in the pursuit of education, learning and research at the highest international levels of excellence.

www.cambridge.org
Information on this title: www.cambridge.org/9781107627468

First published 2013

A catalogue record for this publication is available from the British Library

ISBN 978-1-107-02874-6 Hardback
ISBN 978-1-107-62746-8 Paperback

Additional resources for this publication at www.cambridge.org/9781107627468

For Julie and Leo

and

In memory of Tony Foley

Contents

Preface

It is often said, quite correctly, that to really learn a foreign language one needs to immerse oneself in it. For me, getting to grips with chemical and process engineering requires the same degree of immersion. It means formulating lots of problems using the core engineering skills of material and energy balancing. It also means using key mathematical and computational techniques on a regular basis. Having taught on an interdisciplinary degree programme for more than 20 years, it is clear to me that very few students can cope with doing a relatively small amount of engineering while at the same time studying lots of biology, or some other, less quantitative, subject. For most people, constant practice of engineering skills is required. The late and great golfer, Seve Ballesteros, once said: *To give yourself the best possible chance of playing to your potential, you must prepare for every eventuality. That means practice.* Engineering is similar to golf; the more time students spend solving practice problems, the better they become at *doing* engineering.

This book is all about immersing the reader in the field of *membrane filtration*. To use a football term, it is about encouraging the reader to 'get stuck into' quantitative aspects of membrane process design and analysis. As a consequence, the book is somewhat 'equation heavy' but that is the nature of engineering. In any event, none of the mathematics used is beyond what an undergraduate engineering student should be able to cope with.

In many ways, this is quite a conventional book, albeit heavily problem-based, and requires some effort on the part of the student. There is a tendency nowadays for teaching methods to make life almost too easy for the student. Consequently, there are lots of textbooks available that are beautifully produced and contain many coloured boxes and bulleted lists with headings like '*What you need to know*', '*Learning outcomes*', '*Key points to remember*', etc. While superficially appealing, I think these encourage a very limited and examination-focused way of learning by students. To get the most out of this book, the reader needs to have pen and paper, and a computer, to hand. Active reading is required.

When I told friends and colleagues that I was writing a book on membrane filtration, they usually replied with something like: *How can you write a whole book on that?* This is a reasonable question. Superficially, membrane filtration is a simple process: a feed is presented with a barrier and some of the feed, a solid particle for example, is blocked by the barrier. When one thinks about a day-to-day activity like filtering coffee, things do indeed look very simple and hardly the subject worthy of a whole book. However, I say two things to answer the question. First I say that membrane filtration spans everything from simple particulate filtration to the filtration of very small molecules such as salt ions. The underlying physics is quite different in each case. Second, I point out that engineers need to actually *calculate* things, whether it is to design a filtration system, analyse the performance of a system, or even optimise a system. All of these tasks require having a *mathematical description* of the process and therein lies the scope for a whole book. Problems must be formulated in the language of mathematics and solutions must be implemented.

The scope of the book is membrane filtration of *liquid systems* and includes dead-end filtration, crossflow microfiltration, ultrafiltration and diafiltration, nanofiltration and reverse osmosis. Each of these techniques is at a different stage of development, both from a practical and a theoretical point of view. Thus, a completely even treatment of each topic is impossible. The chapters on ultrafiltration and diafiltration (Chapters 4 to 8) form the heart of the book. I have chosen to have a separate chapter (Chapter 7) dealing with incomplete rejection because I think students learn better when going from the specific to the general rather than the other way around.

The emphasis throughout is on process design and analysis and there is very little coverage of membrane structure and membrane fabrication. In effect, this is very much a chemical engineering book. There are other fine books available that give broader coverage of the area but these can usually be classified as books on membrane *science*.

I always tell my students that solving problems in engineering has two parts; formulating the problem *and* actually doing the calculations. This book, therefore, places considerable emphasis on getting numerical predictions from the process models that are developed. Engineering calculations almost invariably require use of numerical methods. Most problems cannot be solved with neat, analytical solutions and the ability to solve non-linear algebraic and ordinary differential equations is a key skill for engineers. In the past, trial and error and graphical techniques abounded but these are largely obsolete as long as one formulates the problem appropriately. Nowadays, there are many options available for implementing numerical methods on personal computers but it is probably fair to say that MATLAB is the most widely used computation environment by engineers.

Getting to grips with MATLAB requires some effort and it would be preferable if the reader had some knowledge of the basics of this package. However, a lot can be achieved by simply copying the code of others and learning as one goes along. My own style of programming is very much influenced by my training in the classic language, FORTRAN – hence my frequent use of *global* variables (not necessarily best practice), which are essentially the same idea as the *common* statement in FORTRAN. Thankfully, none of the problems in this book is very demanding on computer power so the focus of the reader, at least in the short term, should be to get the correct answer and not worry too much about program elegance. That will come in time as one gets more experienced with using MATLAB.

All the codes in this book were written using MATLAB R2011b, kindly denoted by The Mathworks Inc. I also used the Optimisation and Symbolic Math toolboxes. All of the required functions that are used in the book are available in the Student Version of MATLAB. I would recommend that readers try to code their own solutions to the worked examples before looking at my solution. Writing computer code is a great way to learn attention to detail.

When I originally set out to write this book, my vision was to create a textbook for an undergraduate module on *membrane engineering*. However, as I have progressed, the vision has broadened to providing a comprehensive source of material for anyone teaching membrane engineering at a variety of levels. Therefore, it is unlikely that the book will be used in its entirety for a single module. Rather, instructors can pick and choose from the book, depending on the time they have available and the overall structure of their programme. Thus, the book could quite easily be used at every level from the third year of a four-year engineering programme to master level, including interdisciplinary programmes like biotechnology. As I progressed through the writing of the book, it became clear to me that many of the topics within the book required me to venture into research territory. I needed to formulate process models that have not appeared in either textbooks or the research literature. Therefore, I would hope that the book will also be of interest to PhD students and more experienced researchers.

For the last 25 years, I have worked in the School of Biotechnology in Dublin City University. I would like to thank all my colleagues who have supported me or who have simply helped me to enjoy my 25 years working in education and research. The people in DCU whom I most need to thank, however, are my many students, although they probably had no idea that they were helping me! There is no better way to improve your understanding of a topic than having to teach it. On many occasions I have had flashes of understanding right in the middle of a lecture.

Most of all, I must thank all of my immediate and extended family for their fantastic support and love throughout the years, especially my wife, Julie, and my little boy, Leo. Without the love and support of everyone, especially Julie, this book would never have happened.

Greg Foley
Dublin, Ireland

Abbreviations

CFD	Computational fluid dynamics
CFMF	Crossflow microfiltration
CVD	Constant volume diafiltration
DAE	Differential algebraic equation
DDF	Discontinuous diafiltration
DEF	Dead-end filtration
DF	Diafiltration
LF	Limiting flux
MBR	Membrane bioreactor
MF	Microfiltration
MFCVD	(Crossflow) Microfiltration with constant volume diafiltration
MWCO	Molecular weight cut-off
NF	Nanofiltration
NLAE	Non-linear algebraic equation
ODE	Ordinary differential equation
OP	Osmotic pressure
PDE	Partial differential equation
PEG	Polyethylene glycol
RO	Reverse osmosis
UF	Ultrafiltration
UFCVD	Ultrafiltration with constant volume diafiltration
UFDF	Ultrafiltration with diafiltration
UFVVD	Ultrafiltration with variable volume diafiltration
VOP	Viscosity–osmotic pressure
VVD	Variable volume diafiltration

1 Introduction to membrane filtration of liquids

1.1 Introduction

This book is largely concerned with solving process problems in the membrane filtration of *liquids*. In that sense, it is more a chemical engineering book than a membrane science book. There are many fine books available which provide much more information on membrane synthesis and structure, module design, transport processes in membranes and applications of membrane technologies in industrial and medical processes [1–4].

Nonetheless, a small amount of background on membrane separations is needed before the business of process modelling, analysis and design can begin. So, in this chapter, some general terminology is defined, membranes and modules are described and some of the main characteristics of the various membrane filtration techniques are outlined.

A membrane is simply a physical barrier through which pure solvent can pass while other molecules or particles are retained. In the case of ultrafiltration (UF) and microfiltration (MF) this semi-permeability is largely a result of the relative sizes of the solute/particles and the membrane pores. Solute retention is a little different in reverse osmosis (RO) and is largely determined by charge effects. The RO membrane can be thought of as a matrix through which solvent and solute diffuse at different rates. Nanofiltration (NF) occupies a transition zone between UF and RO and solute retention is complex, involving both size and charge effects.

MF is the filtration of *suspensions* containing particles that are generally less than 10 μm in diameter. For particles greater in size than this, the 'micro' prefix is dropped but there is no significant difference, at least in terms of the basic concepts involved, between filtration and microfiltration. Throughout this book, the terms filtration and microfiltration are used at different times in the knowledge that any equations or models derived are applicable to either process. UF can be thought of as the filtration of solutions of high molecular weight molecules. In the NF region, we are usually dealing with molecules such as peptides, antibiotics and other compounds with molecular weights in the range 100–1000 Da. In RO, the solutions being filtered contain very small ionic species such as Na^+ and Cl^- ions. It is worth mentioning in passing that there are situations where one might use a membrane in a filtration process for which it is not actually designed. For example, it would not be unheard of to use a UF membrane to filter a microbial suspension because such a choice may lead to better long-term performance.

Whether this is UF or MF is a debatable point but it is probably more sensible to classify it as UF.

Microfiltration and ultrafiltration are used widely in industry. MF is used in areas like microbial and animal cell separation, often as part of the downstream processing of bioproducts. It is also used in clarification of beverages such as wine, beer and fruit juice, and in dairy processing and wastewater treatment. UF and the related technique of diafiltration (DF) are used for concentration and purification of solutions of macro-molecules of all kinds, especially proteins. All of the above examples of MF and UF are 'bio' in nature and that is the emphasis throughout this book. The main application of RO is in the desalination of water or the production of ultra-pure water for electronics industries [5]. NF, which is an emerging technology, lies somewhere between RO and UF and has potential for use in wastewater treatment, food processing and textile dye removal. Further details are given in the work of Schafer *et al.* [6], probably the only book available at present which gives a broad and detailed introduction to this relatively new, and complex, area.

All of the techniques mentioned above can be termed *pressure-driven* processes. Table 1.1 shows the approximate pressure ranges required for the various membrane filtration processes.

Table 1.1 Approximate pressure ranges for membrane filtration processes.

Technique	Pressure range (bar)
MF	0.5–3
UF	1–10
NF	7–40
RO	25–100

The variety of pressures required reflects the nature of the suspension/solution in each case and the type of membrane employed. In MF and UF, the membranes have distinct pores and fluid flow theory tells us that smaller pores lead to greater pressure drops across the membrane. In the case of UF, the need for greater pressures, resulting from the smaller pores, is enhanced by the osmotic pressure of the solution adjacent to the membrane. This creates an osmotic 'back pressure' that opposes the applied pressure. In RO, the permeability of the solvent through the dense, pore-free, membrane is low while the osmotic back pressure created by the low molecular weight solutes is much higher than it is for the high molecular weight solutes that arise in UF. The pressures employed in NF fall between those used in UF and RO, reflecting the fact that NF has both UF and RO characteristics.

To understand membrane filtration processes, some basic knowledge of fluid mechanics is required. For MF, one needs to understand some complex ideas relating to the compression of highly concentrated suspensions, or *cakes*, as well as issues concerning the various electrostatic and adhesive interactions that can occur between particles. However, for modelling purposes, much of the complexity of MF is lumped into a small number of experimental parameters, a common chemical engineering strategy when confronted by highly complex problems. Therefore, the equations derived in Chapters 2 and 3 do not involve any intricate theory, but remain at a level that should be accessible to most.

UF is the best understood of the membrane filtration processes. As a general rule, predicting the behaviour of solutions is easier than predicting the behaviour of suspensions. If you are from a chemical engineering background, you only have to think of a subject like distillation where accurate models of the vapour–liquid equilibrium of multi-component systems have been developed, allowing for accurate design of distillation columns. In contrast, describing an apparently simple operation like the settling of solid particles is a very complicated task that often defies a purely theoretical approach. To understand UF, the key engineering subjects required, in addition to fluid flow, are physical chemistry and mass transfer. The latter is a subject that is probably unique to chemical engineering and can involve some quite tricky ideas. However, in the context of UF, it is usually enough to know the basic concepts without getting bogged down in some of the notoriously difficult ideas that are scattered throughout mass transfer. The rigour and detail of Treybal's famous work comes to mind [7]. Physical chemistry, in the form of chemical thermodynamics, is important in UF, not only because it is an inherent part of mass transfer, but because a key thermodynamic concept, osmotic pressure, appears explicitly in the equations.

One aspect of UF that makes it simpler than NF and RO is the fact that the solute rejection process, i.e., the process by which the solute is *rejected* by, or *retained* by, the membrane, is essentially a *sieving* process. This means that the transmission of a solute through the membrane largely depends on the relative size of the solute and the pores. Matters are not so simple in NF and RO. While the behaviour of the fluid adjacent to the membrane can be described with a similar theoretical approach to that used in UF, the transport of solvent and solute through the membrane is more complex, being driven by chemical potential differences rather than simple hydraulic pressure differences. In RO, the rejection of ions is normally complete and determined by charge. In NF, there is a range of rejection mechanisms, from RO-like behaviour at the lower end of the pore size spectrum to UF-like behaviour at the upper end. In between, the mechanism of rejection is a complex mixture of charge and size effects.

1.2 Definitions and terminology

1.2.1 Dead-end and crossflow configurations

Figure 1.1 below illustrates the difference between the dead-end and crossflow configurations using MF as an example. In the dead-end configuration, the feed flows normal (perpendicular) to the membrane while in crossflow it flows parallel, or 'tangential', to the membrane.

The dead-end terminology tends to be used mainly in the bioprocessing field, and most chemical engineers would drop the dead-end term and refer to this configuration simply as filtration or microfiltration. However, given the importance of the crossflow mode of operation, the dead-end prefix has become important for clarity.

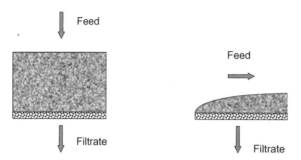

Figure 1.1 Dead-end and crossflow modes of operation illustrating the accumulation of a solid 'cake' on the membrane.

When MF is done in crossflow mode, it is termed crossflow microfiltration (CFMF), at least in this book. In the literature you will sometimes see it referred to as tangential flow filtration (TFF), crossflow microfiltration or crossflow filtration. As UF, NF and RO are normally done in crossflow mode, the 'CF' prefix is generally not used for these operations.

The rationale for using a crossflow mode of operation in CFMF is that it reduces *cake formation*, i.e., the accumulation of a layer of solids on the membrane. Cake formation leads to a reduction in flow through the membrane so it is always desirable that it be minimised. It is important to note that cake formation does not really occur in UF, NF and RO. These processes exhibit a phenomenon referred to as concentration polarisation. This is the formation of a gradient in solute concentration whereby its concentration is highest adjacent to the membrane and lowest in the bulk flow. It is somewhat akin to boundary layer formation in fluid mechanics and is explained in Chapter 4.

An additional and attractive feature of crossflow operation, which follows from the fact that cake formation/concentration polarisation is limited, is that steady state operation is possible, at least in principle. This is in contrast to most dead-end operations, which tend to be batch in nature. In dead-end filtration, cake continues to accumulate until the capacity of the filter is reached. At this point the equipment must be dismantled, the cake recovered or discarded, the system cleaned and a new batch commenced. There are exceptions, of course, such as when a rotary vacuum filter is used for continuous filtration of suspensions. This is explained in Chapter 2.

It is commonplace, particularly when perusing the research literature, to read statements to the effect that crossflow techniques offer significant advantages over dead-end techniques. There is clearly some truth in this but the two techniques are not always comparable. In dead-end filtration, for example, the product of the process can be a cake of particles, whereas in CFMF the product is typically a concentrated suspension. Of course, both techniques can be used to recover a solute in the *filtrate*, also known as *permeate*. This is the liquid that has passed through the membrane. The word, *filtrate*, is generally used for MF processes, whereas *permeate* tends to be used for UF, NF and RO. Both terms are used throughout this book and the reader should take them as being identical.

It is important to mention some terminology that is used throughout the research literature and other textbooks, but which is not used so much in this book. This is *gel formation*, which is the idea, commonly put forward in UF work, that when the solute reaches a very high concentration at the membrane, a gel forms. A gel is a solid, jelly-like substance formed by the action of intermolecular forces, or cross-links, between the solute molecules. Evidence for gel formation in UF is essentially circumstantial as it is based purely on flux measurements. However, UF can be explained without having to invoke this phenomenon. We return to this issue in Chapter 4.

Figures 1.2 to 1.4 show the range of process configurations that are available for crossflow membrane filtration techniques. They can be divided into continuous, batch and fed-batch systems and all are important in practice. The continuous configurations are shown below.

In the feed-and-bleed mode, a portion of the retentate is recirculated as shown above and this requires an additional pump. The presence of the recirculation loop increases the pressure of the fluid entering the module as well as the volumetric flowrate through the module. Both of these changes have the effect of increasing the *filtrate* flowrate. Recirculation can also be used in batch systems as shown in Fig. 1.3.

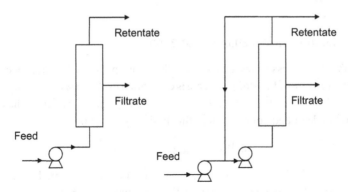

Figure 1.2 Process configurations for single pass and feed-and-bleed continuous membrane filtration.

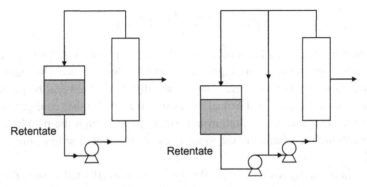

Figure 1.3 Process configurations for batch membrane filtration processes showing simple and recirculation options.

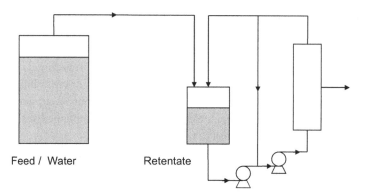

Figure 1.4 Process configurations for fed-batch membrane filtration and diafiltration.

Fed-batch systems are used when the retentate tank has insufficient capacity. In Fig. 1.4, fed-batch operation with a recirculation loop is shown. The volume of solution in the retentate tank is typically kept constant while the feed tank empties. When the feed to the retentate tank is pure water this operation is referred to as constant volume diafiltration (CVD).

1.2.2   Key process parameters in membrane filtration

All the membrane processes described in this book are pressure driven processes. This means that the flowrate of filtrate / permeate is controlled by the *trans-membrane pressure* in the same way that a heat flow is controlled by a temperature gradient in heat transfer processes. In the dead-end configuration, this is simply given by

$$\Delta P_{TM} = P_{feed} - P_{filtrate}, \tag{1.1}$$

where P_{feed} is the feed pressure and $P_{filtrate}$ is the filtrate pressure. In the crossflow configuration, one has to take account of the pressure loss of the feed as it passes along the membrane. Thus, the trans-membrane pressure is typically computed as the difference between the average pressure of the feed and the filtrate pressure, i.e.,

$$\Delta P_{TM} = \frac{P_{feed} + P_{retentate}}{2} - P_{filtrate}. \tag{1.2}$$

The retentate is technically the concentrated fluid that emerges from the membrane module. It is at a lower pressure and a higher concentration than the feed. Only in continuous, single pass configurations is there a significant difference between the particle/solute concentrations at the inlet and exit of the membrane. In all other configurations, there is very little change in concentration in a single pass through the membrane and the exit concentration is essentially the same as the concentration at any point in the flow channel.

Rather than dealing with permeate flowrates, one normally talks about the permeate *flux*. This is simply the permeate flowrate divided by the membrane area and is usually denoted by J, with units of m/s. An important concept that arises in membrane filtration processes is the idea of a *resistance to flow*. If pure water of viscosity, μ, passes through a membrane with a flux, J, the flux is related to the trans-membrane pressure, by the expression

$$J = \frac{\Delta P_{TM}}{\mu R_m},$$ (1.3)

where R_m is termed the *membrane resistance* and has units of m^{-1}. This parameter typically depends on the properties of the membrane, including pore size, porosity, charge, hydrophobicity, etc. The origins of Eq. 1.3 are explained in Chapter 2. The advantage of the resistance idea is that it is easily extended to account for multiple resistances in series. Thus, for an MF process, it is shown in Chapter 2 that

$$J = \frac{\Delta P_{TM}}{\mu(R_m + R_c)},$$ (1.4)

where R_c is the resistance of the cake. As is also seen in Chapter 2, the resistance of a filter cake is a complex function of particle and liquid properties and, usually, the trans-membrane pressure, a phenomenon referred to as *cake compressibility*.

In this book, this resistance-in-series approach is not used for the molecular filtration process of UF, NF and RO. Many researchers and practitioners do indeed adopt this strategy but in Chapter 4, an entirely different language, based on the idea of osmotic pressure, is used to describe molecular filtration processes.

The *tangential* or *crossflow velocity* is a key parameter in all membrane processes. This parameter is simply the average velocity of the solution or suspension as it flows tangentially to the membrane. Suppose, for example, the feed flowrate is Q m^3/s and the membrane module contains N individual tubular membranes of diameter d_t. The crossflow velocity, u, is then simply given by

$$u = \frac{Q/N}{\pi d_t^2/4}.$$ (1.5)

Membrane fouling is a very important phenomenon in determining fluxes and it is important to be clear as to how it is defined. Many people use the term loosely to describe the reduction of flux with time that occurs almost universally with membrane processes. However, in this book we take the term 'fouling' to mean *an increase in membrane resistance*. Thus, concentration polarisation and cake formation are *not* considered as fouling even though they contribute significantly to the reduction in flux from its clean membrane value.

1.2.3 Solute and particle rejection

The whole purpose of membrane filtration is that a solute or particle is retained on the membrane. This phenomenon is usually referred to as *rejection*. In many MF applications, membranes with a pore size of 0.2 μm are employed, and even allowing for the fact that real membranes have a distribution of pore sizes, one can be almost certain that no particles, even bacteria, will pass into the filtrate. In that case, there is *complete rejection*. However, complete rejection does not always occur in molecular filtration processes and the degree of rejection depends on the properties of the solute, the properties of the membrane and the precise operating conditions. Rejection is quantified with a parameter known as the rejection coefficient, σ, defined by

$$\sigma = 1 - \frac{c_p}{c}, \tag{1.6}$$

where c_p is the solute concentration in the permeate and c is its concentration in the retentate. Thus, complete rejection implies $\sigma = 1$. If a membrane presents no barrier to a solute, $\sigma = 0$. In reality, there are some subtleties that must be dealt with when defining rejection coefficients and these are explained in Chapters 7, 8 and 9.

Even in UF, where rejection is a result of the difference between the size of the solute molecules and the pore size, rejection is a relatively complex phenomenon. One reason for this is that a molecule, such as a protein, does not have a definite size, regardless of the solvent characteristics. The precise conformation and size of the molecule usually depend on liquid properties such as pH and ionic strength. Furthermore, molecules are not spheres and whether they pass through a pore, or not, may well depend on their orientation relative to the membrane as they approach the pore entrance. In addition, the pores of a typical membrane have a distribution of diameters. As a consequence, the rejection of macromolecules by UF membranes is something of a probabilistic phenomenon and rather than talking about the size of the molecule, it is generally better to deal with its molecular weight. UF membranes are then defined, not in terms of their pore size, but in terms of their molecular weight cut-off (MWCO). The exact definition of this parameter varies depending on the manufacturer but it is typically defined as the molecular weight of a standard test molecule for which the rejection coefficient has some value between 0.95 and 1.0. Often the standard test molecule used to determine the MWCO is a polysaccharide such as dextran. Characterisation of the membrane involves measuring σ in a stirred cell apparatus under well-defined conditions for a range of molecular weights of this molecule. A typical plot of σ versus solute molecular weight will look something like Fig. 1.5.

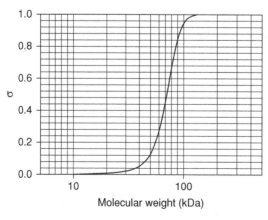

Figure 1.5 Rejection curve for a UF membrane.

It is clear that rejection is not a simple process and, in practice, it is made more complicated by the fact that the solute undergoing UF is usually not the same as the test solute used to characterise the membrane. Users will generally choose a MWCO that is significantly lower than the molecular weight of the solute they are filtering.

1.3   Membranes and their properties

Ideally a membrane should provide as low a resistance to solvent transport as possible. This resistance is determined, at least in the case of MF and UF, by the size of the pores within the membrane and the membrane porosity, i.e., the fraction of the total membrane area taken up by the pores. From basic fluid mechanics, we know that the smaller the pores, the more resistant is the membrane, but the choice of pore size is usually determined by the process specification. For example, UF of a low molecular weight protein, of necessity, requires a membrane with much smaller pores than UF of a high molecular weight protein.

In the case of RO, it is better not to think of a membrane as being a barrier with identifiable pores, but a three-dimensional mesh through which molecules diffuse at different rates, if at all. Charge rather than size is the key determinant of membrane transport in RO. NF is a complex process lying between RO and UF and has characteristics of both of these techniques.

Finally, it is very desirable that a membrane be as resistant to fouling as possible. Many manufacturers produce membranes that are 'low fouling'. Typically this means that solutes, especially proteins, have a low tendency to bind to the membrane.

1.3.1   Membrane materials

Nowadays, most membranes are made from synthetic polymers. For UF and MF applications, polysulphones are often used. Other materials that are used commercially include polyvinylidenefluoride (PVDF), polyacrylonitrile (PAN) and polypropylene (PP). In recent times, interest in the use of ceramic materials has increased. These tend to be very robust and can withstand harsh operating and cleaning conditions. RO membranes are often made of polyamide while NF membranes are constructed from a variety of polymers, including polysulphone and polyimide, and ceramic materials.

1.3.2   Membrane morphology

MF membranes are typically symmetric with a sponge-like structure. This means that the pore size and porosity are the same throughout the depth of the membrane. Ultrafiltration membranes are typically asymmetric i.e., they have a thin region ('skin') of well-defined molecular weight cut-off, supported by a thicker, highly porous region. It is important to note that this asymmetry is an inherent part of the membrane and arises naturally from the chemistry of the membrane synthesis process. It is *not* two separate components stuck together. The less dense part of the membrane provides very little resistance to flow and gives the skin mechanical strength. All the flow and separation characteristics are determined by the membrane skin.

RO and NF membranes are typically classified as thin film composites (TFCs). In contrast to asymmetric UF membranes, TFCs *do* involve separate polymeric layers bonded together, the top layer often being made of polyamide or polyimide and the support layer typically made from polysulphone. The polysulphone layer itself is often supported by a fabric backing, especially in RO membranes.

Further details on membranes and their structure are to be found in the references at the end of this chapter, although there is some further discussion of NF and RO membrane characteristics in Chapter 9.

1.4   Membrane modules

This section describes the types of equipment used in crossflow membrane filtration processes. Discussion of dead-end equipment is left for Chapter 2. The housing in which a membrane is placed is termed a *module*, or in some instances, a *cartridge*. The same basic module types can be used for all membrane processes ranging from CFMF to RO, although specific types of module tend to be preferred for different types of filtration.

1.4.1   Flat sheet and spiral wound modules

Flat sheet modules, shown schematically in Fig. 1.6, use multiple flat sheet membranes in a sandwich arrangement consisting of membrane sheets, attached at their edges to a backing support.

Spacer screens, or meshes, provide the channel for the feed flow and are typically 0.3–1 mm in height. The space between the membrane support and the membrane itself acts as the channel for permeate flow. Several of these membranes and spacers are stacked alternately and held tightly together to form a cassette. Although flat sheet modules are prone to having their flow channels clogged, they have still been used successfully

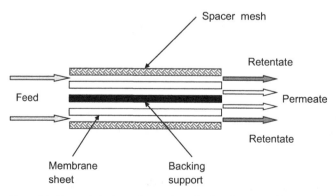

Figure 1.6 Expanded view of a portion of a flat sheet cartridge showing examples of feed splitting into retentate and permeate.

in bacterial and yeast cell separations and find widespread use in UF. Quantifying the channel Reynolds number is an important aspect of understanding the performance of crossflow filters. In flat sheet systems, where the membranes are separated by separator meshes, determining the precise flow regime can be difficult and the use of computational fluid dynamics (CFD) to model the flows in these systems is a growing area of membrane research.

Spiral wound modules are similar to the flat sheet type except that the whole assembly is wrapped up rather like a 'Swiss roll'. The feed flows down the retentate channels, while permeate flows radially through the membrane, spiralling into a central permeate duct. Because of the presence of the spacer screens, the flow channel can become clogged quite easily when used for MF applications. Spiral wound systems are used extensively for UF applications in dairy industries, as well as in NF and RO. Again, the channel flow is complex in these systems and use of CFD is generally required to get a full understanding of the flow characteristics.

1.4.2 Shell-and-tube modules

The shell-and-tube configuration, illustrated in Fig. 1.7 and used extensively in heat exchangers, is also widely used in membranes. The feed enters through the 'tubes', which are porous. The permeate flows radially through the wall of the tubes into the 'shell' to the exit ports. Different types of shell-and-tube module are distinguished by the size of their membrane diameters. In tubular modules, individual membrane tubes with internal diameters > 6 mm are packed in small bundles which are kept in place by two end-plates. The primary advantage of the tubular membrane module design is the turbulent flow which leads to high fluxes. Furthermore, they are not easily plugged, something that makes them suitable for applications involving high solids concentrations. The main disadvantage of tubular modules is their very low membrane area per unit volume.

Hollow fibre modules consist of large numbers of flexible, narrow-bore (≤ 200 μm internal diameter), porous fibres bonded at each end to a common header. A separate

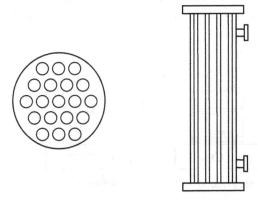

Figure 1.7 Top and side views of a tubular module.

category, the capillary module, is defined for modules with fibre diameters in the range 200 μm to 6 mm. The primary advantage of hollow fibre and capillary modules, in which the flow is laminar, is their high permeate flowrate per unit module volume. The high flowrate per unit volume is a consequence of the large area per unit volume rather than high fluxes. The main disadvantage of the hollow fibre module design is that it is susceptible to particulate plugging. Hollow fibres find widespread use in medical applications, especially haemodialysis.

1.4.3   Stirred cells

Stirred cells (Fig. 1.8) are frequently used for small scale operations such as those required for basic research and process development. The feed is placed in a pressurised cell, typically made of plastic but occasionally of glass.

The cell contains a stirrer, placed very close to the membrane itself, and typically controlled magnetically from a stirrer plate on which the cell is placed. Sometimes, when the feed volume is larger than the cell volume, the cell is fed from a separate reservoir using the arrangement shown in Fig. 1.8. This is a fed-batch operation. The reservoir can also be filled with pure water, meaning that the cell operates at constant concentration. The action of the impeller in a stirred cell creates a tangential flow of fluid across the membrane, which is broadly comparable to that found in the larger scale modules described above. Permeate volume is typically measured using an electronic balance which is best connected to a computer.

1.5   Flux characteristics in membrane filtration

1.5.1   Dead-end filtration

As will be seen in Chapter 2, there are no great mysteries associated with the flux in dead-end filtration processes. Cake accumulates and the flux declines until the capacity of the equipment is reached. The higher the solids concentration in the feed, the more rapidly the cake accumulates and the faster the flux declines. Higher trans-membrane pressures lead to higher fluxes but the relationship is not necessarily a simple linear

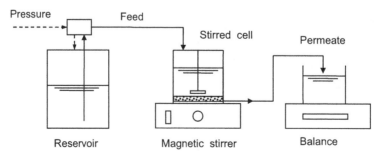

Figure 1.8 Stirred cell apparatus including reservoir for fed-batch operation.

one. This is because the cake resistance term in Eq. 1.4 is usually itself a function of pressure. The precise mathematical form of the flux decline and its dependence on process parameters is derived in Chapter 2.

1.5.2   Crossflow configuration

The crossflow configuration is much more interesting. It is important to state at the outset that it is understood throughout this section that the module is operating with a constant feed concentration. For small scale research or process development operations, this is often achieved by running the unit in *total recycle mode*, where the retentate and permeate are *both* recycled into the feed tank. By doing this, the solute concentration remains essentially constant over time.

In all membrane processes, membrane fouling can cause a decline in flux from the initial, clean membrane value. The rate of this decline depends on the precise mechanism of fouling as well as whether other processes, such as cake formation and/or concentration polarisation are occurring. Membrane fouling processes are very difficult to model with any rigour but some attempts at modelling are described in Chapter 10.

In the absence of fouling, cake formation leads to a decline in flux from the clean membrane value to a steady state value whose precise magnitude depends on a range of suspension properties and operating conditions. This decline in flux usually takes place on a measurably long time scale and thus cake formation dynamics must be incorporated into many models of CFMF. The steady state flux increases with trans-membrane pressure and exhibits saturation-like behaviour at high pressures, i.e., the flux is weakly dependent on pressure in this region. At any given pressure, it declines with increasing particle concentration (often in a power-law fashion) and increases with increasing crossflow velocity. Further detail on these trends is given in Chapter 3.

Qualitatively similar behaviour is observed in molecular filtration processes but there are some key differences. First, when concentration polarisation occurs, steady state is established essentially instantaneously. There is no need to concern oneself with the dynamics of concentration polarisation. Second, the saturation-like behaviour observed at high pressures is much more definite and universally observed, especially in UF where operating pressures are typically in the saturation region. When UF is operated in this region, it is said to operate at the *limiting flux*. RO and NF are typically operated below the limiting flux region. An example of saturation behaviour is shown in Fig. 1.9.

A third difference between the molecular filtrations and CFMF is that in the former, the limiting flux tends to show a logarithmic dependence, as opposed to a power depen-dence, on the solute (retentate) concentration over a reasonably broad range of operating conditions. An example is given in Fig. 1.10.

A final difference is the fact that the *limiting* flux in molecular filtration processes shows precise dependencies on the tangential velocity, u, for laminar and turbulent flows. This is in contrast to CFMF where behaviour is very varied. With laminar flow, the limiting flux in molecular processes scales as $u^{1/3}$, while under turbulent conditions, it scales as $u^{0.8}$. To the chemical engineer, these exponents are an immediate reminder of heat and mass transfer correlations such as the famous Leveque and Dittus–Boelter

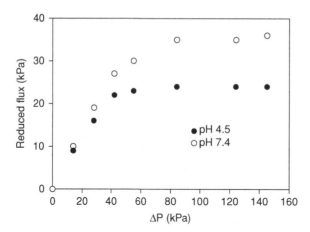

Figure 1.9 Flux versus pressure at two pH values during UF of an IgG solution. Adapted from [8]. Note that the reduced flux in this figure is defined as $\mu R_m J$.

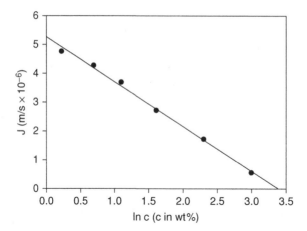

Figure 1.10 Semi-log plot of flux versus concentration during UF of BSA. Adapted from [9].

equations. They strongly imply the importance of mass transfer in molecular filtration processes. Indeed, this is a hint that although the flux characteristics are qualitatively similar in CFMF and the molecular processes, the underlying mechanisms are quite different. In the case of CFMF, the flux is reduced by the formation of an additional resistance to flow while in the molecular processes, mechanisms are a little bit more complex, including the fact that the driving force for filtration is reduced as a result of osmotic pressure effects.

1.6   Conclusions

This chapter has given a brief overview of membrane filtration processes and the reader is referred to the References and Additional reading for more information. There are many excellent books that provide most of the relevant information that one needs to have a good knowledge of membrane filtration processes. However, the aim of this book is to encourage the reader to delve into the subject in a much more computational manner, and thus this chapter is somewhat different from the remainder of the book. All of the theoretical issues touched on in this chapter are covered again in the relevant chapters and the focus will be on using models to tackle problems in process analysis and design. The remainder of the book is reasonably mathematical and it might be a good idea for the reader to glance through the Appendix at this point to get a sense of the level of mathematics and computation required.

References

1. Cheryan, M. (1998). *Ultrafiltration and Microfiltration Handbook*, 2nd Edition. CRC Press, Florida, USA.
2. Baker, R.W. (2004). *Membrane Technology and Applications*, 2nd Edition. Wiley, West Sussex, UK.
3. Zeman, L.J. and Zydney, A.L. (1996). *Microfiltration and Ultrafiltration. Principles and Applications*. Marcel Dekker, New York, USA.
4. Schweitzer, P.A. (Ed.) (1997). *Handbook of Separation Techniques for Chemical Engineers*, 3rd Edition. McGraw-Hill, New York, USA.
5. Kucera, J. (2010). *Reverse Osmosis: Design, Processes, and Applications for Engineers*. Wiley, New Jersey, USA.
6. Schafer, A.I., Fane, A.G. and Waite, T.D. (Eds.) (2005). *Nanofiltration: Principles and Applications*. Elsevier, Oxford, UK.
7. Treybal, R.E. (1979). *Mass Transfer Operations*, 3rd Edition. McGraw-Hill, New York, USA.
8. Wang, Y. and Rodgers, V.G.J. (2008). Free-solvent model shows osmotic pressure is the dominant factor in limiting flux during protein ultrafiltration. *Journal of Membrane Science*, **320**, 335–343.
9. Porter, M.C. (1979). In: *Handbook of Separation Techniques for Chemical Engineers*, Schweitzer, P.A., (Ed.). McGraw-Hill, New York, USA.

Additional reading

Cussler, E.L. (2009). *Diffusion: Mass Transfer in Fluid Systems*, 3rd Edition. Cambridge University Press, New York, USA.

Mulder, J. (1997). *Basic Principles of Membrane Technology*. Kluwer Academic, Dordrecht, The Netherlands.

Strathmann, H. (2011). *Introduction to Membrane Science and Technology*. Wiley-VCH, Weinheim, Germany.

Wang, W.K. (Ed.) (2001). *Membrane Separations in Biotechnology*, 2nd Edition. Marcel Dekker, New York, US.

2 Dead-end filtration

2.1 Introduction to dead-end filtration

In Chapter 1 the various membrane filtration processes were classified and it was seen that filtration and microfiltration involve the separation of particulate matter from liquids. Filtration and microfiltration are essentially the same process but the term microfiltration is usually reserved for filtration of particles with diameters in the range 0–10 µm. Many biological suspensions, such as suspensions of bacteria or yeast, fall within this range. Typically, synthetic membranes ('microfilters') are used with pore diameters less than 1 µm in diameter and many companies manufacture membranes with pore diameters of 0.22 µm and 0.45 µm. Filtration is a technology that has been used extensively in more traditional chemical industries where separation of larger particles such as crystals may be required. For these bigger particles, synthetic membranes are not necessary and filter cloths with considerably large pore sizes are employed. Often, suspensions of smaller particles, such as yeast cells employed in brewing processes, are filtered with filter cloths rather than synthetic membranes. However, this requires use of *filter aids* as discussed later.

The important point to note throughout this chapter is that filtration and microfiltration operate on the same principle and the underlying theory covers both processes. Both terms are used at various times throughout this chapter.

This chapter focuses on filtration performed in dead-end mode as illustrated in Fig. 2.1.

Actually, the dead-end terminology has evolved quite recently and is used mainly in bioprocessing industries to distinguish it from crossflow filtration. Some people refer to dead-end filtration as *normal-flow filtration* or *conventional filtration*.

At this stage it is important to mention some terminology. As indicated in Fig. 2.1, the clear liquid that emerges from the filter is defined as the *filtrate*. The layer of solids that accumulates on the membrane is the filter *cake*. Indeed, this type of filtration is often referred to as *cake filtration* to distinguish it from *depth* filtration. In the latter process, the solids actually penetrate into the filter medium where they are entrapped or captured by adsorption.

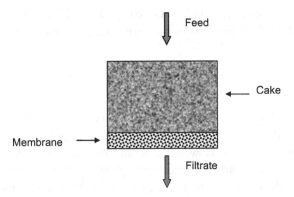

Figure 2.1 Dead-end filtration.

2.2 The filtrate flux equation

Filtration is a pressure-driven process and the following pressures and pressure drops can be defined:

$$\Delta P = P_s - P_f, \tag{2.1}$$

$$\Delta P_c = P_s - P_m, \tag{2.2}$$

$$\Delta P_m = P_m - P_f, \tag{2.3}$$

giving

$$\Delta P = \Delta P_m + \Delta P_c, \tag{2.4}$$

where ΔP is the total pressure drop, ΔP_c is the pressure drop across the cake, ΔP_m is the pressure drop across the membrane, P_s is the feed (suspension) pressure, P_m is the liquid pressure at the membrane (on the feed side) and P_f is the filtrate pressure. In this chapter, ΔP is often referred to as the *operating pressure* or the *applied pressure*.

The goal of this section is to develop a relation between the filtrate flux and ΔP. The filtrate flux is simply the volumetric flowrate of filtrate per unit membrane area. The starting point is to recognise that the flow of filtrate through the cake and membrane combination is, in essentially all cases, laminar. For laminar flows in general, mean fluid velocity is proportional to the pressure drop and inversely proportional to the fluid viscosity. For laminar flow in a pipe, for example, the following well-known Hagen–Poiseuille expression applies [1]:

$$u = \frac{\Delta P}{32\mu L_p/d_p^2}, \tag{2.5}$$

where u is the mean fluid velocity, L_p is the pipe length, d_p is the pipe diameter, ΔP is the pressure drop down the pipe and μ is the fluid viscosity. The combination of variables $32 L_p/d_p^2$ can be viewed as the pipe *resistance to flow*. By analogy, therefore, the following expressions for the filtrate flux, J, can be written:

$$J = \Delta P_m/\mu R_m, \tag{2.6}$$

$$J = \Delta P_c/\mu R_c, \tag{2.7}$$

where μ is the filtrate viscosity, R_m is the resistance to flow provided by the membrane and R_c is the resistance to flow provided by the cake. These resistances have units of m^{-1}. Combining Eqs. 2.4, 2.6 and 2.7 gives

$$J = \frac{\Delta P}{\mu(R_m + R_c)}. \tag{2.8}$$

This equation is of no real practical value since there is no a priori way (i.e., one based on first principles) of predicting the resistance terms unless assumptions are made about the membrane and cake structures. However, a further step can be made in light of Eq. 2.5. This suggests that R_m should be proportional to the membrane thickness and R_c proportional to the cake thickness. The traditional approach in filtration theory has been to retain R_m in the flux equation because there is no computational value in recasting the resistance in terms of a constant membrane thickness. In contrast, it is essential to write R_c in terms of the amount of cake deposited if one is to capture the dynamic character of most filtration processes. Rather than relating R_c to the cake thickness, it turns out to be more useful to relate R_c to the *mass* of solids deposited per unit membrane area, M, through the expression

$$R_c = \alpha M, \tag{2.9}$$

where α is termed the *specific cake resistance* (technically the *mean* specific cake resistance as explained later) with units of m/kg. For microbial suspensions, M is usually the mass of wet cells per unit area but dried cell mass is also sometimes used. In order to understand where Eq. 2.9 comes from, it is worthwhile to consider a cake whose properties are uniform. In this case, the cake mass per unit area is related to the cake thickness, L_c, by the expression

$$M = \rho_s \phi_c L_c, \tag{2.10}$$

where ρ_s is the density of the particles and ϕ_c is the particle volume fraction in the cake. Thus Eq. 2.9 is equivalent to saying that the cake resistance is proportional to the cake thickness.

The complex nature of the flow, in which fluid permeates through the cake, is hidden in the specific cake resistance, α. This parameter plays a central role in filtration theory by quantifying the ease with which a suspension can be filtered. In summary, therefore, the flux equation for cake filtration can be written

$$J = \frac{\Delta P}{\mu(R_m + \alpha M)}. \tag{2.11}$$

The derivation of this equation has been quite simplistic, and far more rigorous developments of filtration theory are available in the literature. These are much more mathematically complex and applying them in practice is difficult and beyond the scope of this text. The best way to think about Eq. 2.11 is to see it as a *definition* of the specific cake resistance parameter used throughout this and other chapters. The next section focuses on this important parameter.

2.3   The specific cake resistance

2.3.1   Cake compressibility

For most materials, the specific cake resistance is found to increase as the pressure drop across the cake increases. We refer to this phenomenon as *cake compressibility*. The precise mechanism of cake compressibility is not always well understood and it can have a number of causes including particle deformation, breakdown of particle aggregates and re-orientation of non-spherical particles [2, 3]. For many microbial suspensions it is probably cell deformation that is the cause of cake compressibility, as microbial cake compression is often found to be reversible. This means that when the pressure applied to a cake is increased, its specific resistance increases, but when the pressure is returned to its initial value, the specific resistance also returns to its initial value. In contrast, suspensions of powders like calcium carbonate tend to show irreversible increases in specific resistance, a fact that is probably due to the breakdown of particle aggregates under the action of the applied pressure [2].

Regardless of the exact mechanism, it is important, from a computational point of view, that the effect of pressure on specific resistance be quantified. The most commonly used expression in this regard is the following:

$$\alpha = a\Delta P^n, \tag{2.12}$$

where a and n are empirical constants, the latter usually having values such that $0 \leq n \leq 1$. If cake compressibility is *defined* as the rate of change of $\ln \alpha$ with respect to $\ln \Delta P$, one gets

$$\frac{d \ln \alpha}{d \ln \Delta P} = n. \tag{2.13}$$

Therefore, the constant, n, is a quantitative measure of the cake compressibility. Equation 2.12 is often found in chemical engineering textbooks but it usually applies only at moderate to high pressures. When microbial suspensions are microfiltered, log–log plots of α versus ΔP are often non-linear, particularly at the low end of the pressure range. Indeed, much yeast filtration data can be correlated more accurately by the simple linear expression [2, 3]

$$\alpha = \alpha_0(1 + k_c\Delta P), \tag{2.14}$$

where α_0 and k_c are empirical constants. A generalisation of Eqs. 2.12 and 2.14 is the following:

$$\alpha = \alpha_0(1 + k_c\Delta P)^m. \tag{2.15}$$

For this equation

$$\frac{d \ln \alpha}{d \ln \Delta P} = \frac{mk_c\Delta P}{(1 + k_c\Delta P)}. \tag{2.16}$$

Thus, the apparent cake compressibility is pressure-dependent. It becomes equal to m at high pressures and dependent on m, k_c and ΔP at low pressures. The use of

Table 2.1 Equation 2.15 used to correlate specific resistance data of some microbial and latex suspensions [4–7].

Suspension	Shape	Diameter (μm)	α_0 (m/kg $\times 10^{12}$)	k_c (Pa$^{-1} \times 10^{-4}$)	m (–)	α at 100 kPa (m/kg $\times 10^{12}$)
[4,5] *S. cerevisiae*	Ovoid	4.8	0.11	1.93	0.47	0.45
[4,5] *B. circulans*	Rod	1.7	1.79	1.41	1.07	32.7
[4,5] *R. spheroides*	Rod	1.6	3.40	1.29	1.01	55.4
[4,5] *E. coli*	Rod	1.2	43.3	5.75	0.69	717
[4] Latex	Spherical	0.55	3.63	0	0	3.63
[4] Latex	Spherical	0.21	23.2	0	0	23.2
[6] *K. marxianus*	Ovoid	n/a	0.30	0.17	1.0	0.81
[6] *K. marxianus*	Filament	n/a	0.21	0.24	1.0	0.71
[7] *S. uvarum*	Ovoid	4.8	0.15	0.24	1.0	0.51
[7] *Z. rouxii*	Ovoid	4.0	0.11	0.18	1.0	0.31
[7] *S. pombe*	Ovoid	4.3	0.24	0.18	1.0	0.67
[7] *S. cerevisiae*	Ovoid	4.9	0.33	0.08	1.0	0.61

Data in [4] are valid for $\Delta P < 200$ kPa. Reference [5] contains the analysis of the data in [4]. Data in [6] are valid for $\Delta P < 500$ kPa and data in [7] are valid for $\Delta P < 200$ kPa. All microbial suspensions are yeasts except for the three rod-shaped cells which are bacteria. The *K. marxianus* filaments had an average length-to-width ratio of 7.3. 'Diameter' is an effective value in all cases.

Eq. 2.15 to describe the compressibility of some microbial and latex suspensions is shown in Table 2.1. Example 2.1 below illustrates the use of non-linear regression, implemented in MATLAB, to arrive at a correlation of the form of Eq. 2.15.

Example 2.1 Correlation of specific resistance with applied pressure

The following specific resistance data have been obtained in a series of constant pressure filtration experiments:

ΔP (Pa $\times 10^5$)	0.2	0.4	0.8	1.2	1.6	2.0	2.5
α (m/kg $\times 10^{12}$)	5.48	5.89	6.59	7.17	7.67	8.12	8.63

Develop a correlation for α based on Eq. 2.15.

Solution. Evaluation of the constants α_0, k_c and m requires non-linear regression. MAT-LAB code for this purpose, employing a least squares method and the **fminsearch** function is shown below. The fit is excellent as shown in Fig. 2.2. The best-fit parameters obtained are $\alpha_0 = 5.0 \times 10^{12}$ m/kg, $k_c = 1.49 \times 10^{-5}$ Pa^{-1} and $m = 5$.

```
function example21
p=[.2 .4 .8 1.2 1.6 2.0 2.5]; % pressure data
alpha=[5.48 5.89 6.59 7.17 7.67 8.12 8.63]; % alpha data
guess=[5.0 1.0 0.5]; % initial guesses of model parameters
params=fminsearch(@fitfun,guess,[],p,alpha)
ppred=linspace(0,2.5); % pressures for computing
```

```
predicted alphas
alphapred=params(1)*(1+params(2)*ppred).^params(3);
%predicted values of alpha
plot(p,alpha,'sk',ppred,alphapred,'-k');
xlabel( '\DeltaP (bar)'); ylabel( '\alpha (m/kg x 10^1^2)');
function sse=fitfun(params,p,alpha); % function to
compute sum of square errors
alpha0=params(1);
kc=params(2);
m=params(3);
err=alpha-(alpha0*(1+kc*p).^m); % difference between
actual alpha and predicted
sqerr=err.^2; %square of difference in each case
sse=sum(sqerr);% sum of squares
```

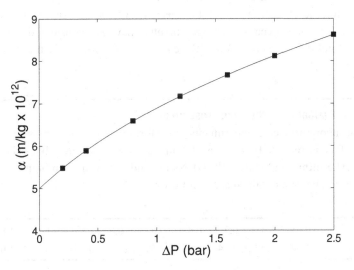

Figure 2.2 Fit of Eq. 2.15 to specific resistance data.

Although it is valid from a pure curve-fitting perspective to correlate specific resistance data with an expression like Eq. 2.15, it is important to note that the real determinant of α is the cake pressure drop, ΔP_c, where Eqs. 2.4 and 2.6 imply that

$$\Delta P_c = \Delta P - \mu R_m J. \tag{2.17}$$

Thus Eq. 2.15 can be re-written more meaningfully as

$$\alpha = \alpha_0(1 + k_c \Delta P_c)^m, \tag{2.18}$$

where, now, α_0 has physical meaning, namely the specific resistance of an unstressed cake. The correlation of specific resistance data with ΔP_c has the practical advantage that the correlation is valid and independent of the precise experimental conditions,

Table 2.2 Parameters for Eq. 2.18 when used to correlate specific resistance data for some powders. Adapted from [8].

Suspension	Range (kPa)	α_0 (m/kg $\times 10^{10}$)	k_c (Pa^{-1} $\times 10^{-4}$)	m (−)	α at 100 kPa (m/kg $\times 10^{10}$)
CaCO$_3$	7–550	5.1	1.43	0.2	5.1
Darco-B	7–275	1.1	5.88	0.4	1.1
Kaolin-Al$_2$SO$_4$	7–415	43	1.43	0.3	4.3
Talc-C	7–1400	4.7	1.82	0.55	4.7
TiO$_2$	7–7000	18	1.43	0.35	1.8

especially the membrane employed. This is technically not the case when a correlation based on ΔP is used, as is explained later. The obvious question, therefore, is: why not use ΔP_c in correlations for α? This has been done in the past, especially for filtration of powder suspensions with filter cloths, and Table 2.2 shows some typical correlations.

However, for microfiltration of microbial suspensions, the ΔP_c approach has generally not been used, perhaps because the experimenters have not considered the difference between ΔP and ΔP_c. Computation of the cake pressure drop is illustrated in Example 2.2 below.

Example 2.2 Evaluation of the cake pressure drop

A series of laboratory scale experiments is performed at different pressures in which the filtrate flux is recorded just as a certain mass of a compressible cake has been deposited. The filtrate viscosity is 0.001 Ns/m^2 and the clean membrane resistance is 5×10^{10} m^{-1}. The data obtained are as follows:

ΔP (Pa $\times 10^5$)	0.05	0.1	0.2	0.4	0.6	0.8	1.0	1.4	1.8
J (m/s $\times 10^{-5}$)	1.34	2.53	4.54	7.48	9.50	11.0	12.1	13.6	14.6

Estimate the cake pressure drop relative to the applied pressure in each case.

Solution. The key equation here is Eq. 2.17. Calculation of the cake pressure drop is a simple issue of 'plugging in the numbers'. Performing the necessary computations, the following results are obtained:

ΔP (Pa $\times 10^5$)	0.05	0.1	0.2	0.4	0.6	0.8	1.0	1.4	1.8
$\Delta P_c/\Delta P$	0.866	0.874	0.887	0.907	0.929	0.931	0.940	0.951	0.959

Clearly the difference between the two pressures becomes more significant at low pressures. A fundamental difficulty with this calculation is that in many practical situations, membrane fouling occurs and, in the absence of any reliable method for separating the effects of membrane fouling from those of cake formation, the cake pressure drop is impossible to compute accurately. This is, perhaps, why one does not seem to encounter correlations for α in terms of cake pressure drop in microfiltration work.

It was mentioned earlier that one of the reasons that it is desirable to have a correlation for α in terms of ΔP_c, is that the correlation is valid regardless of the precise experimental conditions. This is not the case for correlations written in terms of ΔP. To explain why this is, consider an experiment such as that described in Example 2.2, where a fixed mass of solids is deposited on the membrane. Then, combining Eqs. 2.11, 2.17 and 2.18, the following expression is obtained:

$$\Delta P_c = \Delta P - \frac{\Delta P}{1 + \dfrac{\alpha_0 M}{R_m}(1 + k_c \Delta P_c)^m}. \tag{2.19}$$

This is a non-linear algebraic equation for the cake pressure drop which shows that for a fixed mass of cake of a particular solid, ΔP_c depends on the applied pressure *and* the membrane resistance. Therefore, two filtrations performed at the same ΔP but different R_m may involve significantly different cake pressure drops, and the specific resistances in each filtration process may not be the same. This idea is explored in Problem 2.1.

2.3.2 Effects of particle and liquid properties

So far, it has been seen that the specific cake resistance can be a function of operating pressure. However it is important to have a sense of how the properties of the particle and liquid components of a suspension affect the specific resistance. Table 2.1 gives some indication of the role of particle size and shape but to deepen our understanding we must first go back to some of the fundamentals of flow in porous media.

The classic, theory-based, expression that relates the specific cake resistance to particle morphology is the Carman–Kozeny equation which can be written in the filtration context as [9]

$$\alpha = \frac{K a_v^2}{\rho_s} \frac{1 - \varepsilon_c}{\varepsilon_c^3}, \tag{2.20}$$

where K is the Kozeny constant, ρ_s is the particle density, a_v^2 is the particle surface area per unit volume and ε_c is the cake porosity. This equation is based on the assumption that there is point contact between the particles in the cake, i.e., the total contact area between particles and fluid is equal to the sum of the surface areas of the individual particles. It also assumes that the cake properties are uniform. Problem 2.2 takes the reader through a simple derivation of this equation. Neither of the above assumptions is valid when the cake is compressible. Of course a more obvious drawback is that the Kozeny constant is not known a priori. Using the value $K = 5$, which might be reasonable for fluid flow in the packed columns used in gas absorption or distillation [10], is very unlikely to be accurate in filtration where K is often dependent on cake porosity [3]. Ultimately, therefore, the Carman–Kozeny equation cannot be considered as a predictive equation.

Its main advantage is that it provides some insight into the effect of particle size on the specific cake resistance. If the particles in the suspension are spherical and of diameter, d, the Carman–Kozeny equation can be written

$$\alpha = \frac{36K}{\rho_s d^2} \frac{1 - \varepsilon_c}{\varepsilon_c^3}. \qquad (2.21)$$

Indeed, a quick calculation shows that the latex data in Table 2.1 are pretty much consistent with the size-dependence predicted by this equation.

While Eq. 2.21 gives a rough guide as to the (strong) effect of particle size on specific resistance, it is not generally possible to quantify the general effect of particle morphology on the specific cake resistance. The data in Table 2.1 suggest that the 'less spherical' the microbial cell, the more compressible is the filter cake. It is important to realise, however, that the surface properties of the particle are very important in determining the specific resistance and, therefore, it is never possible to make definitive statements about the effects of particle size and shape on specific resistance [11]. Research has been done, however, with a microorganism that can change shape, depending on the growth conditions, and it has been shown that the cake compressibility does increase as the cells become elongated [3]. However, there was no absolute guarantee in that work that the detailed surface properties of the cells were completely independent of shape.

The specific cake resistance of suspensions is often sensitive to liquid properties such as ionic strength and pH [5]. Changes in pH or ionic strength can lead to different states of aggregation. Research with scanning electron microscopy has shown that at certain pH values, microbial cakes, in particular, can contain many cell aggregates and large voids. At other pH values, the cake may be homogeneous and free of aggregates. These changes in cake structure are reflected in changes in the specific cake resistances. It is not possible to predict the effect of liquid properties on the specific resistance and experiment will always be needed. Even if the precise relation between liquid properties and specific resistance could be predicted, one would still be faced with the fact that the specific resistance of microbial suspensions depends in very subtle ways on factors that include the growth medium employed and the time in the growth cycle when the cells are filtered [5]. For the engineer, therefore, filtration of microbial suspensions always presents a big challenge.

2.3.3   Filter aids

Many suspensions, particularly those of a microbial nature, pose problems in filtration. In general, therefore, microbial suspensions must be augmented in some way if they are to be filtered in an efficient manner. Filter aids are solids such as perlite and diatomaceous earth (Kieselgur), which form highly porous, incompressible or weakly compressible filter cakes. They are used for two main reasons: (i) to prevent cells from clogging the filter or passing through it, and (ii) to lower the specific resistance and compressibility of the filter cake.

In the *pre-coat* method, the filter is first coated with a thin cake of filter aid. This cake acts as a secondary membrane, protecting the filter itself and stopping particles from appearing in the filtrate or from clogging the filter pores. This type of protection is usually required when the filter is of the woven fabric type. It should not be necessary with microfilters. The filter aid has a small specific cake resistance and thus no serious reduction in filtrate flux occurs due to the presence of the pre-coat. In the *body feed* method, the filter aid is mixed with the original feed to give a new suspension that produces a filter cake of lower specific resistance. The addition of the filter aid also makes the filter cake less compressible. Filter aids have some disadvantages. Disposal may be difficult, especially if contaminated with microorganisms. In addition, the filter aid may adsorb products such as enzymes, leading to potentially significant losses.

2.4   Analysis and design of batch dead-end filtration

Most dead-end filtration (DEF) processes are carried out in batch mode, as illustrated schematically in Fig. 2.3.

Filtration is initiated and a volume of filtrate is produced before the equipment is dismantled, the cake recovered or discarded, and the equipment cleaned. There is a vast range of equipment for dead-end filtration and there are many books that describe these in detail, including the work of Svarovsky [9]. Even in the new biotechnology industries, new dead-end filters in which containment of all process fluids is essential are beginning to appear. In the spirit of this book, however, the focus here is on the core principles rather than the attributes of different types of filter.

The goal of this section is to devise a general differential equation that provides the tools to answer questions such as:

(i) how long will it take to produce a volume, V_f, of filtrate with a filter of given area?
(ii) how large a filter is needed to produce a given amount of filtrate in a given time? and
(iii) how do the suspension properties affect the filter performance?

In order to answer such questions, Eq. 2.11 must be recast in terms of the filtrate volume, V_f. First, though, it is recognised that, by definition,

$$J = \frac{1}{A}\frac{dV_f}{dt},\qquad(2.22)$$

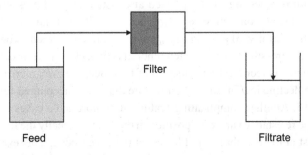

Filter

Feed   Filtrate

Figure 2.3 A batch dead-end filtration process.

where A is the membrane area and V_f is the filtrate volume at time, t. Therefore, using Eq. 2.11, the following differential equation describes the production of filtrate:

$$\frac{1}{A}\frac{dV_f}{dt} = \frac{\Delta P}{\mu(R_m + \alpha M)}. \tag{2.23}$$

A solids balance gives

$$M = \frac{cV_f}{A}, \tag{2.24}$$

which means that Eq. 2.23 is a differential equation in V_f that is easily solved, at least in simple cases. However, Eq. 2.24 is a little unsatisfactory because of the way c is defined, i.e. as the mass of solids deposited *per unit volume of filtrate*, a quantity that would not be known before the filtration has taken place. This is because one does not know in advance how much of the liquid in the feed suspension is retained in the cake. It would be far more convenient, therefore, if c could be defined as the mass of solids *per unit volume of suspension*. This would imply that the fluid retained in the cake is insignificant in comparison with the filtrate volume. Fortunately, in many applications, especially in bioprocessing, it is valid to make this assumption. The precise conditions under which the liquid entrained in the cake can be neglected are explored in Problem 2.3.

The governing differential equation for dead-end cake filtration can finally be written

$$\frac{1}{A}\frac{dV_f}{dt} = \frac{\Delta P}{\mu(R_m + \alpha c V_f/A)}, \tag{2.25}$$

where c is now understood to be the particle concentration in g solids per litre (l) of *suspension*. For the purposes of this chapter, the initial condition for Eq. 2.25 is assumed to be $V_f = 0$ @ $t = 0$. This is something of a simplification as the underlying theory that leads to Eq. 2.25 probably does not apply in the very early stages of filtration, even if the membrane resistance remains constant. The reason for this is that the cake has been assumed to be a continuum, whereas when the cake is very small, this is not valid. This issue is explored further in Problem 2.4, which should be attempted after reading Example 2.3 below.

2.4.1   Analysis and design of batch DEF at constant pressure

Dead-end filtration processes are often operated at constant applied pressure. From a mathematical point of view, constant pressure implies that ΔP is constant in Eq. 2.25 and the integration is straightforward as long as R_m, α and c are considered to be constant. An aspect of cake compressibility that is often ignored is that, even if ΔP is constant, the specific cake resistance is technically a function of time because ΔP_c increases during filtration due to the decline in J in Eq. 2.17, a feature that is not accounted for by either Eq. 2.12 or Eq. 2.15. Another complicating feature of compressible cakes is that their porosity and specific resistance vary with position in the cake. Typically the local specific resistance is highest at the membrane and lowest at the cake's edge. This explains why the specific resistance used here represents a *mean* value for the entire cake.

Given the assumptions outlined above, Eq. 2.25 can be inverted and rearranged to give

$$\int\limits_0^t dt = \int\limits_0^{V_f} \left(\frac{\mu R_m}{A \Delta P} + \frac{\alpha \mu c}{A^2 \Delta P} V_f \right) dV_f. \tag{2.26}$$

This is easily integrated to give the well known result

$$\frac{t}{V_f} = \frac{\mu R_m}{A \Delta P} + \frac{\alpha \mu c}{2 A^2 \Delta P} V_f. \tag{2.27}$$

One use of this expression is that it provides us with a simple methodology for measuring the specific cake resistance. The filtrate volume is recorded as a function of time and a simple plot of t/V_f versus V_f, often referred to as the *Ruth* plot, is made. This should give a straight line from whose slope α can be calculated. This can repeated at different applied pressures to yield information on the cake compressibility. Example 2.3 below illustrates the use of the t/V_f versus V_f method.

Filtration experiments of this type are often conducted in a filter cell. This is simply a cylindrical vessel fitted with a porous support onto which is placed the membrane. The suspension to be filtered is placed into the cell and the lid, which is attached to a nitrogen or air cylinder, is securely tightened. A regulator ensures that filtration is performed at a constant, known pressure. An exit port is provided for the filtrate, which is collected and weighed on a digital balance. For accurate measurements, the balance can be connected to a personal computer. The method simply involves opening the filtrate valve and recording the filtrate volume as a function of time. In effect, the apparatus and measurements are the same as the stirred cell described in Fig. 1.8, except that there is no stirring involved.

Example 2.3 Evaluation of specific cake resistance from constant pressure data

The table below gives experimental data obtained during a bench scale experiment on filtration of suspensions of rehydrated active dry yeast. The membrane area was 0.0017 m², the applied pressure was 60 kPa, the wet cell concentration was 32 g/l and the viscosity of the filtrate can be taken to be 0.001 Ns/m². Determine the specific cake resistance.

t (s)	30	60	120	180	240	270	300	360
V_f (ml)	15.0	21.0	29.6	36.1	41.9	44.2	46.9	51.4

Solution. A simple MATLAB script that employs the **polyfit** function to perform the linear regression and plot the results (Fig. 2.4) is shown below. The same regression can be done interactively within the plot editing window. In the same way, the appearance of the plot can be fine-tuned as done throughout this book.

```
% example 23
t=[30 60 120 180 240 270 300 360];
```

```
Vf=[15 21 29.6 36.1 41.9 44.2 46.9 51.4];
tVf=t./Vf;
a=polyfit(Vf,tVf,1); % a is the vector of polynomial
coefficients
Vfpred=linspace(0,60);
tVfpred=a(2)+a(1)*Vfpred;
plot(Vf,tVf,'ok', Vfpred,tVfpred, 'k-');
xlabel('V_f ( mL )'); ylabel('t / V_f ( s/mL )');
axis([0 60 0 10]);
```

The **polyfit** function gives a slope of 0.138 s/ml^2. Using Eq. 2.27 and making sure to use a consistent set of SI units, we have

$$\alpha = \frac{0.138 \times 10^{12} \times 2 \times (0.0017)^2 \times 60 \times 10^3}{0.001 \times 32} = 1.496 \times 10^{12} \text{ m/kg.}$$

One point worth commenting on is the fact that the linear regression gives a small negative intercept on the t/V_f axis. This is simply an artefact and does not have any physical meaning. If more data had been obtained at low filtrate volumes, non-linearity in the plot would have been observed, giving a positive intercept. The precise location of that intercept would depend on whether there was any membrane fouling. Non-linearity at low pressures can be due to cake compressibility or may simply represent a breakdown of the basic model when the cake is very thin.

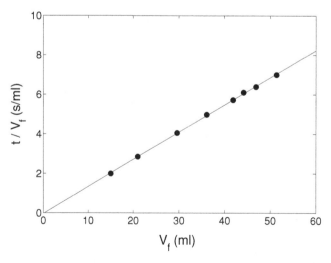

Figure 2.4 Ruth plot of filtration data.

Many filtration practitioners would prefer not to use the t/V_f method described above as there are too many assumptions required for it to be valid, namely constant R_m, α and c. Furthermore, the integration used to arrive at Eq. 2.27 can tend to smooth out effects that would be more visible if one had based the analysis more directly on the original differential equation. Thus, an alternative and potentially better approach is to base the analysis on the inverted form of Eq. 2.25, namely

$$\frac{dt}{dV_f} = \frac{\mu R_m}{A \Delta P} + \frac{\alpha \mu c}{A^2 \Delta P} V_f. \tag{2.28}$$

Now, if one collects V_f versus t data, the left-hand side of this expression can be computed by numerical differentiation and the data interpreted completely objectively. If the plot curves upwards, this could be an indicator of cake compressibility (α increasing with time) while downward curvature is often an indicator of membrane fouling. A third phenomenon that has not been mentioned previously is *cake clogging*, in which the pores of the cake itself become clogged by very small particles that are present along with the dominant larger particles [12]. The use of Eq. 2.28 to compute the specific cake resistance is illustrated in the next example.

Example 2.4 Evaluation of the specific resistance using Eq. 2.28

For the conditions and data of Example 2.3, evaluate the specific resistance using Eq. 2.28.

Solution. The solution to this problem requires numerical differentiation of the experimental data. This can be done quite easily by simply approximating the derivative by the expression

$$\frac{dt}{dV_f} \approx \frac{\Delta t}{\Delta V_f}.$$

Thus the first value of dt/dV_f for plotting Eq. 2.28 is $(60{-}30)/(21{-}15)$. The corresponding value of V_f is taken as the midpoint of the interval, i.e., $(21{+}15)/2$, written equivalently as $15 + (21{-}15)/2$. This approach can be implemented in a MATLAB script using the **diff** command as shown below.

```
% example 2.4
t=[30 60 120 180 240 270 300 360];
Vf=[15 21 29.6 36.1 41.9 44.2 46.9 51.4];
dtdVf=diff(t)./diff(Vf) % approximation to derivative
Vfm=Vf(1:end-1)+diff(Vf)/2 % midpoint of interval
plot(Vfm,dtdVf,'ok');
xlabel('V_f ( mL )'); ylabel('\Deltat / \DeltaV_f (
s/mL )'); axis([0 60 0 15]);
```

Adopting this approach gives the plot shown in Fig. 2.5. The regression was done with the *basic fitting* facility within the plot window giving a slope of 0.262 s/ml². A slope of 0.276 s/ml² (i.e. 0.138×2) would have been in perfect agreement with the slope of the line in Example 2.3. There is a noticeable degree of scatter in the plot and this stems from the errors introduced by numerical differentiation, a notoriously error-prone numerical calculation. Accurate numerical differentiation requires good quality data

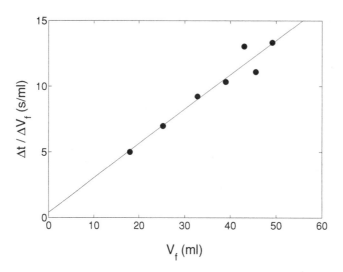

Figure 2.5 Filtration data analysis using Eq. 2.28.

where the difference between data readings is much greater than the precision of the instrument.

Another application of Eq. 2.27 is in computing the filtration time or in estimating the size of a filter required for a given process. In many dead-end filtration processes, it is found that except in the very early stages of filtration, the membrane resistance is negligible in comparison with the cake resistance, and Eq. 2.27 can be simplified to give

$$t = \frac{\alpha \mu c}{2 A^2 \Delta P} V_f^2. \tag{2.29}$$

Use of this equation is illustrated in Example 2.5.

Example 2.5 Calculation of filter area for a constant pressure process

An aqueous suspension forms filter cakes with a specific resistance of 4×10^{11} m/kg. What filter area would be required to produce 40 l of clear filtrate from a 20 g/l suspension in 10 minutes, if the operating pressure is 50 kPa? Take the viscosity of the filtrate to be 0.001 Ns/m^2 and assume that the membrane resistance is negligible.

Solution. Using Eq. 2.29 and rearranging gives

$$A = \left(\frac{\alpha \mu c}{2 t \Delta P} \right)^{1/2} V_f.$$

Ensuring that consistent SI units are used in all cases, the following result is obtained:

$$A = \left[\frac{(4 \times 10^{11})(0.001)(20)}{2(10 \times 60)(50 \times 10^3)} \right]^{1/2} (40 \times 10^{-3}) = 0.46 \text{ m}^2.$$

A simple additional exercise for the reader would be to calculate the area if Eq. 2.27 is used with a membrane resistance of 2×10^{10} m^{-1}.

It is worthwhile now to consider Eq. 2.29 as it applies to a compressible cake. This is most clearly done by using Eq. 2.12 to describe the cake compressibility, giving

$$t = \frac{a\mu c}{2A^2 \Delta P^{1-n}} V_f^2. \tag{2.30}$$

A natural instinct when trying to accelerate a filtration process is to increase the applied pressure. It is clear from Eq. 2.30 that such an approach will be of limited value if $n \to 1$ because the filtration time becomes almost independent of pressure. In practice, therefore, compressible cakes can pose considerable problems and frequently require the addition of filter aids. In some cases, it proves more efficient to use an alternative separation technique such as centrifugation or crossflow microfiltration. The small impact of pressure on filtration time is illustrated in the next example and in Problem 2.5.

Example 2.6 Effect of cake compressibility on filtration time

Filter cakes of *E. coli* are found to have a specific resistance that is described by Eq. 2.12 with $a = 8 \times 10^{10}$ and $n = 0.79$, with ΔP in Pa and α in m/kg. A 20 g/l suspension is to be filtered at 30 kPa in a filter with an area of 0.5 m^2. How long will it take to produce 10 l of filtrate? How long would it take if the pressure were doubled? Take $\mu = 0.001$ Ns/m^2 and ignore the membrane resistance.

Solution. Assuming the membrane resistance is negligible, we can use Eq. 2.30. Therefore, at 30 kPa, we have

$$t = \frac{(8 \times 10^{10})(0.001)(20)}{2(0.5)^2(30 \times 10^3)^{0.21}}(10 \times 10^{-3})^2 = 36724 \text{ s} = 10.2 \text{ hours.}$$

If the pressure is doubled, the time taken is given by $10.2 \times (1/2)^{0.21}$ which works out at approximately 8.8 hours. This represents a reduction in filtration time of only about 14%. It is worth noting that the very long filtration times are due to the extremely high specific resistance of the *E. coli* suspensions employed.

To close this section on constant pressure filtration it would be useful for the reader to do Problem 2.6 now.

2.4.2 Analysis of batch DEF at constant filtrate flux

In this mode of operation, often referred to as *constant rate filtration*, the filtrate flux is kept constant, thus requiring a gradual increase in operating pressure as the cake accumulates. To start, Eq. 2.25 is rearranged to give

$$\Delta P = \mu R_m J + \alpha \mu c V_f J / A. \tag{2.31}$$

For constant flux, J, this becomes

$$\Delta P = \mu R_m J + \alpha \mu c J^2 t, \qquad (2.32)$$

because V_f is just JAt. Clearly, if the cake is incompressible, α is constant and ΔP increases linearly with time. For a compressible cake whose specific resistance is predicted by Eq. 2.15, Eq. 2.32 becomes

$$\frac{\Delta P - \mu R_m J}{(1 + k_c \Delta P)^m} = \alpha_0 \mu c J^2 t. \qquad (2.33)$$

If Eq. 2.18 is used to correlate the specific resistance data, the following equation applies:

$$\frac{\Delta P_c}{(1 + k_c \Delta P_c)^m} = \alpha_0 \mu c J^2 t. \qquad (2.34)$$

As both of the above equations can be defined as non-linear algebraic equations (NLAEs), repeated numerical solution will be required to compute the pressure as a function of time. However, an alternative, possibly simpler, approach is to differentiate Eqs. 2.33 and 2.34 with respect to time, yielding the following ordinary differential equations (ODEs):

$$\frac{d\Delta P}{dt} = \frac{\alpha \mu c J^2}{1 - \mu c J^2 \dfrac{d\alpha}{d\Delta P} t}, \qquad (2.35)$$

$$\frac{d\Delta P_c}{dt} = \frac{\alpha \mu c J^2}{1 - \mu c J^2 \dfrac{d\alpha}{d\Delta P_c} t}, \qquad (2.36)$$

with the initial conditions $\Delta P = \mu R_m J$ @ $t = 0$ and $\Delta P_c = 0$ @ $t = 0$. The details of the differentiations are left as an exercise for the reader. The choice as to which of these equations to use depends on whether specific resistance data are expressed in terms of ΔP or ΔP_c. Use of Eq. 2.36 is illustrated in the next example, where the specific resistance is given by Eq. 2.18. It is worth noting in passing that if Eq. 2.18 is used for α and $k_c \Delta P_c \gg 1$, Eq. 2.34 becomes

$$\Delta P_c = \left(\alpha_0 k_c^m \mu c J^2 t \right)^{\frac{1}{1-m}}. \qquad (2.37)$$

This equation is to be found in most textbook treatments of constant rate filtration where an expression of the form of Eq. 2.12 (with ΔP_c replacing ΔP) is used to correlate the specific cake resistance.

Example 2.7 Calculation of pressure versus time using differential equation approach

Consider a 10 g/l suspension of Talc-C whose specific resistance parameters are given in Table 2.2. The suspension is to be filtered at a constant flux of 2×10^{-4} m/s with a membrane of resistance 1×10^{10} m^{-1}. Assuming a filtrate viscosity of 0.001 Ns/m^2, generate a plot of ΔP versus time using Eqs. 2.36 and 2.37, ending at $\Delta P = 300$ kPa. Note the time at which this pressure is reached. Recalculate this time by numerical solution of Eq. 2.34.

Solution. This problem can be solved easily with a MATLAB function employing MATLAB's 'workhorse' ordinary differential equation solver, **ode45**. Sample code is shown below. The routine is stopped at 3 bar using the **event** facility within MATLAB. The difference between the full numerical solution and the approximate expression, Eq. 2.37, is small, at least for the parameters employed. The final time as computed from the ODE is 1745 s.

```
function example27
global alpha0 kc m u rm j c pf;
pf=3e5; % pressure at which filtration is stopped
alpha0=4.7e10; kc=1.82e-4; m=0.55; u=0.001;rm=1e10;
j=2e-4; c=10;
options=odeset('events', @eventfcn);
[t,pc]=ode45(@constant_rate,[0 7200],0,options);
p=pc+u*rm*j; % p is the total applied pressure
papprox=u*rm*j+(alpha0*kc^m*u*c*j^2*t).^(1/(1-m)); % Eq. 2.37
x=[t,p,papprox];
disp([x])
plot(t,p,'-k',t,papprox,'--k');
xlabel('t ( s )');ylabel('\DeltaP ( Pa )');
legend('Eqn. 2.36','Eqn. 2.37')
function dpcdt=constant_rate(t,pc)
global alpha0 kc m u j c;
alpha=alpha0*(1+kc*pc)^m;
dalphadpc=alpha*kc*m/(1+kc*pc);
dpcdt=alpha*u*c*j^2/(1-u*c*j^2*dalphadpc*t); % Eq. 2.36
function [value, isterminal, direction]=eventfcn(t,pc)
global u rm j pf;
value = pc+u*rm*j-pf ; % stop when final applied
pressure is reached
isterminal = 1;
direction = 0;
```

Results are shown in Fig. 2.6.

To compute the final time using Eq. 2.34, solution of a non-linear algebraic equation is required. This can be done easily with the MATLAB function **fzero** as shown below.

```
function example27a
global alpha0 kc m u rm j c pf;
pf=3e5;
```

```
alpha0=4.7e10; kc=1.82e-4; m=0.55; u=0.001;
rm=1e10;j=2e-4; c=10;
guess=3600;% initial estimate for time taken
time=fzero(@constant_rate_a,guess)
function f=constant_rate_a(t)
global alpha0 kc m u rm j c pf;
pcf=pf-u*rm*j; % cake pressure drop
f=pcf/(1+kc*pcf)^m-alpha0*u*c*j^2*t; %Eq. 2.34
```

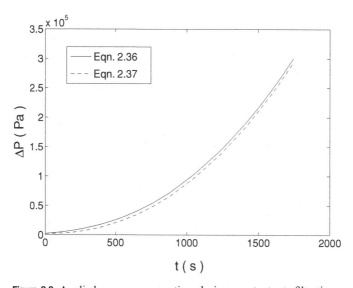

Figure 2.6 Applied pressure versus time during constant rate filtration.

This code returns an answer of 1745 s which is precisely the same as the ODE approach. Obviously, this is a simpler approach if one only has to compute a single time. If the full pressure versus time profile is required, the NLAE must be solved repeatedly and the ODE approach is probably simpler and more efficient. To close this section, the reader should complete Problem 2.7.

2.4.3   Analysis of batch DEF with a centrifugal pump

When a filtration operation is performed with a centrifugal pump, one has to take into account the fact that the flowrate delivered by the pump is related to the pressure at which the fluid is delivered – this is one of the pump's *characteristics*. In the context of Eq. 2.25 this means that both ΔP and the filtrate flowrate, Q_f (dV_f/dt), vary with time. For a given fluid, the pressure–flow characteristic of a typical pump can often be written, in a filtration context, as a quadratic equation of the form

$$\Delta P = \Delta P_{\max} + \kappa_1 Q_f + \kappa_2 Q_f^2, \tag{2.38}$$

where $\Delta P_{\max}$ is the pressure for which the flowrate is zero (the 'shut-off' pressure), κ_1 is a positive constant and κ_2 is a negative constant. The coupling between flowrate and pressure complicates the analysis somewhat. The conventional approach to solving this problem, at least in terms of calculating the filtration time, has been to note that the filtration time is given by

$$t = \int\limits_0^{V_f} \frac{dV_f}{Q_f}. \tag{2.39}$$

The integral is then evaluated graphically by plotting $1/Q_f$ versus V_f. Each coordinate is determined by selecting a value of Q_f, computing ΔP from Eq. 2.38 and evaluating V_f values from Eq. 2.25. This method can be improved by doing the integration numerically with the trapezoidal rule or similar. Here, however, the strategy employed is, once again, to turn the problem into one of solving a system of ordinary differential equations. There already exists a differential equation for V_f (Eq. 2.25). The real challenge is to find an expression for dQ_f/dt, especially when α is a function of ΔP (or ΔP_c) and consequently a function of Q_f. Differentiating Eq. 2.25 with respect to time gives

$$\frac{\mu}{A} \frac{dQ_f}{dt} = \frac{(R_m + \alpha c V_f/A)\dfrac{d\Delta P}{dt} - \Delta P \left(\dfrac{c}{A}\right)\left(\alpha Q_f + V_f \dfrac{d\alpha}{dt}\right)}{(R_m + \alpha c V_f/A)^2}. \tag{2.40}$$

Now using

$$\frac{d\alpha}{dt} = \frac{d\alpha}{d\Delta P}\frac{d\Delta P}{dt} \tag{2.41}$$

and

$$\frac{d\Delta P}{dt} = \kappa_1 \frac{dQ_f}{dt} + 2\kappa_2 Q_f \frac{dQ_f}{dt} \tag{2.42}$$

we ultimately get

$$\frac{dQ_f}{dt} = \frac{-\alpha\mu c Q_f^3/A^2 \Delta P}{1 - \left(\dfrac{Q_f}{\Delta P} - \dfrac{\mu c V_f Q_f^2}{A^2 \Delta P}\dfrac{d\alpha}{d\Delta P}\right)(\kappa_1 + 2\kappa_2 Q_f)}. \tag{2.43}$$

Likewise, the following expression is found when α is expressed as a function of ΔP_c:

$$\frac{dQ_f}{dt} = \frac{-\alpha\mu c Q_f^3/A^2 \Delta P}{1 - \left(\dfrac{Q_f}{\Delta P} - \dfrac{\mu c V_f Q_f^2}{A^2 \Delta P}\dfrac{d\alpha}{d\Delta P_c}\right)(\kappa_1 + 2\kappa_2 Q_f) - \dfrac{\mu^2 R_m c V_f Q_f^2}{A^3 \Delta P}\dfrac{d\alpha}{d\Delta P_c}} \tag{2.44}$$

Detailed derivation of Eqs. 2.43 and 2.44 is left for Problem 2.8. The initial condition for these equations needs a little bit of thought. From Eq. 2.25 we know that at zero time, when no cake is formed, the following expression relating the initial flowrate and the initial pressure applies:

$$Q_{f0} = \frac{A\Delta P_0}{\mu R_m}. \tag{2.45}$$

But the pump characteristic requires

$$\Delta P_0 = \Delta P_{max} + \kappa_1 Q_{f0} + \kappa_2 Q_{f0}^2. \tag{2.46}$$

Solving these two equations simultaneously yields a quadratic which can be solved to give

$$Q_{f0} = \frac{-(\kappa_1 - \mu R_m/A) - \sqrt{(\kappa_1 - \mu R_m/A)^2 - 4\kappa_2 \Delta P_{max}}}{2\kappa_2}. \tag{2.47}$$

Since κ_2 is negative, one always takes the negative root in the quadratic formula. Before going on to Example 2.8, it is worth looking at a special case of Eq. 2.43 and 2.44, namely constant pressure filtration. In that case, we can take $\kappa_1 = \kappa_2 = 0$ and $\Delta P_{max} \equiv \Delta P$, giving

$$\frac{dJ}{dt} = \frac{-\alpha\mu c J^3}{\Delta P}, \tag{2.48}$$

$$\frac{dJ}{dt} = \frac{-\alpha\mu c J^3/\Delta P}{1 - \frac{\mu^2 c V_f R_m J^2}{A\Delta P}\frac{d\alpha}{d\Delta P_c}}, \tag{2.49}$$

where it should be recalled that $J = Q_f/A$. Equation 2.48 suggests an alternative way of evaluating α from measurements of the flux and this is left for Problem 2.9. In Eq. 2.49, the dependence of α on ΔP_c potentially leads to more complex flux behaviour, which is explored in Problem 2.10.

Example 2.8 Calculation of time for centrifugal pump filtration

A centrifugal pump is used to pump a 10 g/l suspension of Talc-C through a filter of area 50 m^2 and resistance 6.4×10^{10} m^{-1}. The pump characteristic constants are $\Delta P_{max} = 2.02 \times 10^5$ Pa, $\kappa_1 = 3.64 \times 10^6$ Pa/m^3 and $\kappa_2 = -1.37 \times 10^9$ Pa/m^6. Taking the viscosity of the filtrate to be 0.001 Ns/m^2 and using specific resistance information from Table 2.2, calculate how long it will take to produce 50 m^3 of filtrate.

Solution. The MATLAB code for this problem is shown below. This uses **ode45** again and returns a time value of 9555 s.

```
function example28
global alpha0 kc m u rm c dpmax k1 k2 A Vffinal;
Vffinal=50; % desired filtrate volume
alpha0=4.7e10; kc=1.82e-4; m=0.55;
u=0.001; rm=6.4e10; c=10; A=50;
dpmax=2.02e5; k1=3.64e6; k2=-1.37e9;
Vf0=0 ; % initial condition
Qf0=(-(k1-u*rm/A)-((k1-u*rm/A)^2-4*k2*dpmax)^.5)/(2*k2)
options=odeset('events', @eventfcn);
[t,y]=ode45(@variable_rate,[0 24000],[Vf0 Qf0],options);
```

```
result=t(end)
function dydt=variable_rate(t,y)
global alpha0 kc m u rm c dpmax k1 k2 A;
% y(1) is Vf, y(2) is Qf
dp=dpmax+k1*y(2)+k2*y(2)^2; % calculate applied
pressure
dpc=dp-u*rm*y(2)/A; % calculate cake pressure drop
alpha=alpha0*(1+kc*dpc)^m; % calculate alpha
dalphadpc=alpha*kc*m/(1+kc*dpc); % calculate
derivative of alpha
Veqn=A*dp/u/(rm+alpha*c*y(1)/A); % ODE for Vf
top=-alpha*u*c*y(2)^3/dp/A^2; % break up Eq. 2.44
into manageable parts
term1=(k1+2*k2*y(2))*(y(2)/dp-
u*c*y(1)*y(2)^2/A^2/dp*dalphadpc)
term2=u^2*rm*c*y(1)*y(2)^2/A^3/dp*dalphadpc
Qeqn=top/(1-term1-term2) %ODE for Qf
dydt=[Veqn;Qeqn];
function [value, isterminal, direction]=eventfcn(t,y)
global Vffinal;
value = y(1)-Vffinal;
isterminal = 1;
direction = 0;
```

The advantage of the ODE approach is that it can be extended quite easily to situations where the pump characteristic is described by a higher order polynomial. The alternative approach, based on numerical integration of the integral in Eq. 2.39, is much more awkward because it is generally impossible to obtain an analytical expression for Q_f in terms of V_f. This is explored in Problem 2.11.

2.5   Continuous filtration

This chapter has so far focused exclusively on batch filtration but it is possible for dead-end filtration to be done on a continuous basis. This requires some mechanism for limiting cake accumulation. The most frequently used technique for doing this is to use rotary vacuum filters, shown schematically in Fig. 2.7.

Here, the filter cloth is on the surface of a rotating drum which is divided into a number of different sectors. A certain fraction of the drum is submerged at any given time. Vacuum is applied to a sector as it passes through the suspension in the trough into which suspension flows continuously. Filtrate is drawn into a central duct while cake forms on the drum surface. The suction in a given sector is switched off to facilitate

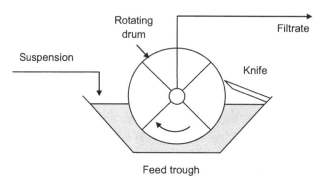

Figure 2.7 A rotary vacuum filter.

continuous cake removal with the 'knife' attachment. Washing and drying steps can also be included.

Rotary vacuum filters are a good example of the important role that mechanical engineering, especially machine design, plays in chemical and biochemical processes. Many process techniques are easy to conceptualise but require cleverly designed equipment to actually implement them in practice.

Rotary vacuum filters operate at constant pressure and therefore Eq. 2.27 can be used, albeit with some modification. Let us consider a single rotation of the drum and denote the time for one revolution to be t_r. Let the fraction of the drum surface that is in contact with the suspension be denoted f_s. (This fraction is known as the *submergence*.) This means that in any one revolution, the actual filtration time is $f_s t_r$. Therefore, the filtrate volume produced in one revolution is given by

$$\frac{\alpha \mu c}{2A^2 \Delta P} V_f^2 + \frac{\mu R_m}{A \Delta P} V_f = f_s t_r. \tag{2.50}$$

Now define the drum speed in revolutions per second to be $N (= 1/t_r)$ and the filtrate flowrate to be $Q_f = V_f/t_r$. Therefore

$$\frac{\alpha \mu c}{2A^2 \Delta P} Q_f^2 + \frac{\mu R_{mc}}{A \Delta P} Q_f N = f_s N, \tag{2.51}$$

where R_{mc} is the resistance of the membrane plus the resistance of residual cake left after the cake cutting step. This might be filter aid and/or some of the cake material itself. If R_{mc} is small relative to the resistance of the cake formed during submergence, Eq. 2.51 reduces to

$$Q_f = \left(\frac{2A^2 \Delta P f_s}{\alpha \mu c} N \right)^{1/2}. \tag{2.52}$$

Therefore, increasing the drum speed increases the filtrate flowrate. Use of Eq. 2.52 is illustrated in Example 2.9.

Example 2.9 Scale-up calculation for a rotary vacuum filter

A rotary vacuum filter with an area of 10 m^2 produces a certain filtrate flowrate at a drum speed of 1.2 rpm. An alternative filter is made available that has an area of 15 m^2. At what speed should this be operated to give the same flowrate?

Solution. This is a simple application of Eq. 2.52. That equation implies the scale-up relation

$$\frac{Q_{f2}}{Q_{f1}} = \frac{A_2}{A_1} \left(\frac{N_2}{N_1}\right)^{1/2}.$$

Inserting the relevant numerical information gives

$$1 = \frac{15}{10} \left(\frac{N_2}{1.2}\right)^{1/2}.$$

Therefore

$$N_2 = 1.2 \times \left(\frac{10}{15}\right)^2 = 0.53 \text{ rpm}.$$

2.6   Modelling product transmission in DEF

Dead-end filtration is a big subject in itself and there are many other topics that could be covered in this chapter. Some of these are touched upon in the problems but to close this chapter, the important issue of product transmission is examined. Product transmission refers to the rate at which the product passes through the membrane and is generally defined as the ratio of the product concentration on the filtrate side relative to its concentration on the feed side. This is an often neglected area of membrane filtration as most work tends to focus on the filtrate flux.

Transmission is a complex problem, especially if the product interacts with the membrane. This interaction causes not only a reduction in transmission but also a reduction in flux. It is a particular problem if the solute tends to form aggregates. A protein that is often used for research studies is bovine serum albumin (BSA), and it is well established that aggregates of this protein can cause severe fouling of the membrane, reducing flux and solute transmission. However, the situation is even more complex when one considers that while the product is passing through the filter, the cake is accumulating. The product might become entrapped in the cake, further reducing the transmission. Paradoxically, however, the formation of the cake can lead to improved fluxes because the cake acts as a filter aid which protects the membrane from being plugged by foulants [13].

In this section, a very simple, somewhat speculative, model of solute transmission is presented. The purpose here is merely to give the reader some insight into how one might go about building an initial model of a poorly understood phenomenon. The key assumptions in the model are: (i) that the solute does not interact with the membrane, and (ii) the solute becomes entrapped in the cake without affecting the specific cake resistance. The basic idea in the model is that the cake acts as a *depth filter*. Depth filters are commonly used for production of drinking water and are a type of packed bed, typically composed of solids like sand, where matter becomes entrapped, resulting in a solid-free filtrate. A simple equation often used in modelling of depth filters is the following [14]:

$$c_s = c_{s0}e^{-\lambda L}, \tag{2.53}$$

where c_s is the solute concentration leaving a depth filter of length, L, c_{s0} is the inlet concentration and λ is an empirical constant. For the purposes of this analysis, it is better to write this in terms of cake mass per unit area, i.e.,

$$c_s = c_{s0}e^{-\lambda M} \tag{2.54}$$

or

$$c_s = c_{s0}e^{-\lambda c V_f/A}. \tag{2.55}$$

Now if it is assumed that filtration is taking place at constant pressure, Eq. 2.27 can be used. Rearranging this gives

$$V_f = \frac{A R_m}{\alpha c} \left[\sqrt{1 + \frac{2\alpha c \Delta P t}{\mu R_m^2}} - 1 \right]. \tag{2.56}$$

Therefore the time-dependence of the solute transmission is given by

$$\frac{c_s}{c_{s0}} = \exp\left(-\frac{\lambda R_m}{\alpha} \left(\sqrt{1 + \frac{2\alpha c \Delta P t}{\mu R_m^2}} - 1 \right) \right). \tag{2.57}$$

Use of this equation is illustrated in Example 2.10 below.

Example 2.10 Modelling the dynamics of solute transmission in DEF

A suspension of a certain solid with a concentration of 25 g/l and constant specific cake resistance of 1×10^{12} m/kg is filtered at an applied pressure of 1 bar with a membrane of resistance 6×10^{10} m^{-1}. The filtrate viscosity can be taken to be 0.001 Ns/m^2 and $\lambda = 0.1$ m^2/kg. Generate a plot of solute transmission (Fig. 2.8) using Eq. 2.57.

Solution. A simple MATLAB script for this problem is shown below.

```
%Example 210
rm=6e10; alpha=1e12; c=25; p=1e5; u=0.001;
lambda=0.1;
```

```
t=linspace(0,3600)
trans=exp(-
lambda/alpha*rm.*((1+2*alpha*c*p/u/rm^2.*t).^0.5-1))
plot(t,trans,'-k');
xlabel('t (s)'); ylabel('Transmission');
```

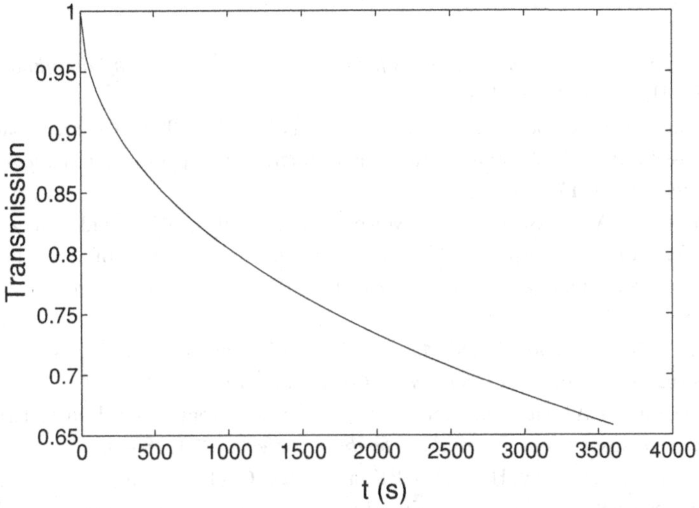

Figure 2.8 Model predictions for product transmission as a function of time.

2.7 Conclusions

In this chapter we have touched on some of the computational issues that arise with very simple models of dead-end filtration. Even allowing for the dependence of specific resistance on cake pressure drop, the equations that result do not really pose any great problems when modern software such as MATLAB is used. As long as one is willing and able to solve non-linear algebraic equations and ordinary differential equations using appropriate numerical techniques, most problems can be solved very rapidly indeed. Of course the big neglected factor in all of the discussion so far has been membrane fouling. In principle, as is seen in Chapter 10, membrane fouling does not pose any great computational problems even when it occurs simultaneously with cake formation. The real challenge is in developing accurate models either through pure theory or via experiment.

As mentioned earlier in this chapter, the theory of filtration presented here is quite simple; much more rigorous theories exist and a good starting point is the recent short paper by Tien and Ramarao [15]. In the next chapter, the related technique of cross-flow microfiltration is discussed and while some of the theory that has been used here can be adapted for the crossflow configuration, it will be seen that quite a few new and interesting concepts emerge. No new computational techniques are required, however.

References

1. Wilkes Q.W. (2005). *Fluid Mechanics for Chemical Engineers with Microfluidics and CFD*. Prentice Hall, New Jersey, USA.
2. McCarthy, A.A., Gilboy, P., Walsh, P.K. and Foley, G. (1999). Characterisation of cake compressibility in dead-end microfiltration of microbial suspensions. *Chemical Engineering Communications*, **173**, 79–90.
3. McCarthy, A.A., O'Shea, D.G., Murray, N.T., Walsh, P.K., and Foley, G. (1998). The effect of cell morphology on the filtration characteristics of the dimorphic yeast Kluyveromyces marxianus var. marxianus NRRLy2415. *Biotechnology Progress*, **14**, 279–285.
4. Nakanishi, K., Tadokoro, T. and Matsuno, R. (1987). On the specific resistances of cakes of microorganisms. *Chemical Engineering Communications*, **62**, 187–201.
5. Foley, G. (2006). A review of factors affecting filter cake properties in dead-end microfiltration of microbial suspensions. *Journal of Membrane Science*, **274**, 38–46.
6. McCarthy, A.A., Conroy, H., Walsh, P.K. and Foley, G. (1998). The effect of pressure on the specific resistance of yeast filter cakes during dead-end filtration in the range 30–500 kPa. *Biotechnology Techniques*, **12**, 909–912.
7. Tanaka, T., Tsuneyoshi, S.I., Kitazawa, W. and Nakanishi, K. (1997). Characteristics in cross-flow filtration using different yeast suspensions. *Separation Science and Technology*, **32**, 1885–1898.
8. Grace, H.P. (1952). Resistance and compressibility of filter cakes – 2. Under conditions of pressure filtration. *Chemical Engineering Progress*, **49**, 367–377.
9. Svarovsky, L. (2000). *Solid-Liquid Separation*, 4th Edition. Butterworth-Heinemann, Oxford, UK.
10. McCabe, W., Smith, J. and Harriott, P. (2005). *Unit Operations of Chemical Engineering*, 7th Edition. McGraw-Hill, Boston, US.
11. Hodgson, P.J., Leslie, G.L., Schneider, R.P., *et al.* (1993). Cake resistance and solute rejection in bacterial microfiltration: the role of the extracellular matrix. *Journal of Membrane Science*, **79**, 35–53.
12. Tiller, F.M. (1981). Filtering coal liquids: Clogging phenomena in the filtration of liquefied coal. *Chemical Engineering Progress*, **77**, 61–68.
13. Guell, G., Czekaj, P. and Davis, R.H. (1999). Microfiltration of protein mixtures and the effect of yeast on membrane fouling. *Journal of Membrane Science*, **115**, 113–122.
14. Rushton, A., Ward, A.S. and Holdich, R.G. (2000). *Solid-Liquid Filtration and Separation Technology*, 2nd Edition. Wiley-VCH, Weinheim, Germany.
15. Tien, C. and Ramarao, B.V. (2008). On the analysis of dead-end filtration of microbial suspensions. *Journal of Membrane Science*, **319**, 10–13.
16. McGuire, I., Coyle, K. and Foley, G. (2010). Specific cake resistances of yeast suspensions determined by dynamic and steady state methods. *Journal of Membrane Science*, **344**, 14–16.

Problems

Problem 2.1   Variation of cake pressure drop during cake formation

Let us look at the following simplified version of Eq. 2.19, where the specific resistance is a *linear* function of cake pressure drop:

$$\Delta P_c = \Delta P - \frac{\Delta P}{1 + \dfrac{\alpha_0 M}{R_m}(1 + k_c \Delta P_c)}.$$

Solve this equation to get an explicit equation for ΔP_c as a function of M. Generate a plot of $\Delta P_c / \Delta P$ versus M for $\Delta P = 25$ kPa, $\alpha_0 = 2 \times 10^{10}$ m/kg, $k_c = 1 \times 10^{-4}$ Pa^{-1}, with M values in the range 0–3.0 kg/m^2. Perform the calculations for $R_m = 1 \times 10^{10}$ m/kg and $R_m = 5 \times 10^{10}$ m/kg. Discuss the significance of your calculations for the correlation of specific resistance data and your understanding of constant pressure filtration.

Problem 2.2   Derivation of the Carman–Kozeny equation

Consider the flow of a fluid in a pipe of length L_p and diameter d_p. By doing a force balance on a cylindrical element of fluid, show that the wall shear stress, τ_w, is related to the pressure drop, ΔP, by the expression

$$\tau_w = \frac{d_p}{4} \frac{\Delta P}{L_p}.$$

Now consider a fluid flowing through a bed of particles where the bed porosity is ε and the surface area per unit volume of each particle is a_v. Ignoring wall effects, show that the shear stress at the surface of the *particles*, τ_b is given by

$$\tau_b = \frac{\varepsilon}{a_v(1 - \varepsilon)} \frac{\Delta P}{L_b},$$

where L_b is the bed length. Now, assuming the Hagen–Poiseuille equation for the mean velocity in pipe flow, i.e.,

$$u = \frac{d_p^2}{32\mu} \frac{\Delta P}{L_p},$$

deduce that the flux (i.e., flowrate per unit cross-sectional area or *superficial velocity*) through the bed can be written

$$J = \frac{\varepsilon^3 \rho_s}{K a_v^2(1 - \varepsilon)} \frac{\Delta P}{\mu M},$$

where M is the mass per unit area of the particles in the bed and ρ_s is the particle density. Confirm that this is consistent with Eq. 2.20.

Problem 2.3   Accounting for the liquid content of the cake

Show that the mass of solids per unit volume of filtrate, c, is related to the mass of solids per unit volume of suspension, c_s, by the expression

$$\frac{1}{c} = \frac{1}{c_s} - \frac{1}{\rho_s} - \frac{m_L/m_s}{\rho_L}$$

where ρ_s is the solids density, ρ_L is the liquid density, m_L is the mass of liquid in the cake and m_s is the mass of solid in the cake.

As an aside, it is worth mentioning that for filtration of powder suspensions, evaluating the mass of liquid and solid in the cake is not especially difficult because the wet cake can be weighed, dried and weighed again. This cannot be done with microbial suspensions as the particles themselves are composed mainly of water. Other approaches must be used [3].

Problem 2.4   Modifying the basic t/V_f versus V_f plot

(a) Change the lower limit on Eq. 2.26 to be $V_f = V_{f0}$ @ $t = t_0$ and repeat the integration.
(b) Given the raw data below, do a plot of t/V_f versus V_f and evaluate the slope of the line. Then, do a plot and regression based on your answer to part (a). Compare the slopes obtained in each case.

t (s)	1.0	4	19	69	151	263	407	582	788
V_f (ml)	5	18	40	70	100	130	160	190	220

Problem 2.5   Effect of applied pressure on constant pressure filtration time

Generate a plot of filtration time versus applied pressure using Eq. 2.27. Assume that the suspension is a filamentous *K. marxianus* suspension (see Table 2.1) with a concentration of 10 g/l. Assume 2 l of filtrate are required. Take the membrane resistance to be 1×10^{10} m^{-1}, the filter area to be 0.02 m^2 and explore pressures up to 300 kPa. Assume the filtrate viscosity is 0.001 Ns/m^2.

Problem 2.6   Cake washing post constant pressure filtration

Washing of filter cakes is an important aspect of filtration and is necessary to clean the solids, if they are the desired product, or to maximise the recovery of a valuable solute that may have become entrapped in the cake. In this problem, the focus is on computing the time required to flush a certain volume of water through the cake. More details on cake washing can be found in [14].

A 25 g/l suspension is filtered at a constant pressure of 50 kPa in a filter of area 0.1 m^2. The membrane resistance is negligible and $\mu = 0.001$ Ns/m^2. Assuming that the specific resistance of the cake can be predicted using the expression $\alpha = 4.0 \times 10^9 \Delta P^{0.3}$, where α is in m/kg and ΔP is in Pa, calculate

(a) the time required to produce 10 l of filtrate;
(b) the cake mass per unit area when 10 l of filtrate has been produced;
(c) the filtrate flux when 10 l of filtrate has been produced;
(d) the time required to wash the cake with 20 l of pure water.

We return to some aspects of this problem in Problem 2.15.

Problem 2.7   Constant pressure filtration followed by constant rate filtration

A 30 g/l suspension is to be filtered in a filter of area 0.002 m^2. The suspension is first filtered at a constant pressure of 50 kPa until 100 ml of filtrate is produced. The filter is then switched to constant rate operation until a *total* of 200 ml of filtrate has been produced.

Calculate

(a) the time at which the operation is changed from constant pressure to constant rate;
(b) the time taken to produce the 200 ml of filtrate;
(c) the total pressure at the end of the process.

Take $\mu = 0.001$ Ns/m^2 and assume that the cake has a constant specific resistance of 4×10^{11} m/kg. You may neglect the filter cloth resistance.

Problem 2.8   ODEs for filtration with a centrifugal pump
Derive Eqs. 2.43 and 2.44.

Problem 2.9   Evaluation of α by direct measurement of the filtrate flux
Explain how Eq. 2.48 can form the basis of a simple method to measure the specific cake resistance at constant applied pressure.

Problem 2.10   Evaluation of specific cake resistance from constant pressure flux data
Consider the constant pressure filtration of a microbial suspension whose specific resistance is given by

$$\alpha = 1.1 \times 10^{11}(1 + 1.93 \times 10^{-4}\Delta P_{\rm c})^{0.47}.$$

The membrane resistance is 1×10^{11} m^{-1}, the membrane area is 0.0017 m^2, the applied pressure is 50 kPa, the suspension concentration is 25 g/l and the filtrate viscosity is 0.001 Ns/m^2. Using Eq. 2.49, compute the flux as a function of time, for times up to 600 s and report your results as a plot of $1/J^2$ versus t. Comment on your results.

Problem 2.11   Calculation of filtration time in centrifugal pump filtration using numerical integration

(a) Show that it is generally impossible to obtain an analytical expression for $Q_{\rm f}$ as a function of $V_{\rm f}$ in centrifugal pump filtration.
(b) Based on your answer to part (a), explain how you could compute the integral in Eq. 2.39 without resorting to graphical integration but employing a data-based method such as the trapezoidal or Simpson's rule. Recalculate the process time from Example 2.8 using this method.

Further problems

Problem 2.12   The steady state method for measuring the specific cake resistance
A convenient way, albeit technically incorrect for very subtle reasons [15], of measuring the specific cake resistance over a range of pressures in one experiment is to use the so-called steady state method. The method involves filtering a volume, $V_{\rm s}$, of suspension of known concentration, $c_{\rm s}$, at a low pressure. The filtrate is collected and passed again through the filter cake at the same low pressure as before. The flux is calculated by

measuring the volume of filtrate collected in a known time. The pressure is increased in a step-wise fashion and as each new pressure is set, the flux is measured.

The following data were obtained in such an experiment. Use the data to generate a plot of specific cake resistance, α, versus pressure, ΔP.

ΔP (kPa)	30	60	100	150	200	300	400
Time (s) to collect 10 ml	146	118	105	102	105	104	103

Take the suspension concentration to be 15 g/l, the filtrate viscosity to be 0.001 Ns/m^2, the suspension volume to be 150 ml, the filter area to be 0.0017 m^2 and the membrane resistance to be 1.4×10^{10} m^{-1}.

As a final comment, it is worth noting that there is some evidence that the steady state method yields a value for α that is greater than the value obtained by the t/V_f versus V_f plot [16].

Problem 2.13   Accounting for the resistance of a filter-aid pre-coat

A yeast suspension is filtered and it is required to produce 100 ml of filtrate. Prior to filtration, the membrane is pre-coated by filtering 50 ml of a 12 g/l suspension of filter aid. The filter aid has a specific resistance of 1×10^{10} m/kg and the membrane has a resistance of 2×10^{10} m^{-1}. The specific cake resistance of the yeast suspension is 8.0×10^{11} m/kg at the relevant operating pressure.

Calculate how long it will take to carry out the *yeast filtration* assuming that the operating pressure is 35 kPa, the filtrate viscosity is 0.001 Ns/m^2, the yeast suspension concentration is 4 g/l and the membrane area is 0.0015 m^2.

Problem 2.14   Optimisation of dead-end filtration cycles

(a) Let us a consider a filtration cycle to consist of (i) the actual filtration of the suspension to produce a certain volume of filtrate, and (ii) a fixed down time, t_d, in which the filter is dismantled, the cake recovered (or discarded) and the membrane cleaned. Show that if filtration is done at constant pressure, the filtrate productivity, P (volume of filtrate produced per cycle time) can be expressed as

$$P = \frac{V_f}{t_d + K_1 V_f + K_2 V_f^2},$$

where K_1 and K_2 are constants. Hence derive expressions for the optimum filtrate volume, the optimum filtration time and the maximum productivity.

(b) Consider now a situation where, after filtration, a cake is to be washed with a volume, V_w, of water. Show that the time required to conduct a washing step can be written as

$$t_w = V_w(K_1 + 2K_2 V_f).$$

Hence show that when the membrane resistance is negligible, the filtration and washing times are related by the simple expression

$$\frac{t_w}{t_f} = 2\frac{V_w}{V_f}.$$

(Actually, this expression was first pointed out to me by one of my third year students, Jonathan Cawley, who used this approach to solve Problem 2.6.)

(c) Define the cycle productivity by the expression

$$P = \frac{V_f}{t_d + t_w + t_f}.$$

Show that if t_d and V_w are fixed, the optimum filtrate volume is given by

$$V_f = \sqrt{\frac{t_d + K_1 V_w}{K_2}}.$$

(d) Finally, repeat part (c) but this time assume that the volume of water is proportional to the volume of filtrate, i.e., $V_w = \beta V_f$, where β is a constant.

Problem 2.15 Centrifugal pump filtration with an incompressible cake

Consider a centrifugal pump filtration process where the cake is *incompressible*. Assume that the pump characteristic is given by Eq. 2.38.

(a) Derive an explicit expression for the filtrate flowrate as a function of filtrate volume. Hence express the filtration time as an integral of filtrate volume only.

(b) A centrifugal pump is used to pump a 10 g/l suspension through a filter of area 50 m^2 and membrane resistance 6.4×10^{10} m^{-1}. The characteristic constants are $\Delta P_{max} = 2.02 \times 10^5$ Pa, $\kappa_1 = 3.64 \times 10^6$ Pa/m^3 and $\kappa_2 = -1.37 \times 10^9$ Pa/m^6. Taking the viscosity of the filtrate to be 0.001 Ns/m^2 and the specific cake resistance to be 2×10^{10} m/s use your answer to part (a), Eqn. 2.39 and the MATLAB function **quad** to calculate how long it will take to produce 50 m^3 of filtrate.

Repeat part (b) but use the ODE approach employed in Example 2.8.

3 Crossflow microfiltration

3.1 Introduction

In Chapter 2 it was seen that dead-end filtration is characterised by the flow of suspension *normal* to the filter. In crossflow microfiltration (CFMF), the suspension flows *parallel* (or *tangential*) to the membrane. The rationale behind the crossflow mode of operation is that the 'scouring' action of the flow parallel to the membrane inhibits the growth of filter cake, thus creating the potential for high filtrate fluxes and steady state operation. CFMF is normally considered when the solids in a suspension prove difficult to separate in a dead-end filter or a centrifuge. Dead-end filtration is difficult when the particles are 'small' and have the tendency to form highly resistant and compressible filter cakes. Problems in centrifuging suspensions arise when the particles are small and have densities that are close to that of the suspending fluid. All of the above characteristics are typical of microbial suspensions. Consequently, crossflow systems have the potential to be economically viable for bacterial cells (as compared to centrifugation) and larger filamentous cells (as compared to pre-coat rotary vacuum filtration). CFMF is also useful for separating shear sensitive animal cells, for clarifying beverages such as beer and fruit juice, in separation of blood cells from plasma (plasmapheresis) and in sterile filtration of pharmaceuticals. Use of crossflow microfiltration as a key component of submerged membrane bioreactors (MBRs) is now commonplace. These devices combine biological waste treatment with membrane filtration, the performance of the latter being improved by the presence of the air bubbles required by the biological reactions. This use of air sparging to improve membrane filtration is discussed later in this chapter.

It is very important to remember that comparing dead-end and crossflow microfiltration is not really comparing like with like. In dead-end processes, a filter cake is recovered. An example of this is use of rotary vacuum filters in production of cakes of baker's yeast used in the bread making industry. In CFMF, however, no substantial cake is formed and the best that can be achieved is to *concentrate* a suspension. Both dead-end and crossflow techniques can, of course, be used for soluble product recovery.

3.2 Modes of operation

CFMF can be operated in any of the modes described in Chapter 1, although the continuous modes tend to be reserved for membrane bioreactors. The batch mode would typically be used in cell harvesting operations during downstream processing of a bioproduct. Here, the aim of the process is to reduce the suspension volume and increase the solids concentration. This might be necessary, for example, before the cell disruption step in the recovery of an intracellular product. A suspension washing (diafiltration) step can be added to remove impurities before performing the cell disruption. Batch operation can also be used for both extracellular and intracellular product recovery from cell and cell debris suspensions. Fed-batch operation is used if the retentate tank, which often comes as a component of a membrane filtration skid, has insufficient capacity for the feed volume. Diafiltration can be used to enhance the recovery of the product in the filtrate but this leads to a dilution of the product.

In this chapter, the basic characteristics of CFMF are described and some theories and models outlined. Then, design and analysis of batch, fed-batch and continuous operations are examined in detail. The diafiltration mode of operation is postponed until Chapter 6.

3.3 Flux characteristics during continuous operation

Figure 3.1 shows some typical laboratory data obtained during an experiment on the continuous CFMF of a yeast suspension with a tubular ceramic membrane. On a laboratory scale, continuous operation can be conveniently implemented using *total recycle operation*, i.e., both the retentate and filtrate are recycled into the feed reservoir.

From Fig. 3.1 it is clear that, in contrast to DEF, where the flux declines indefinitely (or at least until the capacity of the filter is reached), the crossflow flux approaches a non-zero steady state value. The time scale for this flux decline is long in the sense that

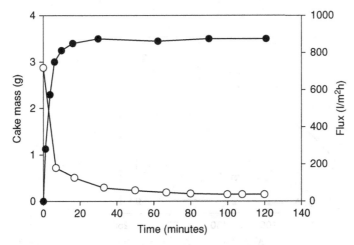

Figure 3.1 Flux and cake mass versus time during CFMF of *K. marxianus*. Solid symbols denote the cake mass per unit area while the open symbols represent the flux (data obtained in author's laboratory).

the dynamics of cake formation must be accounted for in dynamic modelling of CFMF processes. This is in contrast to ultrafiltration where the concentration polarisation layer – somewhat analogous to the cake in microfiltration – forms essentially instantaneously.

Before going on to describe the effects of various process parameters on the steady state flux, it is worth mentioning that in many real systems a true steady state is not reached and a long-term decline in flux is typically observed. This is more than likely due to membrane fouling by plugging of the pores of the membrane itself.

Figure 3.2 shows some typical data for pressure dependence of the flux at various yeast cell concentrations. Note that the simple term 'pressure' is used often in this chapter as shorthand for 'mean trans-membrane pressure', defined in Chapter 1.

Figure 3.2 shows that saturation-like behaviour is observed, especially at high-concentrations. This is similar to what is observed in ultrafiltration. However, with ultrafiltration, the saturation effect is much more definite and a pressure-independent flux is almost universally observed at sufficiently high pressures.

The steady state filtrate flux is generally reported to increase with crossflow velocity in a power law fashion, with exponents in the range 0.5–1.1. This variation partly relates to the fact that in some modules the suspension flow is laminar while in others it is turbulent, leading perhaps to different mechanisms and rates of cake removal. Indeed, the nature of the flow in crossflow microfiltration modules is often difficult to specify because spacer screens in flat sheet and spiral wound modules can promote turbulence even when the flow is nominally laminar.

The apparent dependence of the flux on crossflow velocity is also dependent on the relative importance of the cake resistance (velocity-dependent) and the membrane resistance (velocity-independent). These resistances can be comparable, especially if membrane fouling occurs. In addition, the precise effect of crossflow velocity on flux is dependent on particle properties such as mean size, size distribution and shape. Some of these phenomena are discussed later. All in all, the situation in ultrafiltration is much

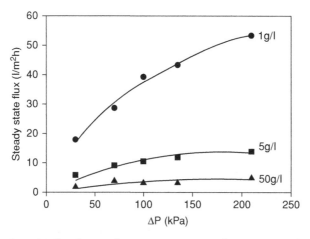

Figure 3.2 Steady state flux versus pressure at three concentrations for a yeast suspension (data redrawn from [1]).

'cleaner'. Indeed, the contrast between CFMF and ultrafiltration is a good example of how it is often possible to create accurate theories of solution behaviour while much greater challenges are posed by suspensions.

The steady state flux is found, not surprisingly, to decrease as the particle concentration, c, increases. Over a limited range of concentrations, it is usually found that the flux is proportional to c^{-s} where s is in the region 0.4–0.6. Inspired by ultrafiltration theory, some practitioners like to relate flux to $\ln c$ but there is no strong theoretical argument for doing this, even if it sometimes yields linear plots over a limited range of concentration.

The overall picture that emerges from studies on CFMF, especially those involving microbial suspensions, is that certain flux characteristics can be expected, albeit with much less universality than is found in ultrafiltration. When it comes to explaining flux data with quantitative, universally applicable models, CFMF lags considerably behind other membrane separation techniques, especially ultrafiltration.

In the next section, selected theories of CFMF are described and it is shown that while they all help to provide insight into the underlying mechanisms of CFMF, they remain a long way from being predictive.

3.4 Theories of crossflow microfiltration

There are a number of theories of crossflow microfiltration, some more rigorous than others. However, none of them can be considered to be of any great predictive use; rather they are a useful aid in trying to interpret experimental and process data. In this section, two theories are described; one is based on the concentration polarisation concept and the other based on an analysis of the forces acting on a single, isolated particle as it deposits on the membrane. Neither approach can be considered good enough to give accurate numerical predictions of fluxes. However, they do provide insight into the fundamental mechanisms of CFMF and give qualitatively accurate predictions of the effects of process variables such as crossflow velocity, particle concentration and particle diameter.

3.4.1 Concentration polarisation and shear-induced diffusion

As will be seen in Chapter 4, ultrafiltration theory, specifically concentration polarisation theory, is firmly rooted in mass transfer. Concentration polarisation is the idea that when a solution undergoes ultrafiltration, convection of solute towards the membrane leads to an accumulation of solute at the membrane leading, in turn, to *back-transport* of solute away from the membrane. A steady concentration gradient is established, essentially instantaneously. Working through the analysis leads to a simple expression for the flux. In UF, the analysis of concentration polarisation is very much couched in the language of mass transfer, incorporating entities like the molecular diffusion coefficient and the mass transfer coefficient.

A key question that arises in CFMF is: can one use the concentration polarisation concept, so successful in ultrafiltration, to understand crossflow microfiltration? The first and obvious point to note is that particles are not molecules, and there could be a problem with the whole concept of a diffusion coefficient. In fact, particles do diffuse due to Brownian motion, and Eq. 3.1 below, the Stokes–Einstein equation, can be used to predict the diffusion coefficient of an isolated particle:

$$D = \frac{k_b T}{3\pi \mu d},$$
(3.1)

where k_b is Boltzmann's constant, T is the absolute temperature, μ is the liquid viscosity and d is the particle diameter. However, if this equation is used in combination with standard concentration polarisation theory, the filtrate flux is predicted to be one to two orders of magnitude lower than typically found in experiments. In addition, the predicted dependence of the flux on particle diameter is completely wrong, i.e., the Stokes–Einstein approach predicts that the flux decreases with increasing particle size, whereas experiments show clearly that the flux increases with increasing particle size.

Early experiments into the phenomenon of *viscous resuspension* [2] suggested that a process other than Stokes–Einstein diffusion might be important in crossflow microfiltration. Viscous resuspension occurs when a settled layer of particles resuspends into the bulk fluid during shear flow. This resuspension of particles can be modelled successfully as a diffusion process involving a new type of diffusion known as *shear-induced diffusion*. The shear-induced diffusion coefficient is a strong function of particle volume fraction, indicating that particle–particle interactions are involved. It is also proportional to the wall shear rate and the square of the particle radius. Using the concept of shear-induced diffusion and following from previous work [3, 4], Davis and Sherwood [5] have derived the following expression for the flux in the (hypothetical) pressure-independent limit in CFMF:

$$J = 0.072\gamma_w \left(\frac{\phi_w r^4}{\phi_b L} \right)^{1/3},$$
(3.2)

where γ_w is the wall shear rate, ϕ_w is the particle volume fraction at the membrane ('wall'), ϕ_b is the bulk particle volume fraction, L is the membrane length and r is the particle radius. In deriving this expression, it is assumed that the suspension is dilute, the cake is incompressible, the cake resistance is very large in comparison with the membrane resistance and the cake thickness is negligible in comparison with the crossflow channel height.

The shear-induced diffusion approach is very elegant and seems to give a reasonably good description of the CFMF of suspensions of uniformly-sized, non-deformable spheres. However, it has not been fully tested on complex microbial suspensions where the cells may be non-spherical and form compressible cakes. Indeed, basic information on the shear-induced diffusion of microbial cells is very rare and very unlikely to be available. It is not an easy parameter to measure. For the present, therefore, shear induced diffusion models are a very valuable contribution to the science of suspension flows but not necessarily of great predictive value for process calculations.

Example 3.1 Flux calculation with the shear-induced diffusion model

Consider a suspension of particles of diameter 5 μm, volume fraction 0.001, being filtered in a module with a single tube of diameter (d_t) 3 mm and length 24 cm. The flowrate (Q) through the tube is 0.5 l/min. The suspension viscosity (μ_s) can be taken to be 0.003 Ns/m². The suspension density (ρ) is 1000 kg/m³. Assuming ϕ_w is 0.58, compute the filtrate flux. Express your answer in the units l/m²h.

Solution. First, the flow regime needs to be established. The Reynolds number can be computed from the expression

$$Re = \frac{4\rho Q}{\pi \mu_s d_t} = \frac{4 \times 1000 \times 0.5 \times 10^{-3}/60}{\pi \times 0.003 \times 3 \times 10^{-3}} = 1179.$$

Thus the flow in the channel is laminar. For laminar flow in a non-porous pipe, the wall shear rate in cylindrical geometry is given by

$$\gamma_w = 32Q/\pi d_t^3,$$

where Q is the tangential flowrate. It is assumed that this expression applies to the membrane, which is obviously porous, but usually this is not a major issue from a computational point of view. Putting numerical values into Eq. 3.2 gives

$$J = 0.072 \left(\frac{32 \times 0.5 \times 10^{-3}/60}{\pi \times (3 \times 10^{-3})^3}\right) \left(\frac{0.58 \times (2.5 \times 10^{-6})^4}{0.001 \times 0.24}\right)^{1/3} = 1.04 \times 10^{-4} \text{ m/s}.$$

Thus the flux in l/m²h becomes

$$J = 1.04 \times 10^{-4} \times 1000 \times 3600 = 371 \text{ l/m}^2\text{h}.$$

3.4.2 The force balance approach

One of the fascinating aspects of science and engineering is that, very often, completely different theoretical approaches can throw light on different aspects of system behaviour. Whereas the shear-induced diffusion model was fundamentally based on a diffusion process that depended upon particle interactions, force balance approaches tend to ignore those interactions completely. Nonetheless, some very useful ideas and predictions emerge from this simple approach. Let us consider a particle depositing on a membrane (or, more likely, depositing on a filter cake that has formed on the membrane) and acted upon by forces as indicated in Fig. 3.3.

In the analysis presented here, adhesive forces between particles and hydrodynamic forces acting normally and away from the membrane are neglected. A typical example of the latter type of force would be the *inertial lift* force which is responsible for the migration of solid particles away from the wall during pipe flow. This type of lift force is generally insignificant in CFMF, i.e., for particles less than about 10 μm in diameter [6].

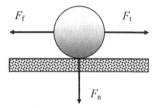

Figure 3.3 Force balance diagram for particle deposition.

Referring to Fig. 3.3, a particle remains deposited unless the tangential hydrodynamic force, F_t, tending to sweep the particle along the membrane exceeds the opposing frictional force, F_f. Thus the critical condition for particle deposition, i.e., the point at which particles begin to be removed by the tangential force, is

$$F_t = F_f. \tag{3.3}$$

The tangential force acting on the particle can be written as

$$F_t = \tau_w \pi d^2, \tag{3.4}$$

where τ_w is the fluid shear stress at the cake surface and d is the particle diameter. From basic physics, the frictional force can be written in terms of the normal force as

$$F_f = \eta F_n, \tag{3.5}$$

where F_n is the normal force 'pulling' the particle onto the membrane and η is the coefficient of friction between the particle and the filter cake or exposed membrane. Assuming the particles are neutrally buoyant (i.e. they have the same density as the pure liquid), the normal force can be calculated using the following modified form of Stokes Law [7]:

$$F_n = k_n \pi \mu d J \varphi^{-2/5}, \tag{3.6}$$

where φ is a dimensionless function of both the particle diameter and the resistance to filtrate flow while k_n is a constant. At the start of operation, when no cake has deposited,

$$\varphi = \frac{L_m}{R_m d^2}, \tag{3.7}$$

where L_m is the membrane thickness and R_m is the membrane resistance. Later in the run, when the cake resistance dominates,

$$\varphi = \frac{1}{K_h d^2}, \tag{3.8}$$

where K_h is the specific resistance of the cake based on the cake thickness (δ) rather than the cake mass per unit area (M), i.e., $K_h \delta = \alpha M$ where α is the specific resistance used in Chapter 2.

Consider, now, the CFMF of a *monodisperse* suspension with particle diameter, d. Monodisperse means that the particles all have the same diameter. At steady state, when the cake resistance dominates and the cake thickness is small relative to the channel width (i.e., τ_w is constant and equal to the clean channel value), the critical condition can be written

$$\eta k_n \pi \mu d J (K_h d^2)^{2/5} = \tau_w \pi d^2. \tag{3.9}$$

Rearranging gives

$$J = \frac{\tau_w d^{1/5}}{k_n \eta \mu K_h^{2/5}}. \tag{3.10}$$

Now, from Eq. 2.21, the Carman–Kozeny equation, the specific cake resistance is known to be proportional to d^{-2}. Therefore, Eq. 3.10 can be written as

$$J = \frac{\tau_w d}{k_n \eta \mu K_{h0}^{2/5}}, \tag{3.11}$$

where K_{h0} is a constant.

Equation 3.11 is the prediction for the steady state flux for the monodisperse suspension in question. Its obvious weakness is the fact that it neglects particle–particle interactions and thus cannot predict the effect of concentration on the flux. It is worth noting, however, that there is a hidden concentration-dependence in Eq. 3.11 in the sense that the wall shear stress is dependent on the suspension viscosity, which is dependent on the suspension concentration.

The force balance approach to modelling is actually of more interest if analysed in a slightly different way. At the beginning of the run, Eq. 3.7 applies and the critical force balance can be written:

$$k_n \eta \pi \mu d J_0 \left(\frac{R_m d^2}{L_m}\right)^{2/5} = \tau_w \pi d^2, \tag{3.12}$$

where J_0 is the initial flux. This means that *no particles will deposit*, even at time zero, if

$$J_0 < \frac{\tau_w d^{1/5}}{k_n \eta \mu} \left(\frac{L_m}{R_m}\right)^{2/5}. \tag{3.13}$$

The right-hand side of this equation is termed the critical flux, J_{crit}. The equation has significant implications. It means that if the initial flux is kept below the J_{crit} value predicted by Eq. 3.13, no particles will deposit on the membrane and the flux will remain constant at the clean-membrane value. This reasoning is what lies behind the influential idea of a *critical flux* [8]. There has been a burgeoning literature on this topic and various modifications made to the basic concept. It is now recognised that a true critical flux is an idealisation and is unlikely to exist for real suspensions which are not monodisperse and may contain particulate matter other than the dominant suspension particles, as well as soluble matter. The former might contribute to cake clogging while both the latter and the former might contribute to membrane fouling. This has led to the concept of *sustainable fluxes* [9], i.e., fluxes that show some decline, but at an acceptable level. The important point in all of this, however, is that for long-term operation it is often better to operate at low fluxes to prevent rapid deterioration of the membrane performance. This idea forms an important part of determining appropriate operating conditions for wastewater treatment processes, especially membrane bioreactors [10].

As a final point, it is important to recognise that the initial flux is determined by the trans-membrane pressure in accordance with the equation

$$J_0 = \frac{\Delta P}{\mu R_{\mathrm{m}}}, \tag{3.14}$$

and therefore the critical pressure for deposition is given by

$$\Delta P_{\mathrm{crit}} = \frac{\tau_{\mathrm{w}} d^{1/5}}{\eta k_{\mathrm{n}}} L_{\mathrm{m}}^{2/5} R_{\mathrm{m}}^{3/5}. \tag{3.15}$$

As long as the pressure is kept below the value predicted by Eq. 3.15, no flux decline will be observed, in theory. Of course, actually computing the critical pressure in practice is another issue altogether.

Example 3.2 Preferential deposition

Suppose a suspension has a range of particle sizes, i.e., it is a *polydisperse* suspension. Use the force balance approach to show that at any given time in a CFMF process, only particles greater than a certain cut-off value will deposit on the membrane. Hence show that the average particle size in the cake decreases with time during constant pressure operation.

Solution. Let us consider Eq. 3.9 in a slightly different way. Writing it as an inequality, we can say that a particle *will* remain deposited on the membrane if

$$k_{\mathrm{n}} \eta \pi \mu d J (K_{\mathrm{h}} d^2)^{2/5} > \tau_{\mathrm{w}} \pi d^2.$$

Therefore, since

$$K_{\mathrm{h}} = K_{\mathrm{h0}} d^{-2},$$

a particle of diameter, d, will deposit only if

$$d < \frac{k_{\mathrm{n}} \eta J K_{\mathrm{h0}}^{2/5}}{\tau_{\mathrm{w}}}.$$

Thus, the deposition process 'selects' the small particles in a suspension. The right-hand side of the above equation can be considered as a *cut-off diameter*, representing the particle size *above which* deposition will *not* occur. As filtration progresses, the flux, J, declines and so does the cut-off diameter. This means that the average particle size in the cake decreases, resulting in an increasing specific cake resistance. Furthermore, increasing the wall shear stress also reduces the cut-off diameter and, consequently, the average particle size in the cake. This raises the possibility that increasing the crossflow velocity will actually reduce the flux, a phenomenon for which there is some evidence, as discussed later.

The above example provides a nice simple explanation for preferential deposition of small particles. It also supplies one explanation for why a steady state is reached in CFMF.

Consider a polydisperse suspension. As CFMF proceeds, the cut-off diameter decreases until eventually it becomes smaller than the smallest particle in the suspension. At this point, according to the force balance model, deposition ceases and the flux remains constant [11].

3.4.3 The spatial dependence of the flux

One aspect of CFMF that has not been mentioned so far is the fact that in this configuration, the flux is actually a function of position within the channel. Indeed, in shear-induced diffusion theory, the flux predicted by Eq. 3.2 is actually a length-averaged flux – hence the explicit mention of the membrane length. In the force balance model, it is taken as a given that J represents an average flux for the membrane. In reality, the thickness of the cake varies with position within the channel. There are two reasons for this. First, mass balance considerations demand that the cake thickness increase with distance from the channel entrance. The basic picture is represented in Fig. 3.4. The particle volume fraction at the wall increases with position down the channel and when it reaches its maximum possible value, a stagnant cake is formed [3].

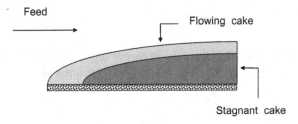

Figure 3.4 Variation of cake thickness with distance down the channel.

Second, there is a pressure drop down the length of the channel, meaning that the local trans-membrane pressure varies. The variation is affected by the precise way that the module is operated. Typically, a crossflow module has two exit ports for the filtrate. Depending on the positions of the valves on these exit ports, one may also have to consider the possibility of a pressure drop on the permeate side. The net result of this complexity is that one really needs to venture into the field of computational fluid dynamics, CFD [12], to fully describe the variation of cake thickness and flux down the channel. CFD is very much beyond the scope of this book. Thus, for the remainder of the book, including chapters on molecular filtration processes, the emphasis continues to be on developing and using models based on the *average flux* for the module.

3.5 A simple kinetic model of CFMF

The shear-induced diffusion and force balance approaches provide considerable insight into the fundamental mechanisms of CFMF, providing elegant explanations for experimental observations. However, it is very desirable to have a model that explicitly incorporates all of the key process parameters, including pressure, wall shear stress or crossflow velocity, particle concentration, membrane properties and cake properties. In particular, the model should be able explain the coupling between these parameters. By this we mean that it should be able to explain why flux behaviour depends not only on individual process parameters but on combinations of these parameters. Ideally, the model should be capable of describing both dynamic (time-dependent) behaviour and steady state behaviour and it should be capable of being modified to include membrane fouling.

These are ambitious targets to set for a new model, and the approach of engineers in dealing with a problem like this is to take a model of a similar, simpler process and modify it so that it applies to the new, more difficult process. In this section, one such model is developed. It is by no means a truly predictive model but it does provide some useful insight. Some of the material presented is a little bit speculative but it does illustrate how use of mathematical models can help one to gain a deeper, albeit incomplete, understanding of a subject through what is essentially a form of mathematical experimentation.

Let us return to a key equation in the analysis of batch DEF. This is Eq. 2.24, the solids balance, which is rewritten here as

$$AM = cV_f, \tag{3.16}$$

where A is the membrane area, M is the cake mass per unit area, c is the suspension concentration and V_f is the filtrate volume at time, t. Differentiating gives

$$\frac{dM}{dt} = cJ. \tag{3.17}$$

This equation states that the rate of accumulation of material per unit area of membrane is equal to the mass flux of material towards the membrane and is valid in the dead-end configuration only. Now, to make this apply to the crossflow configuration, it is modified as follows:

$$\frac{dM}{dt} = cJ - \psi_r, \tag{3.18}$$

where ψ_r is the rate of cake removal, the 'kinetic' term in the model. Taking reaction kinetics as a guide, a simple place to start is to assume a first order cake removal process. The zero order case is left for Problem 3.1. Thus, the following expression is obtained:

$$\frac{dM}{dt} = cJ - k_r \tau_w M, \tag{3.19}$$

where k_r is a cake removal constant and τ_w is the wall shear stress. This basic model has been 'reinvented' on a number of occasions but it probably appeared first in the work of Gutmann [13]. Interestingly, the first order kinetic term, including the dependence on wall shear stress, seems to have appeared originally in work by Kern and Seaton on modelling of heat exchanger fouling [14].

As the main parameter of interest is the filtrate flux, the following filtration equation should be recalled from Chapter 2:

$$J = \frac{\Delta P}{\mu(R_m + \alpha M)}, \tag{3.20}$$

where α is the specific cake resistance discussed extensively in Chapter 2 and μ is the filtrate viscosity. Note that Eq. 3.20 is technically valid for planar geometry. For tubular or hollow fibres, it should really be modified to take account of cake curvature. This is left for Problem 3.2. For the remainder of this chapter, cake curvature is neglected, a reasonable assumption when the cake thickness is small relative to the tube or fibre radius.

Combining Eq. 3.19 and Eq. 3.20 and differentiating while assuming all other parameters are constant gives

$$-\frac{\Delta P}{\alpha\mu J^2}\frac{dJ}{dt} = cJ - k_r\tau_w\left(\frac{\Delta P/\mu J - R_m}{\alpha}\right). \tag{3.21}$$

Tidying up and defining J_{rel} to be the flux at any time relative to the initial flux $(\Delta P/\mu R_m)$ gives

$$\frac{dJ_{rel}}{dt} = -\frac{\alpha c\Delta P}{\mu R_m^2}J_{rel}^3 - k_r\tau_w J_{rel}^2 + k_r\tau_w J_{rel}. \tag{3.22}$$

This is the basic differential equation describing flux dynamics during CFMF at constant pressure and particle concentration.

3.5.1   Steady state behaviour

At steady state, Eq. 3.22 gives

$$F J_{rel}^2 + J_{rel} - 1 = 0, \tag{3.23}$$

where F is a dimensional number defined by

$$F = \frac{\alpha c\Delta P}{\mu R_m^2 k_r\tau_w}. \tag{3.24}$$

It is always desirable that any dimensionless number that arises in engineering analysis should have a definite physical interpretation. The precise meaning of F is not entirely clear and in Section 3.6.2, it is shown that F is best understood as a combination of two dimensionless numbers with physical meanings.

Solving the quadratic that is Eq. 3.23 and doing some algebra gives the following expression for the relative steady state flux:

$$J_{rel} = \frac{2}{1 + \sqrt{1 + 4F}}. \tag{3.25}$$

The details of the algebra are left for Problem 3.3. Thus, the actual steady state flux can be written as

$$J = \frac{2\Delta P/\mu R_m}{1 + \sqrt{1 + 4F}}. \tag{3.26}$$

This is an interesting equation because it contains all the main process parameters and, to some extent, it explains the coupling between these parameters through the dimensionless number, F. This is illustrated in the example below.

Example 3.3 The effect of pressure on the steady state flux

Use Eq. 3.26 to discuss flux behaviour in the low pressure and high pressure regions.

Solution. At low pressures, $F \to 0$ and the flux is a linear function of ΔP. At high pressures, $F \gg 1$ and one gets

$$J = \frac{2\Delta P / \mu R_{\mathrm{m}}}{2F^{1/2}}.$$

Therefore

$$J = \left(\frac{k_{\mathrm{r}} \tau_{\mathrm{w}} \Delta P}{\alpha \mu c} \right)^{1/2}.$$

If the cake is incompressible, this means that $J \propto \Delta P^{1/2}$. If the cake is compressible with specific resistance given by Eq. 2.12, $J \propto \Delta P^{(1-n)/2}$ where n normally has values between 0 and 1. For very compressible cakes with $n \to 1$, the flux becomes independent of pressure at high pressure. Thus, the tendency of the flux to saturate at high pressures represents a combination of two factors; the underlying dynamics of cake formation in which cake removal is coupled to cake formation *and* cake compressibility. The precise pressure dependence that is observed depends in any experiment on the value of F and the compressibility of the cake. Since F incorporates the concentration, wall shear stress and membrane resistance, it becomes clear why such variety is observed in CFMF data.

Discussion of the effect of crossflow velocity, as predicted by the simple kinetic model, is left for Problem 3.4. Discussion of the effect of viscosity on the flux is left for Problem 3.5.

3.5.2 Dynamic behaviour

Equation 3.22 can actually be solved analytically to get an expression for J_{rel} and the integration is best done with a symbolic integrator such as the one employed in MATLAB's symbolic toolbox. This is illustrated in the next example.

Example 3.4 An analytical expression for flux dynamics

Starting from Eq. 3.22, use the MATLAB **Symbolic Toolbox** to derive an analytical expression for the relative flux.

Solution. Rearranging Eq. 3.22 gives

$$t = \int_{1}^{J_{rel}} \frac{d J_{rel}}{-a J_{rel}^3 - b J_{rel}^2 + bJ},$$

where

$$a = \frac{\alpha c \Delta P}{\mu R_m^2},$$

$$b = k_r \tau_w.$$

Simply enter the following commands into the command window:

```
syms a b J
int (1/J/(-a*J^2-b*J+b*J), J)
```

MATLAB returns *Warning: Explicit integral could not be found* and gives a very large 'piecewise expression'. This is really a list of a number of solutions, each valid for different combinations of a and b. Choosing the solution where

```
a <> 0 and b <> 0 and 4*a + b <> 0
```

leads to the following expression for the indefinite integral:

```
log(J)/b - log(a*J^2 + b*J - b)/(2*b) - atan((b + 2*J*a)/
(- b^2 - 4*a*b)^(1/2))/(- b^2 - 4*a*b)^(1/2)
```

After applying the limits on the integral and writing in a more readable format, one gets

$$t = \left(\frac{\ln J}{b} - \frac{\ln(a J^2 + bJ - b)}{2b} - \frac{\tan^{-1}\left(\frac{2a J + b}{\sqrt{-b^2 - 4ab}} \right)}{\sqrt{-b^2 - 4ab}} \right)$$

$$- \left(-\frac{\ln a}{2b} - \frac{\tan^{-1}\left(\frac{2a + b}{\sqrt{-b^2 - 4ab}} \right)}{\sqrt{-b^2 - 4ab}} \right)$$

Apart from being implicit, the above expression is algebraically complex and it is probably simpler to use a numerical approach with the MATLAB function **ode45**. This is illustrated in the example below.

Example 3.5 Numerical simulation of flux dynamics in CFMF

A 30 g/l suspension of a solid with a constant specific resistance of 1×10^{12} m/kg undergoes CFMF at 80 kPa. The membrane resistance is 1×10^{11} m^{-1} and the product, $k_r \tau_w$, is estimated as 0.005 s^{-1}. Take the filtrate viscosity to be 0.001 Ns/m^2. Plot the relative flux versus time and calculate the time required for the flux to drop to 30% of its initial value. Verify that your numerical calculations agree with the predictions of the analytical expression obtained in the last example.

Solution. The problem is solved numerically using **ode45** and sample code is given.

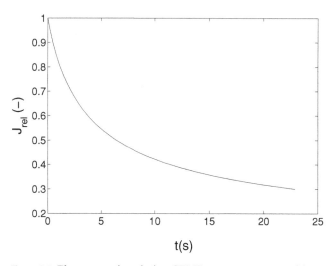

Figure 3.5 Flux versus time during CFMF at constant composition.

```
function example35
global a b;
alpha=1e12;DP=80e3;Rm=1e11;krtauw=0.005;u=0.001;c=30;
a=alpha*c*DP/u/Rm^2;
b=krtauw;
options=odeset('events', @eventfcn);
[t,J]=ode45(@concfmf,[0 1800],1.0, options);
% J is the relative flux in this code
%break up analytical expression for the time
term1=log(J)/b;
term2=-log(a*J.^2+b*J-b)/(2*b);
term3=-atan((b+2*a*J)/(-b^2-4*a*b)^.5)/(-b^2-4*a*b)^.5;
term4=-log(a)/(2*b);
term5=-atan((b+2*a)/(-b^2-4*a*b)^.5)/(-b^2-4*a*b)^.5;
tanal=(term1+term2+term3)-(term4+term5);
disp([t,tanal,J,]);
plot(t,J,'-k');
xlabel('t(s)'); ylabel('J_r_e_l (-)');
function dJdt=concfmf(t,J);
global a b;
dJdt=-a*J^3-b*J^2+b*J;
function [value, isterminal, direction]=eventfcn(t,J)
value = J-0.3;% stop when desired flux is reached
```

```
isterminal = 1;
direction = 0;
```

Figure 3.5 shows the time dependence of the flux and from the program output, the time required to reach 30% of the initial value is 22.911 s according to the analytical method and 22.875 s according to the numerical approach. This is pretty good agreement considering that no effort has been made to optimise the performance of **ode45**. The time involved is probably unrealistically short. Indeed, the prediction of an unrealistically rapid approach to steady state is a feature of this model that hints at some basic flaw.

3.5.3   Constant flux operation

Before going on to examine ways in which the kinetic model can be improved and linked with more rigorous theory, it is worth diverting briefly to have a look at constant flux (rate) operation. This is a mode of operation that is sometimes used in research studies. A useful characteristic of constant flux operation is that the total volume of filtrate that has passed through the membrane is easily computed, and this can be important in understanding membrane fouling.

It was seen in Chapter 2 that when a dead-end process is performed at constant flux, the applied pressure increases indefinitely. The precise shape of the increase depends on cake compressibility. Here, we examine constant flux operation in crossflow mode. The key equation is Eq. 3.19, which can be rearranged to give

$$t = \int_0^M \frac{dM}{cJ - k_r \tau_w M}.$$ (3.27)

Assuming J and τ_w are constant, this is easily integrated, ultimately leading to

$$M = \frac{cJ}{k_r \tau_w}(1 - e^{-k_r \tau_w t}).$$ (3.28)

Thus, unlike dead-end processes, the cake does not accumulate indefinitely but reaches a steady state value given by

$$M_{ss} = \frac{cJ}{k_r \tau_w}.$$ (3.29)

Using Eq. 3.20, the pressure at any given time is given by

$$\Delta P = \Delta P_m + \Delta P_c = \mu R_m J + \frac{\alpha \mu c J^2}{k_r \tau_w}(1 - e^{-k_r \tau_w t}),$$ (3.30)

where ΔP_m and ΔP_c are the pressure drops across the membrane and cake respectively. Therefore, even if the cake is incompressible, i.e., $\alpha = $ constant, the pressure shows a non-linear dependence on time. For compressible cakes where α is a function of ΔP_c, the behaviour is potentially quite complex. If the specific resistance is given by Eq. 2.18, one gets

$$\frac{\Delta P_c}{(1 + k_c \Delta P_c)^m} = \frac{\alpha_0 \mu c J^2}{k_r \tau_w}(1 - e^{-k_r \tau_w t}).$$ (3.31)

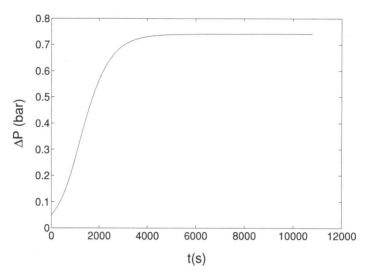

Figure 3.6 Pressure versus time during constant flux CFMF.

This is a non-linear algebraic equation and the time course of the cake pressure drop can be computed by repeatedly solving this equation at various times. However, as shown in Chapter 2, a potentially better approach might be to transform it into an ODE by differentiating with respect to time. Differentiating Eq. 3.31 gives

$$\frac{d\Delta P_c}{dt} = \frac{\alpha \mu c J^2 e^{-k_r \tau_w t}}{1 - \dfrac{\mu c J^2}{k_r \tau_w} \dfrac{d\alpha}{d\Delta P_c}(1 - e^{-k_r \tau_w t})} \tag{3.32}$$

The next example illustrates the predictions of this model.

Example 3.6 Simulation of constant flux operation

Consider a 10 g/l suspension whose specific resistance parameters $\alpha_0 = 4 \times 10^{11}$ m/kg, $k_c = 1.8 \times 10^{-4}$ Pa^{-1} and $m = 0.9$. The suspension is to be filtered at a constant rate of 5.0×10^{-5} m/s with a membrane of resistance 1×10^{11} m^{-1}. Assuming a filtrate viscosity of 0.001 Ns/m^2 and $k_r \tau_w$ to be 0.0015 s^{-1}, generate a plot of ΔP versus time until $t = 3$ hours (see Fig. 3.6).

Solution. MATLAB code to solve Eq. 3.32 is shown below.

```
function example36
global alpha0 kc m u rm j c krtau;
alpha0=4e11; kc=1.8e-4; m=0.9; u=0.001;rm=1e11;
j=5e-5; c=10; krtau=0.0015;
[t,pc]=ode45(@constant_rate,[0 10800],0);
p=(pc+u*rm*j)/1e5;
plot(t,p,'-k');
xlabel('t(s)'); ylabel('\DeltaP (bar)');
```

```
function dpcdt=constant_rate(t,pc)
global alpha0 kc m u j c krtau;
alpha=alpha0*(1+kc*pc)^m;
dalphadpc=alpha*kc*m/(1+kc*pc);
dpcdt=alpha*u*c*j^2*exp(-krtau*t)/(1-
u*c*j^2/krtau*dalphadpc*(1-exp(-krtau*t)));
```

3.5.4 Linking the kinetic model with force balance theory

One of the obvious issues that arise with the simple kinetic model is that it predicts cake formation to be a completely reversible process. Consider a situation where the cake mass has reached a certain value, M_f, and the filtrate valve is closed. This means that Eq. 3.19 becomes

$$\frac{dM}{dt} = -k_r \tau_w M. \tag{3.33}$$

Integrating gives

$$M = M_f e^{-k_r \tau_w (t - t_f)}, \tag{3.34}$$

where t_f is the time at which $M = M_f$. Thus, cake will eventually be completely removed and, in the absence of membrane fouling, the flux will return to its initial value. This is the basis of *crossflushing*, a useful technique that many experimentalists use routinely as part of membrane cleaning protocols [15]. Interestingly, the same exponential decrease in cake mass is predicted if instead of the filtrate valve being closed, the feed suspension is replaced by pure water, thus making $c = 0$ in Eq. 3.19. Most people with experience of working with membranes would agree that this is an unlikely scenario in practice and that cake, once formed, will tend to 'stick' to the membrane if the filtrate valve is not closed. Indeed, even an aggressive combination of crossflushing and air sparging (see Section 3.8.3 below) often cannot fully recover a cake. Some experimental evidence of this is shown in Table 3.1.

There is no doubt that a complete understanding of the irreversibility of cake formation requires some knowledge of particle–particle and particle–membrane interactions, especially the possibility of adhesive forces. However, some insight into irreversibility can be gained without introducing any new forces beyond those described in Section 3.4.2.

Table 3.1 Cake recovery during CFMF of *K. marxianus* suspensions [17].

Growth conditions	Cake recovery (%)
Batch culture / yeast extract medium	85
Batch culture / whey medium	91
Continuous culture / yeast extract medium	24
Continuous culture / whey medium	20

Consider the particles of a suspension depositing on a membrane at an initial flux that is greater than the critical flux. At this stage it is reasonable to suppose that there is no cake removal, and cake formation occurs exactly as in dead-end filtration. Now suppose that the flux declines until it reaches a certain *transition* value, J^*, at which time the tangential force exceeds the friction force. If the cake resistance is dominant at this point, the transition flux can be predicted from Eq. 3.11. Whereas the conventional force balance model assumes that the flux remains constant at this value, it is probably more likely that this point represents the beginning of cake removal. A possible way of advancing the model is to write

$$\frac{dM}{dt} = cJ - k_r\tau_w(M - M^*), \tag{3.35}$$

where M^* is the cake mass per unit area at the transition flux. Thus, the removal mechanism is assumed to act on the cake mass that is in excess of the transition value. Equation 3.35 predicts that if pure water is introduced into the feed at t_f when $M = M_f$ ($> M^*$), one finds

$$M = M^* + (M_f - M^*)e^{-k_r\tau_w(t-t_f)}. \tag{3.36}$$

Thus, M will not return to zero, but to M^*. On the contrary, if the permeate valve is closed, the model predicts that cake mass will still return to zero as there will be no normal force 'holding' the particles onto the membrane while $J = 0$. In effect, $M^* = 0$ when $J = 0$.

Example 3.7 Computing the steady state flux using the transition flux concept

Assuming that Eq. 3.35 applies to the cake removal phase, derive an expression for the steady state flux in terms of the transition flux. Assume that the membrane resistance is negligible at steady state.

Solution. At steady state

$$cJ_{ss} - k_r\tau_w(M_{ss} - M^*) = 0.$$

Assuming the membrane resistance is negligible, the flux is related to the cake mass by the expression

$$J_{ss} = \frac{\Delta P}{\alpha\mu M_{ss}},$$

where it has been assumed that the specific resistance of the 'dead-end' portion of the cake is the same as that of the cake formed during the cake removal phase. Thus

$$\frac{c\Delta P}{\alpha\mu M_{ss}} - k_r\tau_w(M_{ss} - M^*) = 0.$$

This can be rearranged to give the following quadratic:

$$M_{ss}^2 - M^* M_{ss} - \frac{c\Delta P}{\alpha\mu k_r\tau_w} = 0.$$

Solving gives

$$M_{ss} = \frac{M^*}{2} + \frac{1}{2}\sqrt{(M^*)^2 + \frac{4c\Delta P}{\alpha\mu k_r\tau_w}}.$$

This can be shown, after a little rearranging, to give

$$J_{ss} = \frac{2J^*}{1 + \sqrt{1 + \frac{4\alpha\mu c(J^*)^2}{k_r\tau_w\Delta P}}},$$

where

$$J^* = \frac{\Delta P}{\alpha\mu M^*}.$$

The transition flux is, in theory, predicted by Eq. 3.11, thus providing a nice link between the force balance model and the kinetic model. In Problem 3.6, you are asked show that the steady state flux is ultimately related to the *critical flux* by the expression

$$J_{ss} = \frac{2\zeta J_{crit}}{1 + \sqrt{1 + \frac{4\alpha\mu c\varsigma^2 J_{crit}^2}{k_r\tau_w\Delta P}}},$$

where ζ is a constant.

3.5.5 Flux dynamics with a transition flux

Previously, in Example 3.5, the dynamics of the relative flux were examined using the un-modified kinetic model. The following example illustrates use of the modified model.

Example 3.8 Simulation of flux dynamics while incorporating the transition flux

A 30 g/l suspension of a solid with a constant specific resistance of 1×10^{12} m/kg undergoes CFMF at 80 kPa. The membrane resistance is 1×10^{11} m^{-1} and the product, $k_r\tau_w$, is estimated to be 0.002 s^{-1}. Take the filtrate viscosity to be 0.001 Ns/m^2 and assume that the relative transition flux is 0.3, i.e., cake removal begins when the flux has dropped to 30% of its initial value. Report your findings as a plot of $1/J_{rel}^2$ versus time. You may recall from Chapter 2 that in DEF such a plot is linear if the cake is incompressible.

Solution. The MATLAB code for solving this problem is very little different from that used in Example 3.5. The only changes to be made involve inserting an **if** statement to ensure that cake removal begins at the appropriate flux value. Unlike Example 3.5, which deals in fluxes only, this code solves the ODE for the cake mass directly.

```
function example38
global alpha DP Rm u c;
alpha=1e12;DP=80e3;Rm=1e11;u=0.001;c=30;
```

```
[t,M]=ode45(@transflux,[0 180],0)
% J represents the actual flux in this code
J=DP/u./(Rm+alpha*M);
Jrel=J/(DP/u/Rm);
y=1./Jrel.^2;
plot(t,y,'-k');
xlabel('t(s)'); ylabel('1/J_r_e_l^2 (s/m)^2');
function dMdt=transflux(t,M);
global alpha DP Rm krtauw u c;
J=DP/u/(Rm+alpha*M);
Jstar=0.3*DP/u/Rm;
Mstar=Rm/alpha*(DP/(u*Rm*Jstar)-1);
if J > Jstar
krtauw=0;
else
krtauw=0.002;
end
dMdt=c*J-krtauw*(M-Mstar);
```

Running the code gives the result in Fig. 3.7 and shows some departure from linearity due to cake removal.

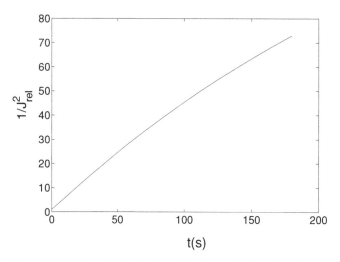

Figure 3.7 Variation of relative flux with time indicating the effect of the cake removal term.

3.6   Process design and analysis

Despite the obvious limitations of the kinetic model, it is useful to use it to examine some real process configurations. The material presented here is more an exercise in mathematical modelling than true process design and analysis. This is in contrast to design and analysis of ultrafiltration systems, which is the subject of Chapter 5. The focus here is on continuous feed-and-bleed, batch and fed-batch systems. The aim is to show how simple modelling can provide valuable insights into system behaviour. Suspension washing (diafiltration) is left for Chapter 6.

3.6.1   Analysis of a continuous feed-and-bleed system

The process layout for a single-stage feed-and-bleed system is shown in Fig. 3.8.
   A solids balance gives

$$Q_0 c_0 = Q_1 c_1, \tag{3.37}$$

where Q_0 represents the volumetric flowrate of the feed, Q_1 represents the flowrate of the retentate while c_0 and c_1 represent concentrations in the relevant streams. A liquid balance gives

$$Q_0 = Q_1 + JA, \tag{3.38}$$

where J is the flux in the system and A is the membrane area. Now, a key feature of feed-and-bleed systems is that they are well-mixed, i.e., the retentate concentration is fully representative of the uniform solids concentration in the module itself. Therefore, in this case, the flux is determined by c_1. In general, it is reasonable to assume that the flux is a power function of the concentration, at least at high pressures, i.e.,

$$J = b c_1^{-n}, \tag{3.39}$$

where b is a constant that depends on trans-membrane pressure and crossflow velocity and n is an empirical constant. In the case of the kinetic model described earlier, n is equal to 0.5 at high pressures. Combining these three equations gives

$$\frac{Q_0}{A} \left(1 - \frac{c_0}{c_1} \right) - b c_1^{-n} = 0. \tag{3.40}$$

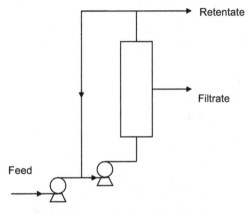

Figure 3.8 Continuous feed-and-bleed CFMF.

Now, for consistency with Chapter 5 where this notation is used extensively, this equation is now written

$$\frac{Q_0}{A}(1 - x_1) - bc_0^{-n}x_1^n = 0,$$

(3.41)

where x_1 is defined by

$$x_1 = \frac{c_0}{c_1}.$$

(3.42)

Thus, analysis of a continuous feed-and-bleed system requires solution of a non-linear algebraic equation which, as we have seen on a number of occasions, does not pose any great computational challenges. This analysis can be extended to multi-stage systems without any difficulty. There, the retentate from one stage forms the feed to the next and, as will be seen in Chapter 5, there are considerable advantages to using the multi-stage approach. In that chapter, solution of the non-linear algebraic equations governing continuous ultrafiltration is covered extensively.

3.6.2 Modelling the dynamics of batch CFMF

The basic layout of a simple batch membrane filtration process is recalled in Fig. 3.9.

In a batch system, the retentate is returned to the feed tank while the permeate is continuously withdrawn. Thus, the retentate volume declines and the particle concentration increases. In this section, the evolution of the flux is modelled and the role of changes in fluid viscosity explored. In particular, the important modelling technique of non-dimensionalisation is demonstrated.

A key part of the modelling of batch systems is the assumption that the particle concentration is the same throughout the system. This is generally a good assumption for batch operation. Consider the retentate volume and particle concentration at any time to be V and c respectively. A particle balance for the system can be written

$$\frac{d(Vc)}{dt} = -A\frac{dM}{dt},$$

(3.43)

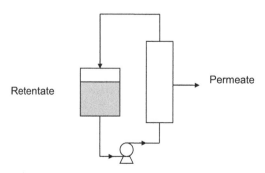

Figure 3.9 Simple batch operation.

where A is the membrane area and M is the cake mass per unit area at time, t. The volume balance for the system can be written

$$\frac{dV}{dt} = -JA. \tag{3.44}$$

For the un-modified kinetic model, cake formation is described by

$$\frac{dM}{dt} = cJ - k_r \tau_w M. \tag{3.45}$$

Combining these three equations gives

$$\frac{dc}{dt} = \frac{A k_r \tau_w M}{V}. \tag{3.46}$$

Now, the wall shear stress can be related to the mean crossflow velocity by the expression

$$\tau_w = \frac{1}{2} \rho u^2 f, \tag{3.47}$$

where f is the Fanning friction factor and u is the mean fluid velocity. For tubular geometry, the wall shear stress is found from basic fluid mechanics to be

$$\tau_w = \frac{8 \rho Q^2 f}{\pi^2 d_t^4}, \tag{3.48}$$

where d_t is the tube diameter, ρ is the fluid density and Q is the tangential flowrate.

Technically, allowance should be made for the fact that the effective diameter of the tube is reduced by the growth of the cake. To keep things simple, this complication is ignored here but explored in Problem 3.8 which is probably best done after doing Example 3.9. Neglecting cake curvature, i.e., assuming a thin cake, the flux is written as before, namely

$$J = \frac{\Delta P}{\mu(R_m + \alpha M)}. \tag{3.49}$$

Rearranging gives

$$M = \frac{\Delta P / \mu J - R_m}{\alpha}. \tag{3.50}$$

Therefore Eq. 3.46 becomes

$$\frac{dc}{dt} = \frac{A k_r \tau_w}{\alpha V} \left(\frac{\Delta P}{\mu J} - R_m \right). \tag{3.51}$$

Combining Eqs. 3.45 and 3.50, as done previously, gives

$$\frac{dJ}{dt} = -\frac{\alpha \mu c}{\Delta P} J^3 - \frac{k_r \tau_w \mu R_m}{\Delta P} J^2 + k_r \tau_w J. \tag{3.52}$$

Thus, the core of the model is a system of three ordinary differential equations in V, c and J, i.e., Eqs. 3.44, 3.51 and 3.52.

Ideally, one would like to set up a model that encompasses all types of flow, from fully turbulent to laminar. In theory, one should account for the roughness of the cake and the porosity of the membrane and cake. To keep things simple, it is assumed that the cake (and membrane) is smooth and equations developed for non-porous pipes can be used. For this scenario, Churchill has developed the following, imposing correlation (*Re* is the Reynolds number) for smooth pipes that applies to all flow regimes, including laminar, transition and turbulent [16]:

$$f = 2 \left[(8/Re)^{12} + \{ (-2.457 \ln(7/Re)^{0.9})^{16} + (37530/Re)^{16} \}^{-3/2} \right]^{1/12}. \quad (3.53)$$

Figure 3.10 is a graphical representation of this equation, the classic Moody diagram. The fact that it captures the transition from turbulent to laminar flow is particularly useful for simulation purposes but it should be stressed that prediction of friction factors in transition flow is somewhat fraught with uncertainty. Transition flows contain laminar and turbulent regions and the precise nature of the mix of flow regimes is very system-specific. Nonetheless, Eq. 3.53 is a very convenient route to predicting wall shear stress in membrane systems and evaluating, in a semi-quantitative way, the effect of changes in fluid rheology on the system performance.

For predicting the friction factor, the Reynolds number can be conveniently written as

$$Re = \frac{4\rho Q}{\pi \mu_s d_t}, \quad (3.54)$$

where μ_s is the *suspension* viscosity and ρ is the suspension density, which is assumed to be the same as the pure liquid throughout. For the purposes of this simulation it is assumed that the suspension viscosity is given by a simple equation of the form

$$\mu_s = \mu e^{\gamma c}, \quad (3.55)$$

where γ is an empirical constant.

When carrying out simulations to explore a new phenomenon, it is often useful to put equations in non-dimensional form. This can provide extra physical insight and reduces the number of numerical parameters that must be specified in simulations. It also provides greater clarity as to the validity of any assumptions that might be made

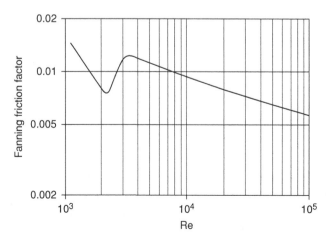

Figure 3.10 Friction factor *versus* Reynolds number for a smooth pipe.

subsequently. Putting a series of differential equations into dimensionless form is usually straightforward but must be done with care. It also requires some judgment as to how precisely to define the various dimensionless variables. The basic task is to recast both the dependent variables and the independent variable in dimensionless form, i.e., as ratios. For example, a logical way of non-dimensionalising a flux is to express it as a ratio of the actual flux to the initial flux, as was done in Section 3.5 when the relative flux was defined. The result of this process is that various parameters are 'left over' and it is always possible to combine these to form dimensionless groups. If the non-dimensionalisation has been done appropriately (and this takes judgement), these groups will have relevant physical meanings.

For this model, the following dimensionless variables are defined:

$$V_{\text{rel}} = V/V_0, \tag{3.56}$$

$$c_{\text{rel}} = c/c_0, \tag{3.57}$$

$$J_{\text{rel}} = J/J_0 = J/(\Delta P/\mu R_{\text{m}}), \tag{3.58}$$

$$t_{\text{rel}} = t/(V_0/A J_0), \tag{3.59}$$

where the '0' subscript denotes initial values. Putting these definitions into the model equations transforms the problem into one of solving the following three differential equations:

$$\frac{dc_{\text{rel}}}{dt_{\text{rel}}} = \left(\frac{k_{\text{r}}\tau_{\text{w}}R_{\text{m}}}{\alpha c_0 J_0}\right)\left(\frac{1}{J_{\text{rel}}} - 1\right) \bigg/ V_{\text{rel}}, \tag{3.60}$$

$$\frac{dJ_{\text{rel}}}{dt_{\text{rel}}} = -\left(\frac{\alpha c_0 V_0}{A R_{\text{m}}}\right)c_{\text{rel}}J_{\text{rel}}^3 - \left(\frac{k_{\text{r}}\tau_{\text{w}}V_0}{A J_0}\right)J_{\text{rel}}^2 + \left(\frac{k_{\text{r}}\tau_{\text{w}}V_0}{A J_0}\right)J_{\text{rel}}, \tag{3.61}$$

$$\frac{dV_{\text{rel}}}{dt_{\text{rel}}} = -J_{\text{rel}}. \tag{3.62}$$

The terms in brackets are dimensionless numbers, the group in Eq. 3.60 being the inverse of the F number defined in Eq. 3.24. However, if it is noted that

$$\left(\frac{k_{\text{r}}\tau_{\text{w}}R_{\text{m}}}{\alpha c_0 J_0}\right) = \left(\frac{k_{\text{r}}\tau_{\text{w}}V_0}{A J_0}\right) \bigg/ \left(\frac{\alpha c_0 V_0}{A R_{\text{m}}}\right), \tag{3.63}$$

it becomes clear that there are actually only two independent dimensionless groups in the model. The dimensionless model equations are now written as

$$\frac{dc_{\text{rel}}}{dt_{\text{rel}}} = \frac{N_2}{N_1}\left(\frac{1}{J_{\text{rel}}} - 1\right) \bigg/ V_{\text{rel}}, \tag{3.64}$$

$$\frac{dJ_{\text{rel}}}{dt_{\text{rel}}} = -N_1 c_{\text{rel}}J_{\text{rel}}^3 - N_2 J_{\text{rel}}^2 + N_2 J_{\text{rel}}, \tag{3.65}$$

$$\frac{dV_{\text{rel}}}{dt_{\text{rel}}} = -J_{\text{rel}}, \tag{3.66}$$

where

$$N_1 = \frac{\alpha c_0 V_0}{A R_{\text{m}}}, \tag{3.67}$$

$$N_2 = \frac{k_{\text{r}}\tau_{\text{w}}V_0}{A J_0}. \tag{3.68}$$

Technically, N_1 is the resistance of a cake formed by deposition of all the particles in the retentate tank, relative to the membrane resistance. The number, N_2, is the cake removal rate if all the particles in the retentate have deposited, relative to the initial particle deposition rate. These interpretations are the subject of Problem 3.7. By their very nature, these groups have wide ranges of possible values, depending on the filtration properties of the particles, the operating conditions and the volume and concentration of the suspension.

The dimensionless number, N_2, varies during CFMF because of the dependence of wall shear stress on suspension viscosity. From Eq. 3.48, the following equation applies at fixed Q and d_t:

$$N_2 = N_{20}\frac{f}{f_0}, \tag{3.69}$$

where the zero subscript denotes the value of the relevant parameter based on the initial viscosity. Changes in friction factor arise due to changes in Reynolds number and this effect can be quantified by

$$Re = Re_0\left(\frac{\mu_{s0}}{\mu_s}\right) = Re_0 e^{-N_3(c_{rel}-1)}, \tag{3.70}$$

where $N_3 = \gamma c_0$.

In the example below, the predictions of this model are explored with special attention being paid to the role of increasing suspension viscosity and the transition from turbulent to laminar flow.

Example 3.9 Dynamic modelling of batch CFMF

A cell suspension is to undergo batch CFMF. The volume is to be reduced by a factor of 20 and the values of the various dimensionless groups are as follows: $N_1 = 10$, $N_{20} = 10$, $N_3 = 0$ or 0.1 and $Re_0 = 6000$. For $N_3 = 0.1$, plot J_{rel} versus t_{rel} and comment on any interesting features. For both values of N_3, note the time taken to complete the operation.

Solution. MATLAB code for solving this problem (for $N_3 = 0.1$) is shown below

```
function example39
options=odeset('events', @eventfcn);
%'rel' notation dropped throughout
[t,y]=ode45(@batchcfmf,[0 50],[1, 1, 1], options);
c=y(:,1);J=y(:,2);
plot(t,J,'-k');
xlabel('t_r_e_l'); ylabel('J_r_e_l ');
function dydt=batchcfmf(t,y)
N1=10;N20=10;N3=0.1;Re0=6000;
% y(1) is c, y(2) is J, y(3) is V
A0=(-2.457*log((7/Re0)^0.9))^16;
```

```
B0=(37530/Re0)^16;
f0=2*(((8/Re0)^12+1/(A0+B0))^1.5)^(1/12);
Re=Re0*exp(-N3*(y(1)-1));
A=(-2.457*log((7/Re)^0.9))^16;
B=(37530/Re)^16;
f=2*((8/Re)^12+1/(A+B)^1.5)^(1/12);
N2=N20*f/f0;
ceqn=N2/N1*(1/y(2)-1)/y(3);
Jeqn=-N1*y(1)*(y(2)^3)-N2*(y(2)^2)+N2*y(2);
Veqn=-y(2);
dydt=[ceqn;Jeqn;Veqn];
function [value, isterminal, direction]=eventfcn(t,y)
%volume to be reduced to V0/20
value = y(3)-0.05;
isterminal = 1;
direction = 0;
```

The output from the program for $N_3 = 0.1$ is shown in Fig. 3.11. The relatively sudden drop in flux close to the end of the run (around $t_{rel} = 1.75$) reflects the reduction in friction factor as the flow passes into the transition flow region. At the very end of the simulation one can begin to see an upturn in the flux, where the friction factor starts to increase as the flow enters the laminar region. The time taken with $N_3 = 0.1$ is 1.897, while when $N_3 = 0$, it is 1.922. This reflects the lower shear stresses that arise under conditions of constant suspension viscosity. Incidentally, when N_3 is zero, there is no sudden drop in flux near the end of the run.

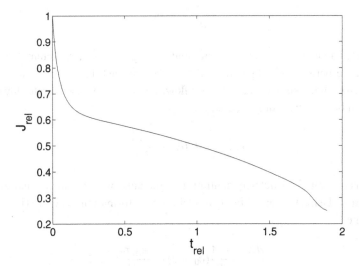

Figure 3.11 Flux versus time in batch CFMF.

3.6.3   Fed-batch operation

The basic layout of fed batch CFMF is recalled in Fig. 3.12.

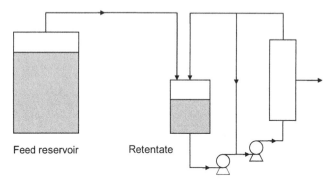

Feed reservoir   Retentate

Figure 3.12 Fed-batch configuration with recirculation loop.

The fed-batch approach can be used if the membrane module is available as part of a skid that has a retentate tank of limited capacity. In fed-batch mode, the feed suspension is pumped into the retentate tank from a larger storage reservoir. In the simplest scenario, the flowrate of feed into the retentate tank exactly equals the filtrate flowrate and thus the volume in the retentate tank is constant. The key difference between the modelling of batch and fed-batch systems is the fact that the concentration and volume balances are altered. The flux equation and the cake formation equation remain unchanged. The non-dimensionalised ordinary differential equation for the retentate volume is simply

$$\frac{dV_{\text{rel}}}{dt_{\text{rel}}} = 0. \tag{3.71}$$

The particle balance states that the accumulation of particles in the retentate is equal to the difference between the input from the external storage tank and the accumulation of particles in the cake. Noting that feed flowrate is equal to the filtrate flowrate, this balance can be written in dimensional form as

$$V\frac{dc}{dt} = JAc_0 - A\frac{dM}{dt}, \tag{3.72}$$

where c_0 is the initial particle concentration in the tank, which is also equal to the feed concentration. Using the same kinetic model for cake formation as before (i.e. Eq. 3.45), this becomes

$$\frac{dc}{dt} = \frac{JA}{V}(c_0 - c) + \frac{Ak_{\text{r}}\tau_{\text{w}}M}{V}. \tag{3.73}$$

Non-dimensionalising as before gives

$$\frac{dc_{\text{rel}}}{dt_{\text{rel}}} = \frac{J_{\text{rel}}}{V_{\text{rel}}}(1 - c_{\text{rel}}) + \frac{N_2}{N_1 V_{\text{rel}}}\left(\frac{1}{J_{\text{rel}}} - 1\right). \tag{3.74}$$

Finally, we need to account for depletion of the external feed, whose volume is denoted by W. This can be expressed by the following:

$$\frac{dW}{dt} = -AJ. \tag{3.75}$$

In dimensionless form it can be written

$$\frac{dW_{\text{rel}}}{dt_{\text{rel}}} = -J_{\text{rel}}, \tag{3.76}$$

where W has been non-dimensionalised with respect to V_0. The following example illustrates the use of this model.

Example 3.10 Dynamic modelling of fed-batch CFMF

A cell suspension is to undergo fed-batch CFMF. The retentate tank contains a dimensionless volume $V_{\text{rel}} = 0.05$ while the feed tank contains a dimensionless volume $W_{\text{rel}} = 0.95$. The process is to be stopped when the feed tank is emptied, thus making this example a fed-batch analogue of Example 3.9 because the suspension volume is ultimately decreased by a factor of 20. Use the same values of N_1, N_{20} as in Example 3.9 and use $N_3 = 0$. Calculate the time required to reach the process objective of emptying the external feed tank. Compare this with the time computed in the previous example.

Solution. MATLAB code for performing the calculation is shown below.

```
function example310
options=odeset('events', @eventfcn);
%'rel' notation dropped throughout
[t,y]=ode45(@fedbatchcfmf,[0 50],[1, 1, 0.05, 0.95], options);
c=y(:,1);J=y(:,2);V=y(:,3);W=y(:,4);
disp([t,J,V,W]);
function dydt=fedbatchcfmf(t,y)
N1=10;N20=10;N3=0.0;
% y(1) is C, y(2) is J, y(3) is V
Re0=6000;
A0=(-2.457*log((7/Re0)^0.9))^16;
B0=(37530/Re0)^16;
f0=2*(((8/Re0)^12+1/(A0+B0))^1.5)^(1/12);
Re=Re0*exp(-N3*(y(1)-1))
A=(-2.457*log((7/Re)^0.9))^16;
B=(37530/Re)^16;
f=2*((8/Re)^12+1/(A+B)^1.5)^(1/12);
N2=N20*f/f0;
```

```
ceqn=y(2)/y(3)*(1-y(1))+ N2/N1*(1/y(2)-1)/y(3);
Jeqn=-N1*y(1)*(y(2)^3)-N2*(y(2)^2)+N2*y(2);
Veqn=0;
Weqn=-y(2);
dydt=[ceqn;Jeqn;Veqn;Weqn];
function [value, isterminal, direction]=eventfcn(t,y)
value = y(4)-0;% stop when feed tank is emptied
isterminal = 1;
direction = 0;
```

Running the simulation gives a dimensionless time of 2.772 which is considerably higher than the time of 1.922 obtained for the batch analogue of this problem. This finding, i.e. of increased times for fed-batch processes, is repeated in ultrafiltration, as shown in Chapter 5.

The above simulations cannot be considered as true process analysis since the underlying flux model is too simplistic. However, mathematical modelling is often a useful exercise even when it cannot produce accurate, quantitative answers to process questions. More often than not, the main use of modelling is in providing physical insight, both in terms of improving our understanding of a process, and as a means of making qualitative decisions about how to run a given process. In these simulations, for example, it has been strongly suggested that a simple batch process will be faster than a fed-batch process. This is not something that would have been obvious before doing the simulations – at least not to this author.

3.7   Cake properties in CFMF

The kinetic model is a useful device for semi-quantitative modelling of CFMF. The lack of rigour in the model has been mentioned before, but there is another important simplification in the model that has not been mentioned yet. This concerns the specific cake resistance, α. In Chapter 2, it was seen that this parameter is, in principle, easy to measure, mainly through the application of the t/V_f versus V_f plot. However, this plot is technically not valid in a crossflow system as its derivation depends crucially on the solids balance, Eq. 3.16, which does not apply in crossflow systems. While the 'transition flux' model developed in this chapter suggests that this plot might well be valid in the early stages of crossflow filtration, it is probably a better approach to assume that the t/V_f versus V_f plot does not apply and other approaches must be used. The usual approach taken to evaluating the specific resistance in crossflow systems is to base the analysis on the basic flux equation, Eq. 3.20. This method involves measuring the steady state cake mass per unit area at a given pressure, measuring the flux and working out α. A variety of approaches to measuring the cake mass has been developed but in no case is it a straightforward task [17]. As a result, crossflow specific resistance data are very rare indeed. This begs the question: can dead-end specific resistance data be used for crossflow calculations? In general, the answer to this question is no. There are two reasons for this. First, the phenomenon of preferential deposition means that the average particle size in a crossflow cake is less than that in a dead-end cake of the same suspension. For narrowly distributed suspensions, this is not such a big issue but it most certainly is for suspensions with a broad range of particle sizes [18]. The second reason is that there are good grounds for believing that the particles in a crossflow cake pack differently from those in a dead-end cake. Tanaka and co-workers have shown, for example, that rod-like bacterial cells in a crossflow cake tend to align parallel to the membrane leading to a tightly packed structure [19]. This effect becomes more pronounced as the crossflow velocity is increased at fixed trans-membrane pressure. However, there is possibly a more fundamental reason for assuming that crossflow filter cakes have a different packing structure from their dead-end counterparts, even if the particles are close to spherical.

When researchers study the rheology of suspensions as a function of particle volume fraction, they find that there is a volume fraction at which the viscosity becomes infinite. This is sometimes known as the maximum packing fraction, ϕ_m. Interestingly this quantity is often found to increase with increasing shear stress [20]. Now, in the shear induced diffusion picture of CFMF, a concentrated layer flows adjacent to the membrane and at some point down the channel, the particle volume fraction at the membrane becomes equal to ϕ_m, i.e., a solid, immobile cake forms. But if ϕ_m increases with increasing shear, that implies that the particle volume fraction in a crossflow cake increases as the crossflow velocity is increased. As a general rule, therefore, the specific resistance of the cake is likely to increase with increasing crossflow velocity. Naturally, therefore, it will be larger in a crossflow setting than a dead-end one. The next example examines this idea more quantitatively.

Example 3.11 Shear stress dependence of the specific cake resistance

The following data have been obtained in a series of rheology experiments [20]:

$\tau\,(N/m^2)$	2.0	9	15	23	33	60	150	260	300
ϕ_m		0.38	0.39	0.43	0.435	0.46	0.475	0.49	0.51	0.515

Assuming the Carman–Kozeny equation, Eq. 2.21, demonstrate graphically the dependence of wall shear stress on specific cake resistance.

Solution. To complete this problem, we need to consider only terms involving volume fraction. Therefore, following from Eq. 2.21 and noting that $\phi_m = 1 - \varepsilon_c$, we simply need to plot α' versus τ where

$$\alpha' = \frac{\phi_m}{(1 - \phi_m)^3}.$$

The results are shown in Fig. 3.13 below. Clearly the specific resistance of this suspension is shear-dependent.

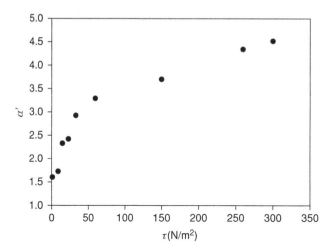

Figure 3.13 Dependence of the specific cake resistance on shear stress.

A final complication worth mentioning in the context of the crossflow specific resistance is the whole issue of cake pressure drop. It was seen in Chapter 2 that the real determinant of specific cake resistance is the cake pressure drop and not the applied pressure. In many dead-end processes, the cake resistance is much larger than the membrane resistance and, therefore, the cake pressure drop and the total applied pressure are essentially the same, except at very early times. In crossflow systems, however, this need not be the case. The cakes in crossflow are very thin and it is possible that the cake resistance is of a similar magnitude to the membrane resistance. This means that the cake pressure drop will be significantly different from the applied pressure. This can lead to further complexity in specific resistance behaviour. McCarthy *et al*. [17], for example, found that when they increased the crossflow velocity at fixed trans-membrane pressure, the specific resistance of a yeast suspension decreased. They were able to show that this was a consequence of the reduced cake pressure drop as the crossflow velocity was increased. The result of all of this is that the specific cake resistance in crossflow systems is a complex and strongly coupled function of suspension properties and operating conditions. As we see in Chapter 10, the possibility, indeed the likelihood, of membrane fouling being important has further significance for specific resistance behaviour.

3.8   Improving the flux

The basic picture of flux behaviour in CFMF is one of a rapidly declining flux, followed by a period of steady state behaviour or maybe a long slow decline due to membrane fouling. The result of this is that fluxes in conventional CFMF can be disappointingly low. The response of some manufacturers has been to develop new modules that vibrate or rotate [21, 22]. The motion of the membrane in each case has the effect of reducing cake formation, and possibly, membrane fouling. However, equipment that involves motion, especially rotation, tends to be more expensive, both in terms of capital cost and running costs. Another approach has been to use helical membranes which promote formation of vortices in the flow channel, again enhancing cake removal. The use of electric fields has also been proposed but one would have to be sceptical of the industrial potential for techniques that involve electric fields in this time of dwindling energy resources. The use of turbulence promoters, such as helical ribbons, in the flow channel, has considerable potential for reducing cake formation but this type of device could potentially promote channel clogging, especially when used with highly concentrated suspensions. Therefore, the approach that is most likely to be taken in industry is to devise novel operational strategies that improve fluxes and maintain them for long periods. In this section, the focus is on some of those strategies.

3.8.1 Backflushing

This is the most important strategy and is used extensively in submerged membrane bioreactors, described later. The basic idea is that the filtrate flux is periodically reversed. The time during which the flow is reversed is typically much smaller than the filtration time. For example, on a laboratory scale, the filtration time might be measured in minutes while the backflushing time will be measured in seconds.

Backflushing has the effect of removing filter cake from the membrane surface and can sometimes remove matter that has penetrated into the membrane pores. Figure 3.14 is a schematic representation of the flux changes that occur with this technique. Note that the backflushing and operating pressures are not necessarily equal in magnitude. The increase in (negative) flux during the backflushing phase is due to removal of cake and/or particles that have penetrated the membrane. It is assumed here that the backflushing returns the membrane to its initial clean state, free of cake on its surface and fouling material within its pores.

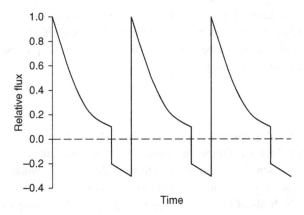

Figure 3.14 Flux changes in backflushing.

Any technique like this poses some obvious questions, such as how often should the permeate flow be reversed and for how long should the flow reversal continue? This is clearly an optimisation problem in which one might want to maximise the average flux over a long period. If this optimisation is to be done theoretically, not only is a good model of cake formation and/or membrane fouling required, but a model of cake removal and fouling reversal is also needed. Realistically, no such models exist and determining the optimum duration of a backflush and the time between backflushes is something that must be determined by experiment.

Backpulsing is an extreme form of backflushing. In this method, the flow is reversed more often and for shorter periods [23]. Here the filtration time is measured in seconds while the backpulsing time is measured in fractions of a second. An approach such as this would be worthwhile if the system were prone to irreversible cake formation and/or membrane fouling. The basic idea is to remove fouling material before adhesive forces have time to bind the material irreversibly to the membrane and/or previously deposited cake. One of the drawbacks of backflushing and backpulsing, however, is that much more equipment, instrumentation and process control is required to enable the filtrate flux to be reversed as required. It should also be pointed out that not all membranes can withstand backflushing or backpulsing so these techniques are by no means a panacea for the improvement of CFMF processes.

3.8.2   Crossflushing

Crossflushing, or zero-flux flushing, is a technique that is often used by researchers when cleaning their membranes between experimental runs [24]. This technique is inspired by the idea that if there is no flux, the forces 'holding' particles onto the membrane should be greatly reduced, or even zero. This concept was raised earlier in Section 3.5.4. The technique involves periodic closing of the filtrate valve. In the zero-flux period, the loosely bound cake is removed. This leads to flux recovery, the degree of recovery depending on the duration of the flushing phase, the extent to which cake can be removed by tangential forces only and the extent to which membrane fouling is significant. The process is illustrated in Fig. 3.15, where the value to which the flux recovers declines gradually because membrane fouling is typically not reversed by the crossflushing action.

Nonetheless, crossflushing remains a useful experimental tool, either for removing cake as part of the initial cleaning process or as part of a cake recovery process in which cake characteristics can be determined.

3.8.3   Air sparging

This is a technique that many researchers will have been aware of before the technique was studied as a potential means of improving fluxes. When a peristaltic pump is allowed to suck air, there is a sudden increase in the turbidity of the retentate when the feed is pure water or cleaning solution. This idea, that the inclusion of air into the feed enhances cake removal, forms the basis of the air sparging strategy [22]. By mixing air with the liquid feed stream, cake removal is enhanced and higher fluxes achieved. As this is now a two-phase flow problem, the relative quantities of air and water are crucially important

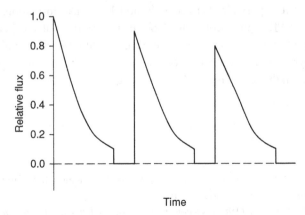

Figure 3.15 Flux patterns during periodic crossflushing with membrane fouling.

for determining the precise flow pattern and the effectiveness of cake removal. The air sparging technique is exploited in submerged membrane bioreactors (MBRs) where the membrane is inserted into an aerobic digester which continually degrades waste products. In these systems the purified water leaves the bioreactor as permeate from the membrane separator. The bubbling of air necessary for aerobic digestion reduces cake formation. By operating at low trans-membrane pressures and using periodic backflushing, sustainable fluxes can be achieved [25]. This is, of course, essential for waste treatment processes that may have to operate for months rather than hours.

3.9　Product recovery in CFMF

So far the emphasis in this chapter has been on the filtrate flux. However, as touched upon briefly in Chapter 2, the purpose of CFMF may not only be to produce filtrate at a rapid rate, but also to separate a soluble product, such as a protein, from the solid particles in the suspension. The transmission of a product, i.e., the ratio of its concentration in the filtrate relative to its concentration in the retentate, is a complex function of time, particle concentration, trans-membrane pressure, crossflow velocity and membrane effects. Le *et al.* [26], for example, showed that the transmission of an enzyme from a cell debris suspension increased rapidly with time before reaching a peak and then declining. At long times, the transmission increased with increasing crossflow velocity and increasing trans-membrane pressure. These effects are not easy to explain and really require an understanding of ultrafiltration, including the concept of solute rejection. We return to this issue in Chapter 7.

3.10　Conclusions

This chapter has shown that understanding CFMF is very much a work in progress and there is still a need to produce realistic models of flux dynamics involving easily measured parameters. It is possible that accurate models of cake formation, possibly involving the use of CFD, will become available in the future, but the goal of the modelling approach, for now, is to provide some insights and qualitative explanations, and predictions, of system behaviour. Of course, just as in any other membrane filtration process, the 'elephant in the room' is membrane fouling.

References

1. Patel, P.N., Mehaia, M.A. and Cheryan, M. (1987). Cross-flow membrane filtration of yeast suspensions. *Journal of Biotechnology*, **5**, 1–16.
2. Leighton, D.T. and Acrivos, A. (1986). Viscous resuspension. *Chemical Engineering Science*, **41**, 1377–1384.
3. Davis, R.H. and Leighton, D.T. (1987). Shear-induced transport of a particle layer along a porous wall. *Chemical Engineering Science*, **42**, 275–281.
4. Romero, C.A. and Davis, R.H. (1988). Global model of crossflow microfiltration based on hydrodynamic particle diffusion. *Journal of Membrane Science*, **39**, 157–185.
5. Davis, R.H. and Sherwood, J.D. (1990). A similarity solution for steady-state crossflow microfiltration. *Chemical Engineering Science*, **45**, 3203–3209.
6. Davis, R.H. (1992). Modelling of fouling of crossflow microfiltration membranes. *Separation and Purification Methods*, **21**, 75–126.
7. Sherwood, J.D. (1998). The force on a sphere pulled away from a permeable half-space. *Physicochemical Hydrodynamics*, **10**, 3–12.
8. Field, R.W., Wu, D., Howell, J.A. and Gupta, B.B. (1995). Critical flux concept for microfiltration fouling. *Journal of Membrane Science*, **100**, 259–272.
9. Bachin, P., Aimar, P. and Field, R.W. (2006). Critical and sustainable fluxes: theory, experiments and applications. *Journal of Membrane Science*, **281**, 42–69.
10. Field, R.W. and Pearce, G.K. (2011). Critical, sustainable and threshold fluxes for membrane filtration with water industry applications. *Advances in Colloid and Interface Science*, **164**, 38–44.
11. Foley, G., Malone, D.M. and MacLoughlin, F. (1995). Modeling the effects of particle polydispersity in crossflow filtration. *Journal of Membrane Sciences*, **99**, 77–88.
12. Ghidossi, R., Veyret, D. and Moulin, P. (2006). Computational fluid dynamics applied to membranes: state of the art and opportunities. *Chemical Engineering and Processing*, **45**, 437–454.
13. Gutmann, R.G. (1977). Design of membrane separation plant. 1. Design of RO Modules. 2. Fouling of RO modules. *Chemical Engineer (London)*, **322**, 510–513 and 521–523.
14. Kern, D.Q. and Seaton, R.E. (1957). A theoretical analysis of thermal surface fouling. *British Chemical Engineering*, **4**, 258–262.
15. Kuberkar, V.T. and Davis, R.H. (2001). Microfiltration of protein-cell mixtures with cross-flushing or backflushing. *Journal of Membrane Science*, **183**, 1–14.
16. Churchill, S.W. (1977). Friction-factor equation spans all fluid flow regimes. *Chemical Engineering*, **84**, 91–92.
17. McCarthy, A.A., Walsh, P.K. and Foley, G. (2002). Experimental techniques for quantifying the cake mass, the cake and membrane resistances and the specific cake resistance during crossflow filtration of microbial suspensions. *Journal of Membrane Science*, **201**, 31–45.

18. Lu, W.M. and Ju, S.C. (1989). Selective particle deposition in crossflow filtration. *Separation Science and Technology*, **24**, 517–540.

19. Tanaka, T., Abe, K. and Nakanishi, K. (1994). Shear-induced arrangement of cells during crossflow filtration of E. coli cells. *Biotechnology Techniques*, **8**, 57–60.

20. Zhou, J.Z.Q., Fang, T., Luo, G. and Lye, P.H.T. (1995). Yield stress and maximum packing fraction of concentrated suspensions. *Rheologica Acta*, **32**, 544–561.

21. Wakeman, R.J. and Williams, C.J. (2000). Additional techniques to improve microfiltration. *Separation and Purification Technology*, **26**, 3–18.

22. Postlethwaite, J., Lamping, S.R., Leach, G.C., Hurwitz, M.F. and Lye, G.J. (2004). Flux and transmission characteristics of a vibrating microfiltration system operated at high biomass loading. *Journal of Membrane Science*, **228**, 89–101.

23. Mores, W.D., Bowman, C.N. and Davis, R.H. (2000). Theoretical and experimental flux maximization by optimization of backpulsing. *Journal of Membrane Science*, **165**, 225–236.

24. Kennedy, M., Kim, S.M., Mutenyo, I., Broens, L. and Schippers, J. (1998). Intermittent crossflushing of hollow fiber ultrafiltration systems. *Desalination*, **118**, 175–187.

25. Le-Clech, P., Chen, V. and Fane, A.G. (2006). Fouling in membrane bioreactors used in wastewater treatment. *Journal of Membrane Science*, **284**, 17–53.

26. Le, M.S., Spark, L.B., Ward, P.S. and Ladwa, N. (1984). Microbial asparaginase recovery by membrane processes. *Journal of Membrane Science*, **21**, 307–319.

Problems

Problem 3.1   Kinetic model with zero order kinetics

Consider Eq. 3.18 with a zero order cake removal rate, i.e., $\psi_r = $ constant. Derive an expression for the steady state flux in this case and explain why it provides no real insight into the factors affecting the steady state flux in CFMF.

Problem 3.2   Filtration in cylindrical geometry

Show that for filtration in cylindrical geometry, the filtrate flux, based on the *inner* tube area is given by

$$J = \frac{\Delta P}{\mu \left[R_m + K_h R \ln \left(\frac{1}{1 - \delta/R} \right) \right]},$$

where K_h is the specific resistance based on the cake thickness, δ, and R is the inner tube radius. As your starting point, you should note that the flux at any radial position in the cake, r, is given by

$$J = -\frac{1}{\mu K_h} \frac{dP}{dr},$$

where P is the fluid pressure at r where r represents the distance from the centre line of the tube or fibre. Finally, simplify your expression for $\delta \ll R$.

Problem 3.3   Derivation of expression for the steady state flux

Derive Eq. 3.25, starting from Eq. 3.23.

Problem 3.4 Effect of crossflow velocity on steady state flux

Using Eq. 3.26, discuss the effect of crossflow velocity on the steady state flux. For laminar flow, assume that the wall shear stress is proportional to $u^{1.0}$ while for turbulent flow, assume it is proportional to $u^{1.75}$.

Problem 3.5 Effect of viscosity on the steady state flux

Using Eq. 3.26 as your basis, discuss the effect of viscosity on the steady state flux.

Problem 3.6 Relation between steady state flux and critical flux

Derive the expression relating the transition flux to the critical flux given in Example 3.7.

Problem 3.7 Physical meaning of dimensionless numbers

Verify the physical meaning of the dimensionless numbers, N_1 and N_2, defined in Eqs. 3.67 and 3.68.

Problem 3.8 Effect of channel narrowing

In a rectangular channel in which the flow is laminar, the wall shear stress τ_w, is related to the initial, clean channel wall shear stress, τ_{w0}, by the expression

$$\tau_w = \frac{\tau_{w0}}{[1 - \delta/H]^2} \frac{\mu_c}{\mu_{c0}},$$

where δ is the cake thickness, H is the channel half-height, μ_c is the suspension viscosity at a concentration, c, and μ_{c0} is the suspension viscosity at the initial concentration. Show that

$$N_2 = \frac{N_{20}e^{N_3(c_{\mathrm{rel}}-1)}}{[1 - N_4(1/J_{\mathrm{rel}} - 1)]^2},$$

where N_4 is a dimensionless number defined by

$$N_4 = \frac{R_{\mathrm{m}}}{\alpha \rho_{\mathrm{p}} \phi_{\mathrm{c}} H}$$

where α is the specific cake resistance, ρ_{p} is the particle density and ϕ_{c} is the particle volume fraction in the cake. What is the physical significance of this number? Now, using the MATLAB code in Example 3.9 as a guide, generate a plot of relative flux versus relative time for $N_4 = N_1 = 10$, $N_{20} = 10$, $N_3 = 0.1$ and $N_4 = 0.1$. End the simulation when the volume has been reduced by a factor of 20.

Further problems

Problem 3.9 Effect of cake formation on pressure loss in channel

This problem is largely concerned with fluid flow and focuses on Eq. 3.48. First derive this equation, starting with Eq. 3.47. Then, by doing a force balance on a cylindrical element of fluid, find an expression for the pressure loss due to friction in terms of the fluid flowrate. You may want to go back to Problem 2.2 to refresh your memory on this.

Hence show that if cake builds up on a tubular membrane of radius, R, the pressure loss due to friction in the channel is given by

$$\Delta P_f = \frac{\Delta P_{f0}}{(1 - \delta/R)^3},$$

where ΔP_{f0} is the initial pressure loss obtained with the clean channel. How could this equation be used as a means of investigating cake formation dynamics in crossflow microfiltration?

Problem 3.10   Analysis of total recycle operation

Total recycle operation is often used to simulate continuous operation. In this mode, both the retentate and the filtrate are returned to the feed tank. In principle, however, one should really take account of changes in bulk particle concentration as a result of cake accumulation. The purpose of this problem is to explore this phenomenon.

Using the kinetic model for CFMF, show that the steady state cake mass per unit area during total recycle operation is obtained by solving the quadratic equation

$$M^2 + \left(\frac{R_m}{\alpha} + \frac{A\Delta P}{\alpha \mu k_r \tau_w V_0} \right) M - \frac{\Delta P c_0}{\alpha \mu k_r \tau_w} = 0.$$

Hence show that total recycle operation is a good approximation to continuous operation when $N_2 \gg 1$ where N_2 is defined by Eq. 3.68.

Problem 3.11   Steady state cake mass with channel narrowing

Let us consider CFMF in a rectangular channel in which the wall shear stress is related to cake thickness, δ, by the expression

$$\tau_w = \frac{\tau_{w0}}{(1 - \delta/H)^2},$$

where τ_{w0} denotes the clean channel wall shear stress and H is the channel half-height. Now, assuming the membrane resistance is negligible, show that the steady state cake mass is given by

$$M = \frac{(c\Delta P/\alpha \mu k_r \tau_{w0})^{1/2}}{1 + \dfrac{(c\Delta P/\alpha \mu k_r \tau_{w0})^{1/2}}{\rho_s \phi_c H}},$$

where ρ_s is the solids density and ϕ_c is the solids volume fraction in the cake.

4 Ultrafiltration flux theories

4.1 Introduction

Modelling, analysis and design of chemical engineering systems generally proceed on two fronts. First, the mass and/or energy balances for the system are formulated. On their own, these balances are rarely sufficient to derive equations with which practical calculations can be performed. This is because the balances typically contain some unknown term, or terms, which quantify the rate at which some key process is occurring. Quantifying this rate is the second step in the analysis. A classic example would be the kinetic term in a model of a batch reactor. Alternatively, an equilibrium relation might provide a crucial relationship between compositions in the liquid and vapour streams of a distillation process, thus providing the missing information required to solve the complete model equations. In this chapter the underlying rate equations for UF design and analysis are established. A number of different equations, of varying complexity, are presented and their main features outlined. The rate equation in this case is simply an equation that relates the permeate flux to measurable process variables, notably the crossflow velocity, the solute concentration and the trans-membrane pressure. In subsequent chapters, each of the rate equations is combined with appropriate mass balances for batch and continuous ultrafiltration and diafiltration systems. As will be seen, the combination of the rate equation and the mass balances leads to a variety of computational challenges. However, with modern software tools, these are easily overcome.

In addition to reviewing the basic theory of UF, this chapter provides the reader with a very brief overview of methods for predicting mass transfer coefficients, viscosities and osmotic pressures for a range of solutions relevant to UF practice. This is by no means an exhaustive review but is intended to give the reader a sense of the issues involved in actually trying to apply the various UF theories. The data provided should also prove useful for performing practice calculations.

In the next section the concept of concentration polarisation is explained and it is shown how it leads to a relationship between permeate flux and the wall concentration of the solute. By 'wall concentration' we mean the concentration of solute at the membrane surface. Two different approaches are then used to show how to compute this wall concentration. Throughout this book, we deal with the *average flux* in the module. More detailed models would require us to account for the fact that in crossflow systems the permeate flux is a function of position within the module. However, such models are considerably more challenging, typically involving partial differential equations, and are beyond the scope of this text.

4.2 Concentration polarisation

Concentration polarisation is the phenomenon whereby there is a gradient in concentration from the channel wall (i.e., the membrane), where the solute concentration is at its highest (c_w) to the bulk solution where it is at its lowest (c), as indicated in Fig. 4.1. This concentration gradient is a natural consequence of the convective flow of solute towards the membrane. There, the solute is completely or partially rejected, leading to accumulation at the membrane surface.

The effect of this concentration polarisation layer and how it impacts on the flux is quite subtle and ultimately requires an analysis based not only on concentration polarisation but also on osmotic pressure. In this section, it is shown how a mass transfer analysis leads to a relationship between the permeate flux, the bulk (retentate) concentration and the wall concentration. Additional information or assumptions are required before the flux can actually be computed.

The approach below is simplified somewhat and the reader is directed to the work of Zeman and Zydney [1] for a more rigorous analysis. The basis of the analysis is that at steady state the rate of transport of solute towards the membrane is equal to the rate of transport of solute away from the membrane ('back-transport'). Transport towards the membrane occurs by convection and, for now, it is assumed that back transport occurs purely by diffusion. To start, let c_y be the concentration of solute at a distance, y, from the membrane and let c_p be the concentration of solute in the permeate. Referring to Fig. 4.1, a solute balance on the region to the left of the dashed line gives

$$Jc_y - Jc_p - \left(-D\frac{dc_y}{dy}\right) = 0, \tag{4.1}$$

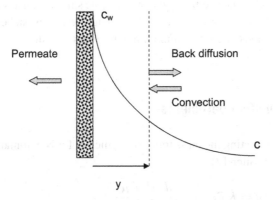

Figure 4.1 Concentration polarisation showing convection and diffusion processes.

where D is the diffusion coefficient. Assuming D is constant, this expression can be re-arranged to give

$$\frac{J}{D} \int_0^\delta dy = - \int_{c_w}^c \frac{dc_y}{c_y - c_p},$$ (4.2)

where δ is the 'film' thickness and c is the bulk concentration. The idea that there is a definite 'film' over which the entire concentration change occurs is obviously a little bit artificial and will be familiar to chemical engineers. However, as seen below, this is just a temporary device which is eliminated from the analysis (at least explicitly) in due course. Integrating, rearranging and applying the limits gives

$$J = \frac{D}{\delta} \ln \frac{c_w - c_p}{c - c_p}.$$ (4.3)

Now, for most of this book (the main exception being Chapter 7), we consider the case of complete rejection of the macrosolute, i.e., $c_p = 0$. Thus, Eq. 4.3 becomes

$$J = \frac{D}{\delta} \ln \frac{c_w}{c}.$$ (4.4)

To avoid the problem of trying to calculate the film thickness, it is recognised that within the context of classic 'film theory' [2], the D/δ term is a *mass transfer coefficient*, typically denoted k. The significance of this definition is that the mass transfer coefficient can, in principle, be predicted with empirical correlations or exact analytical expressions. In a sense, therefore, the identification of D/δ with the mass transfer coefficient allows us to greatly generalise the simple analysis represented by Eq. 4.1 and to incorporate all of the process parameters that are likely to influence the flux in UF. Using the mass transfer coefficient concept, the following result is obtained:

$$J = k \ln \frac{c_w}{c}.$$ (4.5)

Equation 4.5 is only a partial solution to the problem as the wall concentration is unknown. All that has been done, really, is to develop a relationship between the flux and this concentration. To progress, one needs to either make some simplifying assumptions, or to incorporate some additional ideas. Before describing these approaches, a brief survey of the various methods used for computing the mass transfer coefficient is presented.

4.3   Prediction of mass transfer coefficients

The general equation for predicting the mass transfer coefficient for Newtonian fluids in membrane systems can be written [2]

$$Sh = K \, Re^a \, Sc^b \left(\frac{d_h}{L}\right)^c \left(\frac{\mu_c}{\mu_w}\right)^z,$$ (4.6)

where d_h is the module hydraulic diameter, L is the membrane length in the axial direction and K, a, b and c are empirical constants.

The quantity Re is the Reynolds number, defined in terms of tangential flowrate by the expression

$$Re = \frac{4\rho_c Q}{\pi \mu_c d_h},$$ (4.7)

where ρ_c is the fluid density at the bulk fluid concentration, Q is the tangential flowrate (in a single tube, fibre or channel) and μ_c is the fluid viscosity evaluated at the bulk concentration. The Schmidt number, Sc, is defined by

$$Sc = \frac{\mu_c}{\rho_c D_c},$$ (4.8)

where D_c is the diffusion coefficient evaluated at the bulk concentration. The Sherwood number, Sh is defined by

$$Sh = \frac{k d_h}{D_c},$$ (4.9)

where k is the mass transfer coefficient.

Equation 4.6 includes a 'wall correction factor' whereby the mass transfer coefficient is reduced as a consequence of the increased viscosity within the concentration polarisation layer. The parameter, μ_w, represents the viscosity evaluated at the wall concentration and z is an empirical parameter. Choosing the 'correct' value of z is problematic. Theory suggests a value of 0.27 [3] but comparison with a small amount of experimental data suggests that a value of 0.14 might be better [4]. One of the difficulties here is that mass transfer correlations described above have been developed for Newtonian fluids of uniform composition, flowing in a non-porous channel. However, in UF there is a gradient in fluid composition due to concentration polarisation which not only leads to a gradient in physical properties but also, perhaps, to a gradient in rheological characteristics of the fluid. Therefore, an exact description of mass transfer in UF systems is difficult indeed. Some progress has been made in this area by Howell and co-workers [5, 6] but it would be fair to say that the ability to describe mass transfer in non-Newtonian systems, especially in a UF context, is still limited.

In Eq. 4.6, the precise definition of the hydraulic diameter depends on the module geometry. For tubular or hollow fibre modules, it is just the tube or fibre diameter. For rectangular channels, it is twice the channel height, i.e., $2H$. For a channel with a spacer screen, it is given by [1]

$$d_h = \frac{4\varepsilon}{2/H + (1 - \varepsilon s)},$$ (4.10)

where ε is the spacer porosity and s is the specific surface area of the spacer.

Table 4.1 gives values for the constants in Eq. 4.6 for different flow configurations. It should be noted that in turbulent flow, the mass transfer coefficient is independent of L. In laminar flow, the local mass transfer coefficient is actually a function of distance from the channel inlet and the mass transfer coefficient defined by the parameters in Table 4.1

Table 4.1 Exponents in Eq. 4.6 for various flow geometries [1].

Flow type	K	a	b	c
Laminar tube	1.62	1/3	1/3	1/3
Laminar rectangular channel	1.86	1/3	1/3	1/3
Turbulent tube	0.023	0.8	1/3	0
Rectangular channel with spacer	0.664	1/2	1/3	1/2

is an average value over the total length, L. As mentioned above, these correlations are valid for Newtonian fluids only.

In describing mass transfer in UF, where a key aim of the process is to increase the concentration of the solute, it is important to highlight the effect of concentration in the correlations for predicting mass transfer coefficients. A reasonable start is to assume that the fluid density is identical to the pure solvent density. Hence the notation, ρ, is used henceforth for density. This assumption is irrelevant in laminar flow where the mass transfer coefficient is independent of density. Density variations are also likely to be unimportant in turbulent flows. Further investigation of this topic is reserved for Problem 4.1.

To explore the role of concentration further, the mass transfer coefficient can be written in the concise form

$$k = k_c \left(\frac{\mu_c}{\mu_w} \right)^z , \tag{4.11}$$

where k_c is the 'un-corrected' mass transfer coefficient evaluated at the bulk concentration. For laminar flow in a tube where the un-corrected mass transfer coefficient is independent of viscosity (since the viscosity terms in Re and Sc cancel), we can take this a step further and write

$$k_c = k_0 \left(\frac{D_c}{D_0} \right)^{2/3} , \tag{4.12}$$

where k_0 is the infinite dilution value of the un-corrected mass transfer coefficient, D_0 is the infinite dilution value of the diffusion coefficient and D_c is its value at the bulk concentration. Combining Eqs. 4.11 and 4.12 gives

$$k = k_0 \left(\frac{D_c}{D_0} \right)^{2/3} \left(\frac{\mu_c}{\mu_w} \right)^z . \tag{4.13}$$

Equation 4.6 can be rearranged, after taking note of the parameters in Table 4.1, to give the following expression for k_0

$$k_0 = 1.62 \left(\frac{4QD_0^2}{\pi L d_t^3} \right)^{1/3} , \tag{4.14}$$

where d_t is the tube diameter. In a similar manner, analysis of the turbulent flow correlation gives

$$k_c = k_0 \left(\frac{D_c}{D_0}\right)^{2/3} \left(\frac{\mu}{\mu_c}\right)^{0.47},$$
(4.15)

where μ is the solvent ($c = 0$) viscosity. Therefore, in turbulent flow one gets

$$k = k_0 \left(\frac{D_c}{D_0}\right)^{2/3} \left(\frac{\mu}{\mu_c}\right)^{0.47} \left(\frac{\mu_c}{\mu_w}\right)^{z},$$
(4.16)

where the un-corrected, infinite dilution mass transfer coefficient is given by

$$k_0 = 0.023 \frac{D_0}{d_t} \left(\frac{4\rho Q}{\pi \mu d_t}\right)^{0.8} \left(\frac{\mu}{\rho D_0}\right)^{1/3}.$$
(4.17)

Therefore, hidden within the mass transfer coefficient are a number of ways in which the UF flux ultimately depends on concentration. As shown later, the key parameter in this regard is the fluid viscosity and this leads to flux characteristics that are qualitatively different in laminar and turbulent flow modules.

4.3.1 Prediction of diffusion coefficients

The dependence of the mass transfer coefficient on the diffusion coefficient highlights the importance of being able to predict this parameter. However, predicting diffusion coefficients, or even finding relevant diffusion data for polymeric solutions is not easy and could be the subject of a chapter in itself. In this section, some basic ideas are reviewed and some data are given that should prove useful when performing sample calculations and solving problems.

It is important to realise that the D_c term in the above equations represents a *mutual* diffusion coefficient and is a characteristic of the *solution* and not just the solute in question. Unfortunately, it is not possible to quantify the concentration dependence of the mutual diffusion coefficient with universally applicable equations since it depends not only on the *self*-diffusion coefficients (also called *tracer* diffusion coefficients and a measure of the mobility of a molecule) of the solute and solvent, but also on the *solution* thermodynamics [2].

Diffusion coefficients reported in the literature tend to be infinite dilution values and details on the concentration dependence are usually lacking. In practice, therefore, prediction of mass transfer coefficients tends to ignore the dependence of the diffusion coefficient on concentration. Whether this is a good assumption or not is the subject of Problem 4.2. In the event of a lack of infinite dilution diffusion data, it is possible to use one of the many empirical and semi-empirical equations for predicting the infinite dilution diffusion coefficient. As proteins and polysaccharides are probably the most important category of solutes for UF, some of the correlations available to predict diffusion coefficients in these solutions are presented here.

One of the best correlations for proteins is the one developed relatively recently by He and Niemeyer [7]. Their equation can be written

$$D_0 = \frac{6.85 \times 10^{-15} T}{\mu \sqrt{R_G M_w^{1/3}}}, \tag{4.18}$$

where D_0 is in m^2/s, T is the absolute temperature (K), μ is the solvent viscosity in Ns/m^2, R_G is the radius of gyration of the protein in *angstroms* and M_w is its molecular weight (g/mol). A less accurate equation, not requiring radius of gyration data, is the following due to Young *et al.* [8]:

$$D_0 = 8.34 \times 10^{-15} \left(\frac{T}{\mu M_w^{1/3}} \right). \tag{4.19}$$

Some experimental diffusion coefficients of selected dilute protein solutions are given in Table 4.2, along with the parameters used in Eqs. 4.18 and 4.19. In Example 4.1, the accuracies of these equations are compared. In Problem 4.3 you investigate whether Eq. 4.19 can be improved by putting an arbitrary exponent on the molecular weight.

Table 4.2 Diffusion coefficient parameters for selected proteins at 20 °C (adapted from [7]).

Protein	M_w (g/mol $\times 10^3$)	R_G (Å)	$^*D_{exp}$ (m^2/s $\times 10^{-11}$)
Lysozyme	14.4	15.2	11.8
β-Casein	24.1	75.0	6.05
Ovalbumin	38	24.0	7.3
BSA	66	29.8	5.93
Hexokinase	99	24.7	6.0
γ-Globulin	162	70.0	3.7
Catalase	225	39.8	4.1
Fibrinogen	390	142.0	1.86

$^*D_{exp}$ *is the experimental value of the diffusion coefficient.*

Example 4.1 Prediction of diffusion coefficients of proteins

Use Eqs. 4.18 and 4.19 to compute the diffusion coefficient of each of the proteins in Table 4.2. Compare your values with the experimental values and compute the average percentage experimental error defined by $|D_{exp} - D_{calc}|/D_{exp} \times 100$.

Solution. It is probably simplest to do this problem in a spreadsheet package. The numerical results in the table were thus obtained.

Numerical results.

Protein	$^*D_{exp}$	$D_{4.18}$	$D_{4.19}$	%Error$_{4.18}$	%Error$_{4.19}$
Lysozyme	11.8	10.44	10.04	11.5	14.9
β-Casein	6.05	4.31	8.46	28.8	39.8
Ovalbumin	7.3	7.07	7.27	3.2	0.4
BSA	5.93	5.78	6.05	2.5	2.0
Hexokinase	6.0	5.94	5.28	1.0	12.0
γ-Globulin	3.7	3.25	4.48	12.2	21.1
Catalase	4.1	4.08	4.02	0.5	2.0
Fibrinogen	1.86	1.97	3.34	5.9	79.6
			Mean	8.2	21.5

* *Units of diffusion coefficients are $m^2/s \times 10^{-11}$ in all cases.*

Clearly Eq. 4.18 is superior on average. It is worth noting the magnitude of diffusion coefficients of proteins. Typically they are $O(10^{-11})$ m^2/s.

There is even less information available in the literature on diffusion of polysaccharides. However, dextran is a polysaccharide that is used extensively in ultrafiltration studies and there is some diffusion data available. Wu [9] used laser light scattering to derive the following correlation for dextran:

$$D_0 = 1.98 \times 10^{-8} M_w^{-[0.657-0.02 \log M_w]}. \tag{4.20}$$

Of course, many industrial UF processes do not involve a single solute in a well-defined solution. Typically they involve complex mixtures containing more than one macrosolute and possibly many microsolutes. In the UF of dairy feeds such as milk or whey solutions, the feed is a complex mixture of proteins, fats and sugars. In the UF of fruit juices, the feed contains polysaccharide solutes such as pectins as well as sugars and even particulate matter. It is very unlikely that meaningful diffusion coefficients will be available for mixtures such as these. A rigorous analysis of diffusion in such systems would involve using multi-component diffusion theory, something that is not only beyond the scope of this book, but also unlikely to yield any usable equations for process analysis and design. In practice, it is probably a better strategy to approach the problem from the opposite direction, i.e., use a simple UF experiment to evaluate an effective diffusion coefficient for a mixture, as shown later in Example 4.4.

4.3.2 Prediction of solution viscosity

As will be seen in Chapter 5, one of the main applications of UF is concentration of protein solutions. In a batch process, for example, the solution volume is reduced, leading to a simultaneous increase in solute concentration [10]. If the process is performed in a laminar flow module, and the concentration dependence of the diffusion coefficient is neglected, the sole source of the concentration dependence of the mass transfer coefficient is the wall correction factor. In contrast, the mass transfer coefficient in a turbulent flow module is concentration-dependent due to both the wall correction factor and the dependence of the mass transfer coefficient on the bulk viscosity. For both cases, we need to examine how solution viscosity depends on concentration.

Like many areas of bioprocess engineering where physical and transport property data tend to be lacking (at least in comparison with more classic chemical engineering systems), information on the concentration dependence of the viscosity of protein or polysaccharide solutions is scarce, especially at high concentrations. There is, of course, the added complication that at high concentrations, solutions of high molecular weight solutes are likely to be non-Newtonian. Whey protein solutions, for example, exhibit shear-thinning behaviour above about 300 g/l [11]. Many polysaccharides such as xanthan gum, an important food thickening agent, are shear thinning even at low concentrations [5]. For the remainder of this section, non-Newtonian effects are ignored. Not only are they beyond the scope of this text but, with a few exceptions, they have not yet been the subject of comprehensive study, even in the research literature. A complicating feature, for example, is the possibility that the fluid might be Newtonian in the bulk flow but shear thinning close to the membrane, suggesting perhaps that a true understanding of mass transfer in UF requires a much deeper analysis of the underlying fluid mechanics.

It is very difficult to make general statements about the effect of concentration on the viscosity of solutions of proteins or polysaccharides, even if they do remain Newtonian. The precise dependence is affected by morphological parameters such as size, shape, rigidity, presence of side chains, molecular entanglements, etc. However, the small amount of relevant data in the literature suggests that an exponential function of the form [3]

$$\mu_c = \mu e^{\gamma c} \tag{4.21}$$

gives a good description of the concentration dependence of the viscosity of many solutions. Table 4.3 gives values of γ for some selected fluids. The values given have been extracted from a number of publications and should be considered as a rough guide only. This is not only because the data on which the values are based are less than extensive but also because concentrations have been converted to g/l in all cases without accounting for the effect of concentration on the solution density. Note that dextran molecules are assigned a number which represents their average molecular weight in kDa.

Table 4.3 Concentration dependence of viscosity for selected fluids.

Solute	Range (g/l)	γ (l/g)
[12]Whey proteins	0–200	0.011
[11]Dextran T10	0–500	0.0115
[5]Dextran T70	0–500	0.015
[13]Dextran T2000	0–250	0.028

Numbers in square brackets indicate references from which data were obtained.

One of the trends that emerges from even the limited amount of information in Table 4.3 is the fact that the viscosity of a polymer solution is strongly dependent on molecular weight. There is a very large literature devoted to the study of the effect of molecular weight on polymer solution rheology, but from a UF perspective the key point to note is that the higher the molecular weight, the more likely it is that viscosity effects will be important in determining the flux. In contrast it is seen later that the higher the molecular weight, the less significant is the effect of osmotic pressure.

In Example 4.2 the practical consequences of accounting for the effect of concentration on viscosity in calculating the mass transfer coefficient are examined. One of the key findings is that laminar flow and turbulent flow modules are fundamentally different in how the concentration impacts on the mass transfer coefficient.

Example 4.2 Concentration dependence of mass transfer coefficient

A solution of Dextran T10 is concentrated from 10 g/l to 40 g/l in a batch UF process. By what factor is the mass transfer coefficient changed if UF is done (i) in a laminar flow module, and (ii) in a turbulent flow module? Assume that the wall concentration is constant in each case, take $z = 0.14$ and assume the diffusion coefficient is independent of concentration.

Solution
Laminar flow. If changes in the diffusion coefficient are neglected, Eqs. 4.11 and 4.12 imply

$$k = k_0 \left(\frac{\mu_c}{\mu_w} \right)^z .$$

Therefore, if the wall concentration remains constant, and using the viscosity data in Table 4.3, we have

$$\frac{k_2}{k_1} = \left(\frac{\mu_{c2}}{\mu_{c1}} \right)^{0.14} = \exp(0.0115 \times 0.14 \times (40 - 10)) = 1.05.$$

Thus the mass transfer coefficient is actually *increased* by 5% in the laminar flow module. The explanation is that if the wall concentration is constant, an increase in bulk concentration *reduces* the negative effect of the wall correction factor, leading to this counter-intuitive result.

Turbulent flow. If changes in the diffusion coefficient are neglected, Eq. 4.16 becomes

$$k = k_0 \left(\frac{\mu}{\mu_c} \right)^{0.47} \left(\frac{\mu_c}{\mu_w} \right)^{0.14} .$$

Therefore, with a constant wall concentration, we have

$$\frac{k_2}{k_1} = \left(\frac{\mu_{c1}}{\mu_{c2}} \right)^{0.33} = \exp(0.33 \times 0.0115 \times (10 - 40)) = 0.89.$$

Thus, the mass transfer coefficient is *decreased* by 11% in the turbulent flow module. Note that using an exponent of 0.5 instead of 0.47 in the turbulent flow correlation gives

$$\frac{k_2}{k_1} = \left(\frac{\mu}{\mu_c}\right)^{0.5} \left(\frac{\mu_c}{\mu_w}\right)^{0.14} = \exp(0.36 \times 0.0115 \times (10 - 40)) = 0.88.$$

Therefore, to keep equations as neat as possible, an exponent of 0.5 rather than 0.47 is used on the viscosity term in all subsequent calculations of the mass transfer coefficient.

This section has shown that, in principle, it is possible to predict the mass transfer coefficient in UF systems, although the experimental correlations summarised by Eq. 4.6 are of only moderate accuracy. However, for the purposes of this textbook, we now have some of the necessary predictive tools and sufficient physical property data to help solve a wide range of process analysis and design problems in later chapters. Furthermore, when we incorporate phenomena like the effect of concentration on the mass transfer coefficient, we see that UF provides a rich source of interesting system behaviour and some reasonably challenging computational problems.

4.4   The limiting flux (LF) model

As mentioned earlier, a fundamental limitation of concentration polarisation theory is that the wall concentration, c_w, is unknown. A well-known 'solution' to this problem is the so-called gel polarisation model (a terminology that is used sparingly in this book), perhaps first introduced by Blatt and co-workers [14]. The inspiration for this model is the observation that at high pressures the flux becomes independent of trans-membrane pressure, as was mentioned in Chapter 1. This observation is rationalised within concentration polarisation theory by assuming that the wall concentration increases with pressure but eventually becomes independent of pressure at high pressures. This pressure-independent concentration has typically been referred to as the 'gel' concentration, denoted c_g. The terminology owes its origin to the idea that at very high concentrations a 'gelation' process is assumed to occur, and thus c_g reflects a maximum possible concentration of the solute in question. One of the problems with this explanation, however, is the frequent observation that c_g values for the same solution are quite variable and seem to depend on precise process conditions, as shown in Table 4.4.

Thus, c_g does not really seem to represent an inherent property of a solution. Furthermore, 'gel' concentrations seem to have values consistent with the solution being in a liquid state and not some solid-like state. As is seen later, the tendency of the wall concentration to reach an apparently *limiting* value can be explained without any supposed gelation effects. Thus, throughout this book, the still popular c_g notation is replaced with the c_{lim} notation to reflect the fact that although the wall concentration appears to be constant under certain conditions, this is probably *not* as a result of gelation effects. Indeed, we dispense entirely with the 'gel polarisation model' terminology and replace it

Table 4.4 Variation in c_g values for various solutions [10].

Solution	c_g (wt% protein)
BSA	20–44
Skim milk	20–25
Whole milk	9–11
Cheddar cheese whey	20–28.5
Gouda cheese whey	18–20
Gelatin	20–30

with the term *limiting flux model*. In conclusion, therefore, the limiting flux (LF) model is encapsulated in the following expression:

$$J_{lim} = k \ln \frac{c_{lim}}{c}, \qquad (4.22)$$

where c_{lim} must be determined experimentally and k can be computed using the correlations described in Section 4.3. In Problem 4.4, you are asked to explain how this equation is consistent with the observed behaviour of the limiting flux as described in Chapter 1. In Example 4.3, it is demonstrated how the mass transfer coefficient and the limiting concentration can be computed from experimental flux data using this equation.

Example 4.3 Determination of k and c_{lim} from flux data

The following limiting flux data have been obtained during ultrafiltration of a certain protein solution. Verify that these data agree with the LF model, with constant mass transfer coefficient, and compute k and c_{lim}.

c (g/l)	25	50	75	100	125	150	175	200
J_{lim} (m/s $\times$ 10^{-6})	12	9	7.1	5.6	4.5	3.6	2.8	2.1

Solution. We note that Eq. 4.22 can be written as

$$J_{lim} = k \ln c_{lim} - k \ln c.$$

A simple MATLAB script to plot the data in a suitable form (Fig. 4.2) is as follows:

```
%Example 43
c=[25 50 75 100 125 150 175 200];
Jlim=[12 9 7.1 5.6 4.5 3.6 2.8 2.1];
x=log(c);
plot(x,Jlim,'ok');
xlabel('ln c');
ylabel('J_l_i_m (m/s x 10^-^6)');
```

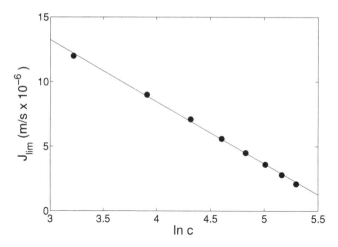

Figure 4.2 Fit of limiting flux model to experimental data (c in g/l).

Performing the linear regression within the MATLAB figure window gives

$$J_{\text{lim}} = 27.61 \times 10^{-6} - 4.79 \times 10^{-6} \ln c.$$

Therefore $k = 4.79 \times 10^{-6}$ m/s and

$$\ln c_{\text{lim}} = \frac{27.61 \times 10^{-6}}{4.79 \times 10^{-6}} = 5.77,$$

from which we get $c_{\text{lim}} = 319$ g/l.

In the earlier discussion on diffusion coefficients, the point was made that calculating diffusion coefficients in complex mixtures is not only difficult in practice, but the precise meaning of a diffusion coefficient in this context is somewhat murky. One way to address this issue is to turn the problem on its head and rather than using a diffusion coefficient to predict a mass transfer coefficient, one can use UF experiments to determine an effective diffusion coefficient for a complex mixture. While this approach may be somewhat dubious for the purist who is interested in studying diffusion as a phenomenon in itself, it may perhaps yield a parameter that is actually more useful, precisely because it will yield a good prediction of the mass transfer coefficient. The basic idea, described in Example 4.4, is that by measuring a mass transfer coefficient under well-defined process conditions, we can work backwards from the correlations defined by Eq. 4.6 to evaluate an effective diffusion coefficient. The correlation can then be used to predict a mass transfer coefficient under different process conditions.

Example 4.4 Determination of effective diffusion coefficient of a complex feed

The data in Example 4.3 were obtained at a total retentate flowrate of 5 l/min in a hollow fibre module containing 200 fibres of diameter 1 mm and length 0.25 m. Assuming that the fluid density is constant and equal to 1000 kg/m^3, estimate the effective diffusion coefficient of this feed.

Solution. The flowrate in m^3/s through an individual fibre is given by

$$Q = \frac{5 \times 10^{-3}/60}{200} = 4.17 \times 10^{-7} \text{m}^3/\text{s}.$$

Therefore the Reynolds number is given by

$$Re = \frac{4\rho Q}{\pi \mu d} = \frac{4000 \times 4.17 \times 10^{-7}}{\pi \times 0.001 \times 0.001} = 531.$$

In the previous example it was found that the mass transfer coefficient was constant over the range of concentrations studied. Denoting this as k and the effective diffusion coefficient as D, one gets, after rearranging Eq. 4.14,

$$D = \left(\frac{\pi L d^3}{4Q}\right)^{1/2} \left(\frac{k}{1.62}\right)^{3/2}.$$

Therefore

$$D = \left(\frac{\pi \times 0.25 \times (0.001)^3}{4 \times 4.17 \times 10^{-7}}\right)^{1/2} \left(\frac{4.79 \times 10^{-6}}{1.62}\right)^{3/2} = 11.03 \times 10^{-11} \text{m}^2/\text{s}.$$

The last two examples have involved a system in which the mass transfer coefficient turns out to be constant and independent of concentration. However, as shown in Section 4.3, the mass transfer coefficient, in general, depends on concentration, mostly via the dependence of the bulk viscosity on concentration (in turbulent flow) and on the wall correction factor (laminar and turbulent flow). The concentration dependence of the diffusion coefficient is also a factor but appropriate data are rarely available. The result of the viscosity dependence is that at high concentrations, a semi-log plot of flux versus concentration may exhibit non-linearity. This is the subject of the next example.

Example 4.5 Non-linear semi-log plots of flux versus concentration

Consider UF of a dextran solution where the wall concentration is at a limiting value of 250 g/l. Generate semi-log plots of dimensionless flux versus concentration in the range 1–250 g/l for (i) laminar flow, and (ii) turbulent flow, with a viscosity dependent mass transfer coefficient in each case. Take $z = 0.14$ and $\gamma = 0.015$ l/g.

Solution. The key equations here are Eqs. 4.13, 4.16 and 4.21. For laminar flow, the following equation applies:

$$J_{\text{lim}} = k_0 e^{0.14\gamma(c - c_{\text{lim}})} \ln \frac{c_{\text{lim}}}{c},$$

while for turbulent flow

$$J_{\lim} = k_0 e^{0.14\gamma(c-c_{\lim})} e^{-0.5\gamma c} \ln \frac{c_{\lim}}{c}.$$

(Recall that the relevant exponent has been assigned the value of 0.5 rather than the technically more nearly correct value of 0.47.)

A MATLAB script for plotting these equations is as follows:

```
%Example 45
clim=250; c=linspace(1,250); x=log(c);
Jlam=exp(0.015*0.14*(c-clim)).*log(clim./c)
Jturb=exp(-0.5*0.015*c).*exp(0.015*0.14*(c-
clim)).*log(clim./c)
plot(x,Jlam,'-',x,Jturb,'--');
xlabel('ln c');
ylabel('J / k_0');
axis([3.5 6 0 1.5])
```

From Fig. 4.3 it is clear that including the viscosity dependence of the mass transfer coefficient introduces non-linearities into the semi-log plots of flux versus concentration. The downward curvature in the laminar plot is a consequence of the fact that the mass transfer coefficient *increases* with concentration at fixed wall concentration. The upward curvature in the turbulent plot is due to the *reduction* in mass transfer coefficient under turbulent flow conditions. These non-linearities may cause problems for determining the value of $c_{\lim}$ from experimental data. This is the subject of Problem 4.5.

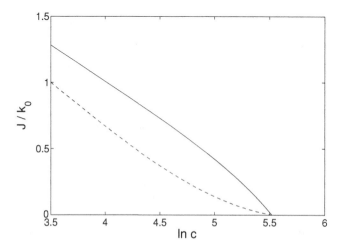

Figure 4.3 Limiting flux versus log concentration (g/l) for laminar flow (solid curve) and turbulent flow (dashed curve).

4.5   The osmotic pressure (OP) model

At this stage, it would be quite reasonable for the reader to be slightly troubled as to why there has been no mention of trans-membrane pressure in the analysis to date. Ultrafiltration is, after all, a pressure driven process and it would seem intuitive that any credible theory of UF should have pressure at its centre. Furthermore, the lack of any mention of the membrane seems to be a significant weakness in the approach. These comments are valid and a more comprehensive model of UF should incorporate these two parameters. While the LF model is adequate when the system is operating under limiting flux (high pressure) conditions, it is very important to have a general theory that is valid over the full range of pressures. This is where the OP model comes in. In this approach, the flux is explicitly related to the trans-membrane pressure, ΔP, through an expression of the form

$$J = \frac{\Delta P - \sigma_o \Delta \pi}{\mu R_m},$$
(4.23)

where μ is the permeate viscosity, R_m is the membrane resistance, σ_0 is the *osmotic reflection coefficient* and $\Delta \pi$ is the osmotic pressure difference across the membrane. This is simply the difference between the osmotic pressure of the solution at the wall concentration and the osmotic pressure of the solution at the permeate concentration. Thus, this model says that accumulation of solute at the membrane creates an osmotic back pressure that must be overcome by the applied trans-membrane pressure. This effect is essentially a thermodynamic one and reflects the fact that in a UF process, water is being forced from a region of low water concentration (on the retentate side) to a region of high water concentration (on the permeate side). The thermodynamic 'tendency', so to speak, is for the water to diffuse in the opposite direction, i.e., from the permeate side to the retentate side. The osmotic reflection coefficient is quite a mysterious concept in some ways, and is, for the most part, beyond the scope of this book. We return to it briefly in Chapter 8. It is true, however, that when there is complete rejection the osmotic reflection coefficient approaches one. Thus, for the case of complete rejection, Eq. 4.23 becomes

$$J = \frac{\Delta P - \pi}{\mu R_m},$$
(4.24)

where π is simply the osmotic pressure of the solution at the wall concentration. Now comparing this with Eq. 4.5, consistency requires

$$k \ln \frac{c_w}{c} = \frac{\Delta P - \pi}{\mu R_m}.$$
(4.25)

Typically, the osmotic pressure is given, to a good degree of accuracy, by a virial equation of the form

$$\pi = \sum_{i=1}^{3} a_i c_w^i.$$
(4.26)

Although the basic form of this type of equation can be derived from statistical thermo-dynamics, the constants, a_i, typically must be determined from experiment. Assuming virial data are available, c_w can be computed (numerically) by solution of the equation

$$k \ln \frac{c_w}{c} - \frac{\Delta P - \sum_{i=1}^{3} a_i c_w^i}{\mu R_m} = 0. \tag{4.27}$$

Once c_w is known, the flux can then be computed using either Eq. 4.5 or Eq. 4.24.

Application of this model obviously requires knowledge of the membrane resistance as well as osmotic pressure data. The resistance of a membrane depends on its material of construction, the mean pore diameter, membrane thickness, membrane porosity and other factors such as hydrophobicity and charge. It would be normal practice in UF (or MF) applications to determine the membrane resistance before beginning a run by measuring the flux of pure water through the membrane. The flux in that case is simply given by

$$J = \frac{\Delta P}{\mu R_m}. \tag{4.28}$$

A plot of J versus ΔP should yield a straight line with slope equal to $1/\mu R_m$. However, in some cases, membrane compression can occur and non-linearities can be observed. Values for pure-water membrane resistances are varied and can be as high as 1×10^{13} m^{-1} for membranes close to the nanofiltration range and as low as 1×10^{10} m^{-1} for membranes with cut-offs in the 10^6 Da range. The term 'pure-water' resistance reminds us that in real processing environments, the effective membrane resistance is likely to be higher than the pure-water value and probably time-dependent. The tendency for the membrane resistance to increase during UF and MF is referred to as *membrane fouling* and is usually irreversible in that cleaning agents are required to return the membrane to its 'pure-water' resistance state. Fouling is examined in detail in Chapter 10.

As well as membrane resistance data, the OP model obviously requires osmotic pressure data. In the past, accumulation of such data has required specialised equipment but, as shown in Problem 4.6, a relatively simple UF technique can be used to measure this important parameter. Table 4.5 provides osmotic pressure data for a range of solutes. Recall that the number used with dextran and polyethylene glycol (PEG) solutes represents the average molecular weight in kDa.

It is worth pointing out at this point that other mathematical forms for the osmotic pressure could be used. One that is sometimes used is the expression

$$\pi = B c_w^n, \tag{4.29}$$

where B and n are empirical constants. Example 4.6 involves a comparison of the accuracy of Eqs. 4.26 and 4.29.

Example 4.6 Comparing equations for correlating osmotic pressure data

For a set of whey concentration values (g/l) of [1, 10 50, 100, 150, 200, 250, 300] investigate whether Eq. 4.29 gives a good approximation to the virial equation for the osmotic pressure of whey given in Table 4.5.

Table 4.5 Virial coefficients for selected solutes (adapted from Table 1.6 of [10]).

Solute	Range (g/l)	a_1 (Pa l/g)	a_2 (Pa l^2/g^2)	a_3 (Pa l^3/g^3)
BSA pH7.4	0–450	378.7	−2.98	0.0100
BSA 5.5	0–450	56.33	−0.28	0.0026
BSA 4.5	0–475	75.39	−0.49	0.0019
Dextran T70	0–110	139	1.10	0.0032
Dextran T500	0–200	8.67	0.30	0.0090
Fibrinogen	0–80	9.948	−0.21	0.0028
β-lactoglobulin	0–250	26.99	1.31	0.0001
Polystyrene (90K)	0–150	56.98	0.83	0.0200
*Whey proteins	0–500	454.3	−0.176	0.0082
*Dextran T10	0–500	1152	−5.07	0.0260
**PEG6000	0–160	−270	15.0	0

Adapted from [11] **Adapted from* [15]

Solution. The osmotic pressure of the whey protein solutions using the information in Table 4.5 is

$$\pi = 454.3c - 0.176c^2 + 0.0082c^3.$$

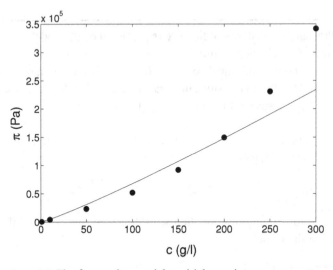

Figure 4.4 Fit of power law model to virial equation.

The MATLAB script below generates the required osmotic pressure 'data' and fits Eq. 4.29 to it.

```
%Example 46
c=[1 10 50 100 150 200 250 300];
osm=454.3*c-0.176*c.^2+0.0082*c.^3;
p=polyfit(log(c),log(osm),1);
```

```
n=p(1);
B=exp(p(2));
disp([B n]);
cpred=linspace(1,300);
osmpred=B*cpred.^n;
plot(c,osm,'ok',cpred, osmpred,'-');
xlabel('c (g/L)');
ylabel('\pi (Pa)');
```

The regression analysis gives $B = 369.1$ (Pa/(g/l)$^{1.13}$) and $n = 1.13$. The quality of the fit is shown in Fig. 4.4

4.5.1   Flux characteristics predicted by the osmotic pressure model

It was seen in Chapter 1 that two of the key characteristics of ultrafiltration are (i) the tendency for the flux to become independent of pressure at high pressures, and (ii) the finding that a semi-log plot of limiting flux versus concentration is linear over a wide range of concentration. In the next example the pressure dependence of the flux as predicted by the OP model is examined, and in the subsequent example the concentration dependence of the flux at a fixed pressure is investigated.

Example 4.7 Pressure dependence of the flux as predicted by OP model

A 20 g/l Dextran T10 solution is ultrafiltered in a module where the mass transfer coefficient, k, is constant and equal to 5×10^{-6} m/s, the membrane resistance is 6.0×10^{11} m^{-1} and the fluid viscosity can be taken to be 0.001 Ns/m^2. Generate a plot of dimensionless flux versus pressure in the range 0–8 bar.

Solution. Rearranging Eq. 4.27 slightly and putting in numerical values, including virial coefficients from Table 4.5, gives

$$\ln \frac{c_w}{20} - \frac{\Delta P - (1152 c_w - 5.07 c_w^2 + 0.026 c_w^3)}{3000} = 0.$$

MATLAB code for solving this equation with the **fzero** function, and the resulting plot (Fig. 4.5), are shown below.

```
function example47
global DP n;
for n=1:16;
DP(n)=n*0.5e5
guess=250;
cw(n)=fzero(@opmodel,guess)
flux(n)=log(cw(n)/20)
end;
plot(DP,flux,'ok')
xlabel('Pressure (Pa)');ylabel('J/k');
```

```
function f=opmodel(cw)
global DP n;
f=log(cw/20)-(DP(n)-(1152*cw-
5.07*cw^2+0.026*cw^3))/3000;
```

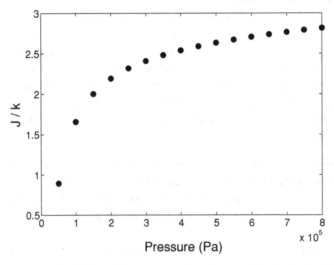

Figure 4.5 Prediction of the pressure dependence of the flux by the OP model.

It is clear from the results shown in Example 4.7 that the flux is predicted to be weakly dependent on pressure at high pressures but it appears that it is never truly independent of pressure. It is worth considering why no truly limiting flux is predicted. The technical definition of a limiting flux is:

$$\frac{dJ}{d\Delta P} = 0. \tag{4.30}$$

Using the chain rule, this implies

$$\frac{dJ}{dc_w}\frac{dc_w}{d\Delta P} = 0. \tag{4.31}$$

Now, implicit differentiation of Eq. 4.25 with constant mass transfer coefficient and bulk concentration gives

$$\frac{dc_w}{d\Delta P} = \frac{1}{k\mu R_m/c_w + d\pi/dc_w}. \tag{4.32}$$

Noting that Eq. 4.5 implies

$$\frac{dJ}{dc_w} = \frac{k}{c_w}, \tag{4.33}$$

the criterion for a technically limiting flux is

$$\frac{k/c_w}{\mu R_m k/c_w + d\pi/dc_w} = 0. \tag{4.34}$$

This can only be true if

$$\frac{d\pi}{dc_{\mathrm{w}}} \to \infty \qquad (4.35)$$

at some finite value of c_{w}. This is *not* mathematically possible, for finite c_{w}, if the osmotic pressure is a polynomial function of c_{w}.

Another interesting feature of Fig. 4.5 is the fact that the plot does not pass through the origin. In fact, the pressure at which the curve crosses the horizontal axis corresponds to the osmotic pressure of a 20g/l solution (the bulk concentration), or about 0.21 bar. Pressures lower than this are not large enough to overcome the osmotic pressure and thus there is no permeate flux. If you have not started Problem 4.6 yet, the last sentence should be a useful hint as to how to go about solving it. In the next example, the concentration dependence of the flux at a fixed pressure is investigated and compared with the predictions of the LF model.

Example 4.8 Concentration dependence of the flux at fixed pressure

Repeat Example 4.7 but fix the pressure at 4 bar and vary the bulk concentration in the range 16–240 g/l. Generate a semi-plot of J/k versus concentration.

Solution. Referring to the previous example, the equation to be solved can be written as

$$\ln \frac{c_{\mathrm{w}}}{c} - \frac{4 \times 10^5 - \left(1152 c_{\mathrm{w}} - 5.07 c_{\mathrm{w}}^2 + 0.026 c_{\mathrm{w}}^3\right)}{3000} = 0.$$

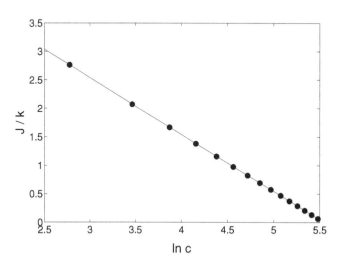

Figure 4.6 Prediction of concentration dependence of flux by OP model. Solid line represents a linear regression on the data; c in g/l.

MATLAB code for solving this problem is shown below and the results are shown in Fig. 4.6.

```
function example48
global c;
```

```
for n=1:15;
c=n*16;
guess=250;
cw=fzero(@opmodel,guess);
flux=log(cw/c);
x(n)=log(c); y(n)=flux;
end;
plot(x,y,'ok')
xlabel('ln c');ylabel('J/k');
function f=opmodel(cw)
global c;
f=log(cw/c)-(4e5-(1152*cw-
5.07*cw^2+0.026*cw^3))/3000;
```

In the above example at least, it is clear that the osmotic pressure model predicts flux versus concentration behaviour that is indistinguishable from the predictions of the LF model. Thus, there is no need to invoke any gelation or precipitation mechanism to explain the *effective* constancy of the wall concentration. The origins of the effectively constant wall concentration are interesting. In purely mathematical terms, it emerges naturally from the non-linear algebraic equation. In more physical terms, the wall concentration approaches a constant value because as the bulk concentration increases, the concentration at the wall also increases. This, in turn, leads to increased osmotic pressure which tends to reduce the flux, thus reducing the tendency for the wall concentration to increase. Ultimately, therefore, a balance is reached and the wall concentration remains essentially constant.

4.6   The osmotic pressure model with viscosity effects

The big assumption in the OP model as presented in the previous section is the neglect of the wall correction factor and, indeed, the concentration dependence of the un-corrected mass transfer coefficient. In this section these effects are included. Thus Eq. 4.27 is modified to give

$$k_c \left(\frac{\mu_c}{\mu_w}\right)^z \ln \frac{c_w}{c} - \frac{\Delta P - \sum_{i=1}^{3} a_i c_w^i}{\mu R_m} = 0. \tag{4.36}$$

We refer to this model as the viscosity–osmotic pressure (VOP) model. Assuming an equation of the form of Eq. 4.21 for the concentration dependence of the viscosity, this becomes

$$k_c e^{\gamma z(c-c_w)} \ln \frac{c_w}{c} - \frac{\Delta P - \sum_{i=1}^{3} a_i c_w^i}{\mu R_m} = 0. \tag{4.37}$$

As before, the best way to show the characteristics of this model is to do some numerical calculations. The following two examples are essentially modified versions of Examples 4.7 and 4.8. Example 4.9 looks at the pressure dependence of the flux with the aim of discovering whether the flux becomes independent of pressure at high pressures. Example 4.10 searches for linearity in a semi-log plot of flux versus concentration.

Example 4.9 Pressure dependence of the flux as predicted by VOP model

Repeat Example 4.7, but account for the wall correction factor using the viscosity parameter for Dextran T10 from Table 4.3 and let $z = 0.14$. Take k_c to be 5×10^{-6} m/s.

Solution. The equation employed in Example 4.7 is modified in this case to give

$$e^{0.0115 \times 0.14(20-c_w)} \ln \frac{c_w}{20} - \frac{\Delta P - \left(1152 c_w - 5.07 c_w^2 + 0.026 c_w^3\right)}{3000} = 0.$$

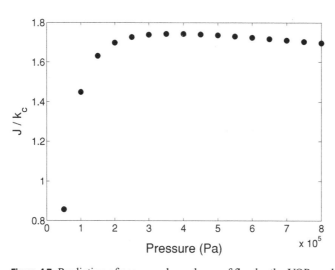

Figure 4.7 Prediction of pressure dependence of flux by the VOP model.

MATLAB code is shown below and the model predictions are shown in Fig. 4.7.

```
function example49
global DP;
for n=1:16;
DP=n*0.5e5;
guess=250;
cw=fzero(@opmodel,guess);
flux=exp(0.14*0.0115*(20-cw))*log(cw/20);
x(n)=DP; y(n)=flux;
end;
plot(x,y,'ok')
xlabel('Pressure (Pa)');ylabel('J/k_c');
```

```
function f=opmodel(cw)
global DP;
f=exp(0.14*0.0115*(20-cw))*log(cw/20)-(DP-(1152*cw-
5.07*cw^2+0.026*cw^3))/3000;
```

Clearly the fluxes computed in Example 4.9 are lower than those obtained in Example 4.7 where the wall correction factor was neglected. Interestingly, the model is found to predict a flux curve that is considerably 'flatter' at high pressures and the flux is even predicted to decline slightly with increasing pressure at high pressures. This is a phenomenon for which there is some evidence in the literature [11].

So, is it possible to determine an analytical expression for the limiting flux in this case? First recall Eq. 4.31. If there is no osmotic source for the limiting flux, this implies that the criterion for a limiting flux becomes

$$\frac{dJ}{dc_w} = 0. \tag{4.38}$$

Including the wall correction factor and assuming an exponential dependence of viscosity on concentration as before gives

$$\frac{d}{dc_w}\left[k_c e^{\gamma z(c-c_w)} \ln \frac{c_w}{c}\right] = 0. \tag{4.39}$$

Differentiating and recognising that the solution to this equation is the limiting concentration, c_{lim}, gives [4]

$$\frac{1}{c_{lim}} - \gamma z \ln \frac{c_{lim}}{c} = 0. \tag{4.40}$$

Thus, for any given concentration there is a limiting concentration that is solely determined by the viscosity characteristics of the solution. This is a remarkable result and is in contrast to the original gel polarisation model which implied that the limiting concentration reflected some sort of gelation process. This is a paradigm shift. Before discussing this idea of a limiting concentration being due to viscosity effects, some more of the features of the VOP model are explored.

Example 4.10 Concentration dependence of flux at fixed pressure – laminar flow

Repeat Example 4.9, but fix the pressure at 4 bar and vary the concentration in the range 10–240 g/l. For laminar flow we can assume that k_c is independent of concentration, i.e., $k_c = k_0$ which we take to be 5×10^{-6} m/s.

Solution. The equation to be solved thus becomes

$$e^{0.0115\times0.14(c-c_w)} \ln \frac{c_w}{c} - \frac{4 \times 10^5 - \left(1152c_w - 5.07c_w^2 + 0.026c_w^3\right)}{3000} = 0.$$

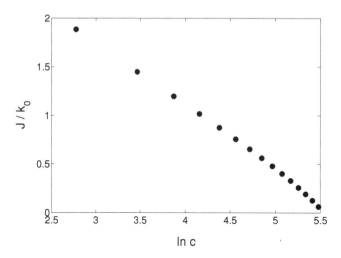

Figure 4.8 Flux versus log concentration (g/l) under laminar flow conditions.

MATLAB code for solving this problem is as follows:

```
function example410
format long g
global c;
for n=1:15;
c=n*16;
guess=250;
cw=fzero(@vopmodel,guess);
flux=exp(0.14*0.0115*(c-cw))*log(cw/c);
x(n)=log(c); y(n)=flux;
end;
plot(x,y,'ok')
xlabel('ln c'); ylabel('J/k_0');
function f=vopmodel(cw)
global c;
f=exp(0.14*0.0115*(c-cw))*log(cw/c)-(4e5-(1152*cw-
5.07*cw^2+0.026*cw^3))/3000;
```

As found with the limiting flux model, downward curvature is seen at high concentrations, see Fig. 4.8.

The next example is a repeat of Example 4.10 but the flow is now assumed to be turbulent, thus making the un-corrected mass transfer coefficient concentration-dependent.

Example 4.11 Concentration dependence of flux at fixed pressure – turbulent flow

Repeat Example 4.9 but assume also that k_c depends on bulk concentration via its dependence on the bulk viscosity and take $k_0 = 5 \times 10^{-6}$ m/s.

Solution. The equation to be solved now becomes

$$e^{-0.5 \times 0.0115 c} e^{-0.14 \times 0.0115 (c - c_w)} \ln \frac{c_w}{c} - \frac{4 \times 10^5 - (1152 c_w - 5.07 c_w^2 + 0.026 c_w^3)}{3000} = 0.$$

MATLAB code for this problem is as follows:

```
function example411
global c;
for n=1:15;
c=n*16;
guess=250;
cw=fzero(@vopmodel,guess);
flux=exp(-0.5*0.0115*c)*exp(0.14*0.0115*(c-cw))*log(cw/c);
x(n)=log(c); y(n)=flux;
end;
plot(x,y,'ok')
xlabel('ln c');ylabel('J/k_0');
function f=vopmodel(cw)
global c;
f=exp(-0.5*0.0115*c)*exp(0.14*0.0115*(c-
cw))*log(cw/c)-(4e5-(1152*cw-
5.07*cw^2+0.026*cw^3))/3000;
```

The resulting plot is shown in Fig. 4.9.

Here we have the same upward curvature as was obtained with the limiting flux model. Of course, this is not surprising given that in the osmotic pressure model, the wall concentration is effectively constant at a fixed, high pressure.

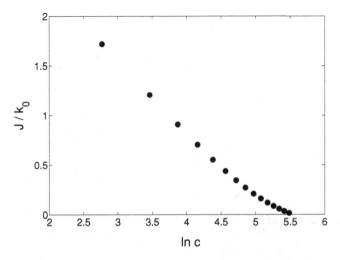

Figure 4.9 Flux versus log concentration (g/l) under turbulent flow conditions.

The above examples have shown that when the viscosity correction factor is included, a departure from linear behaviour is observed at high concentrations. Furthermore, the departure in turbulent flow is qualitatively different from that observed in laminar flow, as seen earlier in the context of the LF model. It is worth noting that even more complex behaviour can be observed if turbulent-to-laminar transitions occur or if the flow is non-Newtonian [5].

4.6.1   Is the VOP model 'correct'?

If you do Problem 4.7 now, it should give you some idea of where this section is headed. Let us go back to Eq. 4.40. This equation predicts that there is a limiting concentration which is a function of the bulk concentration. It is not really a constant at all. For example, with $z = 0.14$ and $\gamma = 0.01$ l/g, one gets the relationship shown in Fig. 4.10 (see Problem 4.8). The $c_{\lim}$ values appear to reach very unrealistically high levels. This fact suggests that there is a fundamental flaw in the VOP model as an explanation for the limiting flux.

The results obtained in Fig. 4.10 lead to the plots of limiting flux versus ln c shown in Fig. 4.11. These were obtained in the same manner as in Example 4.5, the key difference being that $c_{\lim}$ is not constant in this case.

Thus, 'upward curvature' is predicted in both cases. But what exactly is this limiting flux? Let us go back to the osmotic pressure expression for the flux and rearrange it to give

$$\Delta P = \mu R_{\mathrm{m}} J + \pi. \qquad (4.41)$$

Thus, the limiting pressure, i.e., the pressure at which $dJ/d\Delta P = 0$, is given by

$$\Delta P_{\lim} = \mu R_{\mathrm{m}} J_{\lim} + \pi_{\lim}. \qquad (4.42)$$

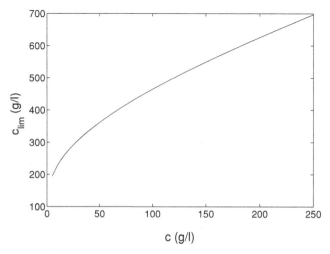

Figure 4.10 Limiting concentration versus bulk concentration for $\gamma = 0.01$ l/g and $z = 0.14$.

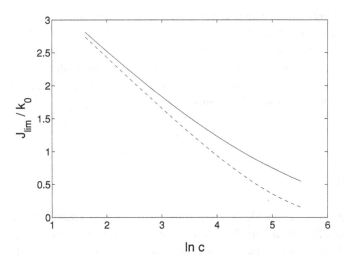

Figure 4.11 Limiting flux versus log concentration (g/l). k_0 is not the same for both flows. Solid curve denotes turbulent flow, dashed line denotes laminar flow.

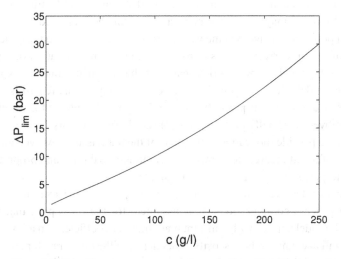

Figure 4.12 Limiting pressure versus concentration. Osmotic pressure data for whey, $\mu = 0.001$ Ns/m^2, $R_m = 1 \times 10^{11}$ m^{-1}, $k_0 = 5 \times 10^{-6}$ m/s, turbulent flow.

Figure 4.12 is a plot of ΔP_{lim} versus c using osmotic parameters for whey (see Problem 4.9).

The key point here is that the limiting flux predicted from the VOP model is something of an abstraction. The model predicts that in *all* cases, the flux will exhibit a *maximum* with respect to pressure, declining with pressure at pressures beyond this maximum. Furthermore, it predicts that if the bulk concentration is increased, the limiting flux occurs at a *higher* pressure. This is contrary to what is typically observed in experiments. The usual observation is that the limiting pressure, i.e., the pressure at which the flux becomes independent of pressure, becomes easier to attain (i.e., occurs at a lower pressure) as the bulk concentration is increased.

The conclusion from all of this is that the VOP model is an improvement on the OP model but it clearly does not tell the full story. It helps to quantify the effect of viscosity on the mass transfer coefficient but it is probable that the pressure independent flux that occurs almost universally in UF systems is due to something other than the viscosity dependence of the mass transfer coefficient. We return to this issue in Chapter 10.

4.7 Conclusions

In this chapter we have provided a brief summary of the best known and easily implemented theories for calculating the flux in ultrafiltration. In the case of the LF model, calculation of fluxes is easy but the limiting concentration must be treated as a purely empirical parameter to be determined by experiment. The equations for the osmotic pressure models are more complex and generally require numerical solution of a non-linear algebraic equation to compute the flux. In all cases, knowledge of the mass transfer coefficient is required and this may involve more than simple application of an experimental correlation. It may be necessary to account for the increased viscosity in the concentration polarisation layer via the wall correction factor and also the concentration dependence of the un-corrected mass transfer coefficient. Incorporation of all of these ideas into a single model uncovers a rich seam of behaviour that cannot be explained by the simple LF model. However, a major practical difficulty in this regard is the dearth of good physical and transport property data for the type of solutions for which ultrafiltration is employed, especially proteins and polysaccharides. In this chapter an attempt has been made to provide the reader some sense of the main issues involved and provide him/her with sufficient theoretical knowledge, computational tools and property data so as to be able to perform analysis and design calculations.

In the next chapter, we move on from theory to practice and examine a range of ultrafiltration process configurations. It is assumed, for the most part, that the basic form of the LF model applies with constant mass transfer coefficient. In a few instances, however, we will account for the viscosity dependence of the mass transfer coefficient. In Chapter 8 the use of the OP and VOP models in process analysis and design is explored. For all of these models it will be seen that most process scenarios require numerical solution of the relevant design equations.

References

1. Zeman, L.J. and Zydney, A.L. (1996). *Microfiltration and Ultrafiltration. Principles and Applications*. Marcel Dekker, New York, USA.
2. Cussler, E.L. (2009). *Diffusion: Mass Transfer in Fluid Systems*, 3rd Edition. Cambridge University Press, New York, USA.

3. Field, R.W. (1990). A theoretical viscosity correction factor for heat transfer and friction in pipe flow. *Chemical Engineering Science*, **45**, 1343–1347.

4. Field, R.W. and Aimar, P. (1993). Ideal limiting fluxes in ultrafiltration: comparison of various theoretical relationships. *Journal of Membrane Science*, **80**, 107–116.

5. Pritchard, M., Howell, J.A. and Field, R.W. (1995). The ultrafiltration of viscous fluids. *Journal of Membrane Science*, **102**, 223–235.

6. Howell, J., Field, R.W. and Wu, D. (1996). Ultrafiltration of high-viscosity solutions: theoretical developments and experimental findings. *Chemical Engineering Science*, **51**, 1405–1415.

7. He, L. and Niemeyer, B. (2003). A novel correlation for protein diffusion coefficients based on molecular weight and radius of gyration. *Biotechnology Progress*, **19**, 544–548.

8. Young, M.E., Carroad, P.A. and Bell, R.L. (1980). Estimation of diffusion coefficients of proteins. *Biotechnology and Bioengineering*, **22**, 947–955.

9. Wu, C. (1993). Laser light scattering characterization of the molecular weight distribution of dextran. *Macromolecules*, **26**, 3821–3825.

10. Cheryan, M. (1998). *Ultrafiltration and Microfiltration Handbook*, 2nd Edition. CRC Press, Boca Raton, FL, USA.

11. Jonsson, G. (1984). Boundary layer phenomena during ultrafiltration of dextran and whey protein solutions. *Desalination*, **51**, 61–77.

12. Gonzalez-Tello, P., Camacho, F., Guadix, E.M., Luzon, G. and Gonzalez, P.A. (2009). Density, viscosity and surface tension of whey protein concentrate solutions. *Journal of Food Process Engineering*, **32**, 235–247.

13. Tirtaatmadja, A., Dunstan, D.E. and Boger, D.V. (2001). Rheology of dextran solutions. *Journal of Non-Newtonian Fluid Mechanics*, **97**, 295–301.

14. Blatt, W.F., Dravid, A., Michaels, A.S. and Nielsen, L. (1970). Solute polarization and cake formation in membrane ultrafiltration: causes, consequences, and control techniques. In: *Membrane Science and Technology*, Flinn, J.E. (Ed.): Plenum Press, NY, USA.

15. Money, N.P. (1989). Osmotic pressure of aqueous polyethylene glycols. *Plant Physiology*, **91**, 766–769.

16. Gill, W.N., Wiley, D.E., Fell C.J.D. and Fane, A.G. (1988). Effect of viscosity on concentration polarization in ultrafiltration. *AIChE Journal*, **34**, 11563–11567.

Problems

Problem 4.1   The significance of a concentration dependent density

The data below give densities as a function of concentration for BSA and Dextran T70 solutions as well as a whey protein solution [12, 16]. Use these data to draw some conclusions as to the importance of accounting for the concentration dependence of the density in predicting mass transfer coefficients in these systems.

BSA		Dextran T70		Whey proteins	
c (wt%)	ρ_c (kg/m^3)	c (wt%)	ρ_c (kg/m^3)	c (wt%)	ρ_c (kg/m^3)
0.1	1003	0.5	1000	0	998
1.0	1015	1.0	1002	5.0	1013
5.0	1028	5.0	1017	10.0	1026
10.0	1040	10.0	1037	20.0	1060

Problem 4.2   The significance of a concentration dependent diffusion coefficient

The data below give diffusion coefficients as a function of concentration for a BSA solution and a Dextran T70 solution [16].

BSA		Dextran T70	
c (wt%)	D (m^2/s $\times 10^{-11}$)	c (wt%)	D (m^2/s $\times 10^{-11}$)
0.1	7.093	0.5	4.108
1.0	7.026	1.0	4.188
5.0	6.740	5.0	5.810
10.0	6.402	10.0	7.812
15.0	6.082	15.0	8.063
20.0	5.782	20.0	8.079
40.0	4.745		

Use these data to discuss the importance of the concentration dependence of the diffusion coefficient for predicting the mass transfer coefficient in these systems.

Problem 4.3   Improving the prediction of diffusion coefficients

We saw in Example 4.1 that Eq. 4.19 is only moderately accurate for predicting the diffusion coefficients of the solutes in Table 4.2. Generalise this equation by allowing the power on the molecular weight to be an adjustable parameter, n, instead of being fixed at $-\frac{1}{3}$. Use a regression analysis to work out the best fit value of n and compare the accuracy of your equation with that of Eq. 4.18.

Problem 4.4   Verification of limiting flux model

Give a short explanation of why Eq. 4.22 is consistent with observations of limiting flux behaviour described in Chapter 1.

Problem 4.5   Errors in estimating c_{lim}

Consider the plots generated in Example 4.5. By considering data obtained at concentrations up to 100 g/l only, estimate c_{lim} in each case by linear regression. Compare your answer with the values given for c_{lim} in Example 4.5.

Problem 4.6   Measuring osmotic pressures using UF experiments

Consider a stirred cell ultrafiltration apparatus containing a known concentration of solute. Now, imagine an experiment in which the pressure is initially set to zero and then very slowly increased. Pursue this idea and describe how it leads to a method for measuring the osmotic pressure of the solution.

Problem 4.7   Pressure dependence of the flux at different concentrations

Repeat Example 4.9 for a range of bulk concentrations and plot your results on a single graph. How is the location of the optimum affected by concentration according to the model?

Problem 4.8   Concentration dependence of the limiting concentration and limiting flux

Solve Eq. 4.40 to reproduce Figs. 4.10 and 4.11.

Problem 4.9   Effect of concentration on the limiting pressure

Solve Eq. 4.40 and use Eq. 4.41 to reproduce Fig. 4.12.

Further problems

Problem 4.10   Variation of wall concentration with pressure

Show that if the wall correction factor is included and the bulk concentration is fixed

$$\frac{dc_w}{d\Delta P} = \frac{1}{\mu R_m k/c_w + \mu R_m \dfrac{dk}{dc_w} \ln \dfrac{c_w}{c} + \dfrac{d\pi}{dc_w}}.$$

Problem 4.11   Flux versus log concentration at low pressures

In this problem we re-visit Example 4.8. Plot J/k versus $\ln c$ with $R_m = 6.0 \times 10^{12}$ and $\Delta P = 0.5$ bar and assuming all other parameters are the same as in that example. Discuss your results and illustrate your answer with a plot of c_w versus pressure.

Problem 4.12   Limiting concentration when the viscosity is a power function of concentration

Equation 4.40 is based on the assumption that the viscosity is an exponential function of concentration. Derive a similar equation for the case where the viscosity is related to concentration by a polynomial expression of the form

$$\mu_c = \mu \left(1 + \sum_{i=1}^{N} a_i c^i \right)$$

where the a_i are constants.

5 Ultrafiltration process analysis and design at the limiting flux

5.1   Introduction

In this chapter, mathematical models for analysis, design and optimisation of a variety of continuous, batch and fed-batch UF systems are developed. In all cases the system is assumed to be operating under limiting flux conditions, i.e., Eq. 4.22 applies. It should be recalled that this equation implies complete rejection of the solute. It is seen that, despite the relative simplicity of the models, the presence of the logarithmic term in the flux expression requires one to routinely use numerical techniques to solve the model equations. In addition, it is shown how a graphical method can be useful for certain types of calculations and also it is demonstrated that by employing *special functions*, numerical solution can be avoided in solving certain batch and continuous problems.

In the next section, continuous feed-and-bleed systems are examined, where the basic computational problem is to solve systems of non-linear algebraic equations (NLAEs). First, process analysis calculations are examined where the goal is to calculate the exit concentration for a given system. In process design calculations, the required area for a given process specification is computed. In process optimisation calculations, the problem of how to minimise the required area in multi-stage systems is examined. Subsequent sections examine batch, fed-batch and single pass continuous systems.

5.2   Continuous feed-and-bleed ultrafiltration

Consider the example of continuous concentration of a macromolecular solution by the feed-and-bleed mode illustrated in Fig. 5.1.

In the feed-and-bleed configuration, a portion of the product stream (retentate) is recycled back into the feed. This causes the system to operate at a sufficiently high pressure for the flux to be at its limiting value. The increased flowrate through the module as a result of the recirculation loop also increases the mass transfer coefficient, thus increasing the limiting flux. Typically, the flowrate through the module greatly exceeds the feed flowrate, and thus the system can reasonably be assumed to be well mixed, i.e., the retentate concentration can be taken to be the same as the mean concentration in the module itself. This is in contrast to single pass operation, which

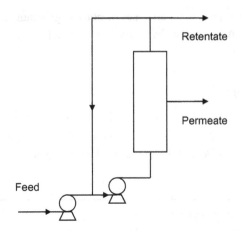

Figure 5.1 Continuous feed-and-bleed operation.

is examined in Section 5.5, where there is no recycle and the variation of flowrate and concentration with distance along the membrane must be accounted for in the analysis.

Given that the module is assumed to be well mixed and operating at the limit, the limiting expression for the flux can be written as

$$J_1 = k \ln \frac{c_{\lim}}{c_1}, \tag{5.1}$$

where J_1 is the *limiting* flux in the module and c_1 is the *exit* concentration. Unless otherwise stated, the mass transfer coefficient is assumed here, and in subsequent calculations, to be independent of concentration. The solute balance for the system, assuming complete rejection (i.e., no solute in the permeate), can be written

$$Q_0 c_0 = Q_1 c_1, \tag{5.2}$$

where Q_0 is the feed volumetric flowrate, c_0 is the feed concentration and Q_1 is the retentate flowrate. An overall balance gives

$$Q_0 = Q_1 + J_1 A, \tag{5.3}$$

where A is the membrane area. Combining Eqs. 5.1 to 5.3 (Problem 5.1) gives the following governing equation of a single-stage system:

$$\frac{Q_0}{kA}(1 - x_1) - \ln x_1 - \ln \frac{c_{\lim}}{c_0} = 0, \tag{5.4}$$

where the dimensionless concentration, x_1, is defined by

$$x_1 = \frac{c_0}{c_1}. \tag{5.5}$$

Equation 5.4 is a key equation for subsequent analyses of both single-stage and multi-stage systems. In the next section its use in the analysis of a single-stage system is illustrated.

5.2.1   Analysis of a single-stage system

In this type of problem, it is assumed that the membrane area, the feed flowrate, the feed concentration and the mass transfer coefficient are known. The goal therefore is to calculate the exit concentration using Eq. 5.4. Example 5.1 demonstrates the solution of a problem of this type.

Example 5.1 Analysis of a single-stage system

Consider a 1 l/min feed of a protein solution which enters a single-stage continuous feed-and-bleed UF system. The feed enters at 10 g/l and c_{lim} is 300 g/l. The mass transfer coefficient is 3.5×10^{-6} m/s and the membrane area is 2.7 m^2. Calculate the exit concentration.

Solution. Putting the relevant parameters into Eq. 5.4 and remembering to convert the flowrate to m^3/s gives

$$\frac{1 \times 10^{-3}/60}{3.5 \times 10^{-6} \times 2.7}(1 - x_1) - \ln x_1 - \ln \frac{300}{10} = 0.$$

Working to three decimal places, this becomes

$$1.764\,(1 - x_1) - \ln x_1 - 3.401 = 0.$$

This a non-linear algebraic equation, which, as seen in previous chapters, provides no great challenges for MATLAB. Code for the solution of this equation is shown below. The use of the 'x' parameter means that the initial guess must be between zero and one.

```
function example51
guess=0.5;
x=fzero(@singlestage,guess);
function f=singlestage(x);
f=1.764*(1-x)-log(x)-3.401;
```

Running the program gives $x_1 = 0.1495$. Hence

$$c_1 = \frac{c_0}{x_1} = \frac{10}{0.1495} = 66.89 \text{ g/l.}$$

5.2.2   Analysis of a multi-stage system

Extension of the above analysis to multi-stage systems is straightforward and involves no new concepts. In a multi-stage system, the retentate from one stage forms the feed to the next, as shown schematically in Fig. 5.2.

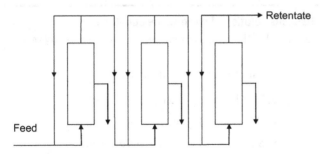

Figure 5.2 Three-stage feed-and-bleed system.

It is easy to show (Problem 5.2) that the governing equation for stage i can be written

$$\frac{Q_0}{kA}(x_{i-1} - x_i) - \ln\frac{c_{\lim}}{c_0} - \ln x_i = 0, \tag{5.6}$$

where it is assumed that the area and mass transfer coefficient are the same for each stage. For a system with an arbitrary number of stages, N, a simple way to solve these equations is to tackle them sequentially, as shown for a three-stage system in Example 5.2 where the MATLAB routine **fzero** is used. In Example 5.3, the same equations are solved simultaneously using MATLAB's **fsolve**.

Example 5.2 Sequential solution of multi-stage equations

In this example, 1 l/min of a protein solution is fed to a 3-stage continuous feed-and-bleed ultrafiltration system. The feed again enters at 10 g/l and $c_{\lim}$ can be taken to be 300 g/l. The mass transfer coefficient is 3.5×10^{-6} m/s in each stage and the area of each stage is 0.9 m^2, thus giving the same total area as in Example 5.1. Use **fzero** to compute the concentration leaving the third stage.

Solution. With the numbers supplied, Eq. 5.6 becomes for the first stage:

$$5.291(1 - x_1) - \ln x_1 - 3.401 = 0.$$

Solving this equation using the **fzero** function within MATLAB, with an initial guess of 0.5, as shown in the previous example, gives $x_1 = 0.4915$. The equation for the second stage thus becomes

$$5.291(0.4915 - x_2) - \ln x_2 - 3.401 = 0.$$

Using the same initial guess, we get $x_2 = 0.1765$ and thus the equation for the third stage becomes

$$5.291(0.1765 - x_3) - \ln x_3 - 3.401 = 0.$$

Solving this equation gives $x_3 = 0.0613$. Therefore, the exit concentration from the final stage is $10/0.0613 = 163.1$ g/l, which is considerably greater than the 66.89 g/l achieved with a single-stage system of the same total area as found in Example 5.1.

Example 5.3 Simultaneous solution of multi-stage equations

Repeat Example 5.2 but solve the equations simultaneously using **fsolve**.

Solution. Using Eq. 5.6 and putting in numerical values gives

$$5.291\,(1 - x_1) - \ln x_1 - 3.401 = 0,$$
$$5.291\,(x_1 - x_2) - \ln x_2 - 3.401 = 0,$$
$$5.291\,(x_2 - x_3) - \ln x_3 - 3.401 = 0.$$

MATLAB code for solving these equations with **fsolve** is shown below. The choice of initial guesses accounts for the fact that the x_i values must get smaller as i increases. Indeed, a real solution is not found if the initial guess for each variable is set at 0.5. This illustrates one of the computational difficulties with NLAEs, namely that convergence is often dependent on the quality of the initial estimates.

```
function example53
x=fsolve(@multistage,[0.5,0.2,0.1])
function f=multistage(x)
x1eqn=5.291*(1-x(1))-log(x(1))-3.401;
x2eqn=5.291*(x(1)-x(2))-log(x(2))-3.401;
x3eqn=5.291*(x(2)-x(3))-log(x(3))-3.401;
f=[x1eqn,x2eqn,x3eqn];
```

Reassuringly, the answers are precisely the same as those obtained in Example 5.2. In Problem 5.3 you are asked to solve a similar problem but with unequal areas.

5.2.3 Analysis of a multi-stage system using ordinary differential equations

The last three examples have used the obvious approach of solving the non-linear algebraic equations describing the system. Here is another way that involves use of ordinary differential equations. It will only work in a multi-stage system if the stages have equal areas. The 'trick' is to differentiate Eq. 5.6 to get

$$\frac{dx_i}{dA} = \frac{\dfrac{Q_0}{k}\dfrac{dx_{i-1}}{dA} - \ln \dfrac{c_{\lim}}{c_0} - \ln x_i}{Q_0/k + A/x_i}. \tag{5.7}$$

This is a useful exercise in implicit differentiation that the interested reader might like to check. Now, an ODE solver can be used to solve this system of equations, noting that $x_i = 1$ @ $A = 0$. Code for re-solving Examples 5.2 and 5.3 is shown below, where it should be noted that the endpoint of the calculations is $A = 0.9$ m^2, i.e., the area for each stage.

```
function odesforuf
[A,x]=ode45(@contuf,[0 0.9],[1, 1, 1, 1])
function dxdA=contuf(A,x)
```

```
clim=300; c0=10; Q0=1e-3/60; k=3.5e-6;
xieqn(1)=0;% eqn for x0
for i=2:4;%stages 1 to 3
xieqn(i)=(Q0/k*xieqn(i-1)-log(clim/c0)-
log(x(i)))/(A/x(i)+Q0/k);
end
dxdA=[xieqn(1);xieqn(2);xieqn(3);xieqn(4)];
```

The results obtained are the same as the previous example as one would expect. The advantage of using the ODE approach is that there is no need to make any initial estimates for the answers. While making good initial guesses is not a major difficulty for the problems tackled in this book, it can sometimes be a troublesome aspect of solving NLAEs. The ODE approach to solving problems that are more obviously formulated in terms of NLAEs does not only apply to membrane systems but can also be used in, for example, chemical reaction engineering. This is illustrated in Problem 5.4.

5.2.4   Analysis of a multi-stage system using a graphical method

Yet another way of solving problems of this type is to use a graphical approach. Chemical engineers will be very familiar with a host of ingenious graphical techniques, developed in the pre-computer era, for solving multi-stage problems in areas like distillation, liquid–liquid extraction and gas absorption. While somewhat obsolete as calculation tools, these methods retain some pedagogical value for some people. Here, a simple graphical technique for solving continuous UF analysis problems is presented. Let us consider the first stage of a multi-stage system and recall that the flux can be written, by expanding out Eq. 5.1, in terms of the x parameter as

$$J = k \ln \frac{c_{\lim}}{c_0} + k \ln x, \tag{5.8}$$

where, here, x represents the ratio c_0/c and c is the exit concentration whose precise value for the module in question (c_1) is to be determined. Similarly, we can write

$$J = \frac{Q_0}{A} (1 - x). \tag{5.9}$$

Equation 5.8 is somewhat analogous to the *equilibrium curve* typically encountered in equilibrium stage operations such as distillation, liquid–liquid extraction and absorption. Equation 5.9 represents a straight line and can thus be thought of as the *operating line* of the module. Thus, feed-and-bleed ultrafiltration can actually be described with much of the same language, and using many of the same methods, as the more conventional unit operations well known to chemical engineers.

Equation 5.9, the operating line, has an x-axis intercept of 1.0 and a y-axis intercept of Q_0/A. The concentration leaving the stage, i.e., x_1, can be found as the point of intersection of the flux curve and the operating line. The value of this approach is the ease with which it can be extended to multi-stage systems. Carrying out a similar analysis for the second stage, one can write the following expression for the flux in the second stage:

$$J = \frac{Q_0}{A}(x_1 - x), \tag{5.10}$$

where equal areas have been assumed. Assuming the mass transfer coefficient is unchanged in the second stage, Eq. 5.8 can be applied as before. Now, Eq. 5.10 is a straight line with x-intercept $= x_1$ and y-intercept $= Q_0 x_1 / A$. Clearly, therefore, Eqs. 5.9 and 5.10 represent parallel lines and the concentration from the second stage can be evaluated precisely as done for the first stage. Example 5.4 illustrates the use of this technique for the same problem posed in Examples 5.2 and 5.3.

Example 5.4 Graphical solution of multi-stage analysis equations

Repeat the previous example but use the graphical method described above.

Solution. For this problem, Eq. 5.8 becomes

$$J = 1.19 \times 10^{-5} + 3.5 \times 10^{-6} \ln x.$$

From Eq. 5.9 the operating line for the first module is

$$J = 1.852 \times 10^{-5}(1 - x).$$

Plotting these two expressions leads directly to the construction shown in Fig. 5.3, and x_3 can be rapidly determined. Reading from the graph gives $x_3 = 0.06$. Therefore $c_3 = c_0/0.06 = 167 \text{g/l}$, in close agreement with the numerical solution obtained previously. In Problem 5.5, you are asked to solve a two-stage problem with unequal areas.

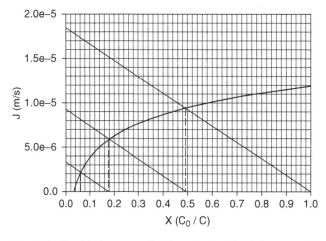

Figure 5.3 Graphical construction for solution of multi-stage problem.

5.2.5   Design of a single-stage system

In this type of problem, the exit concentration is specified and it is the area that must be calculated. For a single-stage system, therefore, Eq. 5.4 becomes after a little rearranging

$$A = \frac{Q_0(1 - x_1)/k}{\ln(c_{\text{lim}}/c_0) + \ln x_1}. \tag{5.11}$$

In this case, calculation of the required area is trivial as long as the mass transfer coefficient is known. But what if the mass transfer coefficient is not known? Recall from Chapter 4 that the mass transfer coefficient depends on a number of factors, including operating conditions, module geometry and fluid properties. Suppose, for example, that the module to be sized is of the hollow fibre type in which the flow is laminar. Then, using Eq. 4.14 one gets (see Problem 5.6)

$$k = 2.57 \left(\frac{D}{d_t} \right)^{2/3} \left[\frac{Q_{\text{in}}}{A} \right]^{1/3}, \tag{5.12}$$

where Q_{in} is the *total* flowrate of fluid actually entering the module, i.e. the feed flow plus the flow in the recirculation loop. Recall that D is the diffusivity of the solution and d_t is the fibre diameter. Combining Eqs. 5.11 and 5.12, the following expression is obtained for the required area:

$$A = 0.24 \frac{d_t}{D} Q_0 \left(\frac{Q_0}{Q_{\text{in}}} \right)^{1/2} \left(\frac{(1 - x_1)}{\ln(c_{\text{lim}}/c_0) + \ln x_1} \right)^{3/2}. \tag{5.13}$$

Use of this equation is illustrated in the example below.

Example 5.5 Single-stage design with unknown mass transfer coefficient

A membrane company produces ultrafiltration hollow fibres of diameter 1.0 mm. Pump and mechanical limitations mean that the inlet flowrate to the fibres cannot exceed 50 l/min. Calculate the minimum area required for a module containing these fibres to concentrate 1.0 l/min of a 10 g/l protein solution to 40 g/l. Take the limiting concentration to be 280 g/l and the diffusivity for this protein solution to be 9.0×10^{-11} m^2/s. If the fibres are made with a length of 0.5 m, how many of these fibres would be needed in the module?

Solution. Inserting the relevant numbers and ensuring that S.I. units are used, the area is given by

$$A = 0.24 \left(\frac{0.001}{9.0 \times 10^{-11}} \right) \left(\frac{1 \times 10^{-3}}{60} \right) \left(\frac{1}{50} \right)^{1/2} \left(\frac{(1 - 1/4)}{\ln(280/10) + \ln(1/4)} \right)^{1.5} = 1.50 \text{ m}^2$$

Now, the cartridge area is given by

$$A = N\pi d L_t = N\pi \times 0.001 \times 0.5,$$

where L_t is the fibre length. Therefore

$$N = \frac{1.50}{\pi \times 0.5 \times 0.001} = 955.$$

5.2.6   Design of a multi-stage system with equal areas

For this type of problem, Eq. 5.6 can be used. This represents N equations where the unknowns are the x_i for $1 \le i \le N-1$ and A, the area of each stage. Technically the area term can be eliminated by rearranging of any one of the equations, thus reducing the problem to solving $N-1$ equations for the intermediate compositions. However, the computational advantages in doing this are small and the resulting $N-1$ equations are a little more cumbersome. Solution of the design problem is illustrated in Example 5.6, where the MATLAB routine **fsolve** is used.

Example 5.6 Design of a three-stage equal area system using fsolve

In this example, 1 l/min of a protein solution is fed to a three-stage, equal area, continuous feed-and-bleed ultrafiltration system. The feed enters at 10 g/l and c_{lim} can be taken to be 300 g/l. The mass transfer coefficient is 3.5×10^{-6} m/s in each stage and the retentate concentration leaving the third stage is 100 g/l. Find the area of each stage and the total area.

Solution. Applying Eq. 5.6 and using the relevant data (including $x_3 = 10/100 = 0.1$) gives the following equations to be solved:

$$\frac{4.762}{A}(1 - x_1) - 3.401 - \ln x_1 = 0,$$

$$\frac{4.762}{A}(x_1 - x_2) - 3.401 - \ln x_2 = 0,$$

$$\frac{4.762}{A}(x_2 - 0.1) - 3.401 - \ln(0.1) = 0.$$

The appropriate MATLAB code is shown below.

```
function example56
x=fsolve(@multistage,[0.5,0.2,1])
function f=multistage(x)
x1eqn=4.762/x(3)*(1-x(1))-log(x(1))-3.401;% NB x(3) is A
x2eqn=4.762/x(3)*(x(1)-x(2))-log(x(2))-3.401;
Aeqn=4.762/x(3)*(x(2)-0.1)-log(0.1)-3.401;
f=[x1eqn,x2eqn,Aeqn];
```

Running the code with the initial guesses shown in the code gives $x_1 = 0.574$, $x_2 = 0.264$ and $A = 0.713$ m^2, thus giving a total area of 2.138 m^2.

5.2.7   Optimisation of multi-stage systems

In the last section, it was shown how the use of a multi-stage system is superior to a single-stage system. However, there is no reason why the area in each stage should be the same, although, in practice, the use of equal areas is probably the most likely choice given the limited range of membrane modules produced by manufacturers. For a system with N stages, the goal, therefore, is to find the values of x_i that minimise the total area, A_t, for fixed x_N where, using Eq. 5.6, we can write

$$A_t = \sum_{i=1}^{N} A_i = \frac{Q_0}{k} \sum_{i=1}^{N} \frac{(x_{i-1} - x_i)}{\ln(c_{lim}/c_0) + \ln x_i}, \tag{5.14}$$

with $x_0 = 1$. The optimum is found by applying the following conditions for $1 \leq i \leq N - 1$:

$$\left(\frac{\partial A_t}{\partial x_i}\right)_{x_{j \neq i}} = \frac{\ln(c_{lim}/c_0) + \ln x_i + \frac{x_{i-1}}{x_i} - 1}{(\ln(c_{lim}/c_0) + \ln x_i)^2} - \frac{1}{\ln(c_{lim}/c_0) + \ln x_{i+1}} = 0. \tag{5.15}$$

This was actually a tricky differentiation, which the interested reader should attempt. Rearranging Eq. 5.15 gives

$$\ln x_{i+1} = \frac{(\ln(c_{lim}/c_0) + \ln x_i)^2}{\ln(c_{lim}/c_0) + \ln x_i + \frac{x_{i-1}}{x_i} - 1} - \ln \frac{c_{lim}}{c_0}. \tag{5.16}$$

Therefore, the values of the x_i that lead to a minimum area are found by simultaneous solution of the $N - 1$ equations represented by Eq. 5.16. It is worth noting that in the limit where $x_i \to 1$, these equations can be simplified somewhat. In that case, one can use a Taylor series expansion to write

$$x_{i-1} \approx 1 + \ln x_{i-1}. \tag{5.17}$$

Applying this relation to x_i and using Taylor series again, one gets

$$\frac{1}{x_i} \approx 2 - x_i = 1 - \ln x_i. \tag{5.18}$$

This gives

$$\ln x_{i+1} = \frac{(\ln(c_{lim}/c_0) + \ln x_i)^2}{\ln(c_{lim}/c_0) + \ln x_i + (1 + \ln x_{i-1})(1 - \ln x_i) - 1} - \ln \frac{c_{lim}}{c_0}. \tag{5.19}$$

Neglecting $(\ln x)^2$ terms gives

$$\ln x_{i+1} + \ln \frac{c_{lim}}{c_0} = \frac{(\ln(c_{lim}/c_0) + \ln x_i)^2}{\ln(c_{lim}/c_0) + \ln x_{i-1}}. \tag{5.20}$$

Now cross multiplying and again neglecting the $(\ln x)^2$ terms leads to

$$\ln x_{i+1} + \ln x_{i-1} = 2 \ln x_i, \tag{5.21}$$

and therefore

$$x_i = \sqrt{x_{i-1}x_{i+1}}, \tag{5.22}$$

which implies

$$\frac{c_{i+1}}{c_i} = \frac{c_i}{c_{i-1}}, \tag{5.23}$$

which is the well known relation due to Rautenbach and Albrecht [1]. In Example 5.7, the exact optimum area for a three-stage system is found and compared with the Rautenbach and Albrecht approximation and with the result obtained previously for a three-stage, equal area system. Problem 5.7 involves using this approximation to relate the intermediate compositions to the exit composition.

Example 5.7 Optimisation of a three-stage system

Consider 1 l/min of a protein solution fed to a three-stage continuous feed-and-bleed UF system. The feed enters at 10 g/l and $c_{\lim}$ can be taken to be 300 g/l. The mass transfer coefficient is 3.5×10^{-6} m/s in each stage. Find the minimum area required to achieve a concentration of 100 g/l in the third stage.

Solution. The unknowns in this problem are x_1 and x_2 ($x_3 = 0.1$), and the two equations represented by Eq. 5.16 can be written after a little rearranging as

$$(3.401 + \ln x_2)\left(3.401 + \ln x_1 + \frac{1}{x_1} - 1\right) - (3.401 + \ln x_1)^2 = 0,$$

$$1.098\left(3.401 + \ln x_2 + \frac{x_1}{x_2} - 1\right) - (3.401 + \ln x_2)^2 = 0.$$

Appropriate MATLAB code is shown below and gives: $x_1 = 0.465$ and $x_2 = 0.209$.

```
function example57
x=fsolve(@optuf,[0.5,0.2])
function f=optuf(x)
x1eqn=(3.401+log(x(2)))*(3.401+log(x(1))+1/x(1)-1)-
(3.401+log(x(1)))^2;
x2eqn=1.098*(3.401+log(x(2))+x(1)/x(2)-1)-
(3.401+log(x(2)))^2;
f=[x1eqn,x2eqn];
```

Therefore, using the individual terms in Eq. 5.14, one gets

$$A_1 = \frac{1 \times 10^{-3}/60}{3.5 \times 10^{-6}} \frac{(1 - 0.465)}{3.401 + \ln(0.465)} = 0.967\,\text{m}^2,$$

$$A_2 = \frac{1 \times 10^{-3}/60}{3.5 \times 10^{-6}} \frac{(0.465 - 0.209)}{3.401 + \ln(0.209)} = 0.665\,\text{m}^2,$$

$$A_3 = \frac{1 \times 10^{-3}/60}{3.5 \times 10^{-6}} \frac{(0.209 - 0.1)}{3.401 + \ln(0.1)} = 0.471\,\text{m}^2.$$

This gives a total area of 2.103 m^2. This compares with 2.138 m^2 in the previous example, where equal areas were used. The Rautenbach and Albrecht approximation gives (see Problem 5.7) $x_1 = x_3^{1/3} = 0.464$ and $x_2 = x_3^{2/3} = 0.215$, both of which values are very close to the exact answers.

5.2.8 Design with viscosity dependent mass transfer coefficient

So far, all the analyses have assumed that the mass transfer coefficient is independent of viscosity. However, in multi-stage systems, there can be a significant change in concentration from the inlet to the exit and, therefore, it is possible that one should account for the fact that the mass transfer coefficient changes as one advances through the stages. To illustrate, let us consider a multi-stage, equal area system. Here it is assumed that the tangential flow is turbulent in each stage. The laminar flow case is kept for Problem 5.8. If one accounts for the effect of viscosity on the mass transfer coefficient, Eq. 5.6 generalises to

$$\frac{Q_0}{k_i A}(x_{i-1} - x_i) - \ln \frac{c_{\lim}}{c_0} - \ln x_i = 0, \tag{5.24}$$

where k_i is the mass transfer coefficient in each stage. Now, neglecting the effect of concentration on the diffusion coefficient, the following expression for the mass transfer coefficient can be used:

$$k_i = k_0 \left(\frac{\mu}{\mu_{c_i}}\right)^{0.5} \left(\frac{\mu_{c_i}}{\mu_{\lim}}\right)^z. \tag{5.25}$$

where $\mu_{\lim}$ is the solution viscosity at the limiting concentration, c_i is the retentate concentration in stage i and k_0 is the infinite dilution, un-corrected mass transfer coefficient. It is assumed here that the tangential flowrate is the same in each module, a reasonable assumption in feed-and-bleed systems where the flowrate actually entering the module is dominated by the recirculation flow. Now, assuming that the viscosity is related to concentration by an exponential function as before (Eq. 4.21), the concentration dependence of the mass transfer coefficient can be written as

$$k_i = k_0 e^{-0.5\gamma c_i} e^{\gamma z(c_i - c_{\lim})}. \tag{5.26}$$

In terms of the x variable, this becomes

$$k_i = k_0 e^{-0.5\gamma c_0/x_i} e^{\gamma z c_0(1/x_i - c_{\lim}/c_0)}. \tag{5.27}$$

Thus, Eq. 5.24 becomes

$$\frac{Q_0}{k_0 A e^{-0.5\gamma c_0/x_i} e^{\gamma z c_0(1/x_i - c_{\lim}/c_0)}}(x_{i-1} - x_i) - \ln \frac{c_{\lim}}{c_0} - \ln x_i = 0. \tag{5.28}$$

In the next example, the calculation of Example 5.6 is redone but Eq. 5.28 is used for each stage.

Example 5.8 System design with viscosity dependent mass transfer coefficient

In this example, 1 l/min of a solution is fed to a three-stage, equal area, continuous feed-and-bleed ultrafiltration system. The feed enters at 10 g/l and c_{lim} can be taken to be 300 g/l. The infinite dilution un-corrected mass transfer coefficient is 3.5×10^{-5} m/s in each stage and the retentate concentration leaving the third stage is 100 g/l. Take z to be 0.14 and $\gamma = 0.01$ l/g and assume turbulent flow. Find the area of each stage and the total area.

Solution. The three equations to be solved can be coded in MATLAB as shown below:

```
function example58
x=fsolve(@vmultistage,[0.5,0.2,1])
function f=vmultistage(x)
g=0.01; c0=10;Q0=1e-3/60;k0=3.5e-6;clim=300;
%Note x(3) is A
v1=exp(-0.5*g*c0/x(1))*exp(g*0.14*c0*(1/x(1)-clim/c0));
v2=exp(-0.5*g*c0/x(2))*exp(g*0.14*c0*(1/x(2)-clim/c0));
v3=exp(-0.5*g*c0/0.1)*exp(g*0.14*c0*(1/0.1-clim/c0));
x1eqn=Q0/k0/x(3)/v1*(1-x(1))-log(x(1))-log(clim/c0)
x2eqn=Q0/k0/x(3)/v2*(x(1)-x(2))-log(x(2))-log(clim/c0)
Aeqn=Q0/k0/x(3)/v3*(x(2)-0.1)-log(0.1)-log(clim/c0)
f=[x1eqn,x2eqn,Aeqn];
```

Running the code gives $x_1 = 0.537$, $x_2 = 0.237$ and $A = 1.292$ m^2. This compares with a stage area of only 0.716 m^2 when viscosity effects are neglected. In Problem 5.8, the task is to modify the above calculation for the case of laminar flow.

5.3   Batch ultrafiltration

The basic configurations of batch ultrafiltration systems are shown in Fig. 5.4.

In a batch process, the solution is fed to a membrane module where ultrafiltration takes place and the retentate is returned to the feed tank while permeate is removed. Consequently, the volume of feed solution decreases over time and the concentration of the macromolecule increases. The simpler configuration in Fig. 5.4 tends to be used for smaller scale operations, especially in research applications. The addition of the recirculation pump boosts the pressure at the membrane inlet and increases the flowrate through the module, thus increasing the mass transfer coefficient.

For both configurations, the key assumption from a process modelling perspective is that the solute concentration is uniform throughout the system. Thus, the concentration in the retentate tank is the same as the concentration in the module at any given time.

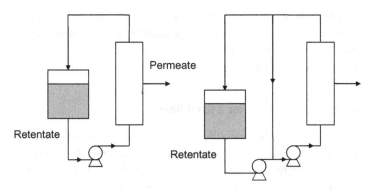

Figure 5.4 Batch ultrafiltration with and without recirculation loop.

Two types of calculation are performed in this section. First, the final concentration (or volume) for a given process time is computed. Then, the process time for a given reduction in solution volume (or increase in solute concentration) is calculated. These are process analysis calculations. The process design calculation, i.e., calculation of the area required to achieve a certain volume reduction in a given time, is essentially the same as the process time calculation.

To begin, the overall balance for the system can be written

$$\frac{dV}{dt} = -AJ, \tag{5.29}$$

where V is the volume at time t, and A is the membrane area. The solute balance can be written for complete solute rejection as

$$Vc = V_0 c_0, \tag{5.30}$$

where c_0 and V_0 are the initial values of c and V respectively. Now assuming operation at the limiting flux, the following governing differential equation is obtained:

$$\frac{dV}{dt} = -kA \ln \frac{c_{\text{lim}}}{c}. \tag{5.31}$$

In passing it should be noted that this approach assumes that the concentration polarisation layer is established instantaneously and that the flux is at its limiting value at all times, including time zero. Using Eq. 5.30, the governing ODE can be written in terms of volumes only as

$$\frac{dV}{dt} = -kA \left(\ln \frac{c_{\text{lim}}}{c_0} + \ln \frac{V}{V_0} \right), \tag{5.32}$$

or in terms of concentrations only as

$$\frac{dc}{dt} = \frac{kA}{V_0} \frac{c^2}{c_0} \ln \frac{c_{\text{lim}}}{c}. \tag{5.33}$$

Essentially, any problem in batch UF, including those with concentration dependent mass transfer coefficient, can be solved by numerical solution of any of the above three equations (in combination with Eq. 5.30 if required). An example of this type of calculation is shown in the next section.

5.3.1   Calculation of final conditions for a fixed time

In this type of calculation, the time and area available to carry out batch UF are specified and the task is to determine the concentration and/or volume at the end of the process. The basic approach here is to solve the governing differential equation numerically. This is illustrated in Example 5.9 below. In addition to enabling one to find the final conditions for a given time, this approach means that full details on the time course of volume and/or concentration can be obtained.

Example 5.9 Computation of final conditions in batch UF

A solution with an initial volume of 60 l and an initial concentration of 14 g/l is to be ultrafiltered in batch mode for one hour in a membrane module of area 1.3 m^2. The limiting concentration is assumed to be 250 g/l and the mass transfer coefficient is 3.0×10^{-6} m/s. Generate plots of relative concentration and relative volume versus time and calculate the final volume and concentration.

Solution. MATLAB code for solving this problem is shown below:

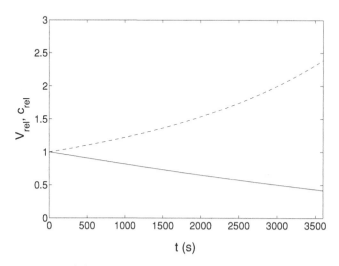

Figure 5.5 Relative concentration and volume versus time in batch ultrafiltration. Dashed curve denotes macrosolute, solid curve volume.

```
function example59
global c0 V0
[t,c]=ode45(@batchuf,[0 3600],[14])
```

```
V=c0*V0./c
crel=c/c0; Vrel=1./crel;
plot(t,Vrel,'-', t, crel, '--');
xlabel('t (s)'); ylabel('V_r_e_l, c_r_e_l');
axis([0 3600 0 3]);
function dcdt=batchuf(t,c)
global c0 V0
k=3e-6; c0=14; V0=60e-3; clim=250; A=1.3;
dcdt=k*A/V0*c.^2/c0.*log(clim./c);
```

The concentration and volume profiles are as shown in Fig. 5.5. The final concentration was 33.44 g/l and the final volume was 25.12 l.

The calculation presented in Example 5.9 assumes a constant mass transfer coefficient, but in batch UF significant changes in solute concentration can occur. It is important, therefore, to be able to modify the above analysis to account for the concentration dependence of the mass transfer coefficient. For laminar flow and accounting for viscosity effects only, the usual approach can be employed, yielding

$$\frac{dV}{dt} = -k_0 e^{\gamma z(c - c_{\lim})} A \ln \frac{c_{\lim}}{c}. \tag{5.34}$$

The analogue of this equation for turbulent flow is

$$\frac{dV}{dt} = -k_0 e^{-0.5\gamma c} e^{\gamma z(c - c_{\lim})} A \ln \frac{c_{\lim}}{c}. \tag{5.35}$$

The next example shows the solution of a batch UF problem with viscosity dependent mass transfer coefficient and laminar flow. In Problem 5.9 you are asked to solve a similar problem where the tangential flow is turbulent.

Example 5.10 Batch simulation with viscosity dependent mass transfer coefficient
For the same conditions as Example 5.9, calculate the final concentration and volume but assume laminar tangential flow and account for the effect of concentration on the mass transfer coefficient by employing Eq. 5.34 and take $\gamma = 0.005$ l/g.

Solution. MATLAB code for solving this problem is almost identical to that used in the previous example, the only difference is that the mass transfer coefficient is specified by the two lines

```
k0=3e-6;
k=k0*exp(g*z*(y(2)-clim));
```

Solving gives $V_f = 29.49$ l and $c_f = 28.48$ g/l. Thus, when compared with the results of Example 5.9, the final volume is higher and the final concentration is lower than when viscosity effects are neglected. This reflects the lower fluxes achieved when viscosity effects are included.

5.3.2   Calculation of batch time by solution of governing ODE

The other type of process analysis calculation that we are interested in is computation of the batch time. The usual approach to this type of problem is to recast the governing ODE as an integral and to solve the integral numerically. This is a perfectly valid approach, although some more algebra can be required to get the integral in final form, and it is used in Chapters 6 and 7. You are also asked to use it in Problem 5.10. However, given that ODEs are so easily solved routinely, and given that a number of MATLAB codes have been created previously, it is a valid strategy to stick with the ODE formulation for now. The basic difference between the batch time calculation and final volume (or concentration) calculation is that one needs to insert an *event* into the MATLAB code to make it stop at the appropriate point. This is illustrated below.

Example 5.11 Calculation of batch time using the governing ODE
A protein solution is to be concentrated from an initial concentration of 14 g/l to a final concentration of 56 g/l. The limiting concentration is assumed to be 250 g/l. Calculate the time required to carry out this process if the initial feed volume is 60 l, the membrane area is 1.3 m^2 and the mass transfer coefficient is constant and equal to 3.0×10^{-6} m/s.

Solution. Solution of this problem simply involves modifying the code used in Example 5.9 by including an *event function* which tells the code when to stop. More precisely, the algorithm actually passes the target and interpolates to get the exact time required to reach that target.

```
function example511
options=odeset('events', @eventfcn);
[t,c]=ode45(@batchuf,[0 10000], [14],options)
function dcdt=batchuf(t,c)
k=3e-6; c0=14; V0=60e-3; clim=250; A=1.3;
dcdt=k*A/V0*c.^2/c0.*log(clim./c);
function [value, isterminal, direction]=eventfcn(t,c)
value = c-56;% stop when desired concentration is reached
isterminal = 1;
direction = 0;
```

Running the code gives a batch time of 5072 s.

5.3.3   Calculation of batch time using special functions

In the last section it was shown how the batch time can be computed by solution of the governing ODE. In this section, an alternative solution, based on recasting the problem as an integral, is developed. While this might seem to contradict what was said in the last section, namely that it is simpler to just work with the original ODEs, the difference here is that the integral does not have to be solved numerically. Essentially, it is solved analytically. We say 'essentially' because the analytical solution is expressed as a special function. Special functions are functions that cannot be expressed as combinations of elementary functions, i.e., powers, logs, exponentials, trigonometric functions, etc. They are discussed further in the Appendix. With software such as MATLAB, the special function approach developed here requires no more effort than would be involved in computing a logarithm. First, Eq. 5.31 is rearranged to give

$$t = \frac{1}{kA} \int_{V_f}^{V_0} \frac{dV}{\ln(c_{\lim}/c)}, \tag{5.36}$$

where V_f is the final volume. Now, a new variable, s, is defined by

$$s = \frac{c_{\lim}}{c} = \frac{c_{\lim}V}{c_0 V_0}. \tag{5.37}$$

Therefore

$$dV = \frac{c_0 V_0}{c_{\lim}} ds. \tag{5.38}$$

Equation 5.36 becomes

$$t = \frac{V_0}{kA} \frac{c_0}{c_{\lim}} \int_{c_{\lim}/c_f}^{c_{\lim}/c_0} \frac{ds}{\ln s}, \tag{5.39}$$

where c_f is the final concentration.

The key step now is to note the existence of the logarithmic integral, denoted $Li(y)$, a special function defined by the expression [2]

$$Li(y) = \int_0^y \frac{ds}{\ln s}. \tag{5.40}$$

Now, this is more than just a simple change of notation because $Li(y)$ can be computed using the following series [2]

$$Li(y) = \xi + \ln(\ln y) + \sum_{n=1}^{\infty} \frac{(\ln y)^n}{n! n}, \tag{5.41}$$

where ξ is the Euler–Mascheroni constant. A special function that is closely related to the logarithmic integral is the exponential integral, defined by the expression [2]

$$Ei(y) = \int_{-\infty}^{y} \frac{e^t}{t} dt, \tag{5.42}$$

and computable using the series

$$Ei(y) = \xi + \ln y + \sum_{n=1}^{\infty} \frac{y^n}{n!n}. \tag{5.43}$$

The fact that the logarithmic and exponential integrals are defined as an infinite series makes them essentially no different from a trigonometric function. Comparing Eq. 5.41 with 5.43, it is seen that

$$Li(y) = Ei(\ln y). \tag{5.44}$$

Going back to Eq. 5.39, this means that the batch time can be computed from the following simple expression

$$t = \frac{V_0}{kA} \frac{c_0}{c_{\lim}} \left(Ei \left(\ln \frac{c_{\lim}}{c_0} \right) - Ei \left(\ln \frac{c_{\lim}}{c_f} \right) \right). \tag{5.45}$$

Use of this expression is illustrated in the example below.

Example 5.12 Calculation of batch time using the exponential integral

Repeat Example 5.11 but use the exponential integral approach.

Solution. Below is the simple MATLAB script to solve this problem. Without getting bogged down in mathematical niceties, it is important to note that the MATLAB function **expint** actually computes a slightly different exponential integral, denoted $E_1(y)$ by mathematicians and related to $Ei(y)$ by

$$Ei(y) = -E_1(-y) - i\pi,$$

i.e.,

$$Ei(y) = expint(-y) - i\pi.$$

Thus, the MATLAB syntax below should be noted carefully and can be entered directly at the command window if desired.

```
x1=log(250/14);
x2=log(250/56);
t=0.06/3e-6/1.3*14/250*(-expint(-x1)+expint(-x2))
```

The imaginary terms obviously cancel when taking differences. Running the script gives a time of 5072 s, which is precisely the same as the answer obtained with the ODE method. This approach is very simple and it is hard to see why one would resort to the more conventional, numerical approach if one has access to software like MATLAB.

Even with simple spreadsheet software this approach is not very daunting as it is straightforward to compute a series with a spreadsheet [3]. In Problem 5.11, the effect of c_{lim} on batch time is examined.

5.3.4 Approximate methods for calculating the batch time

While the exponential integral provides an easy route to computing the batch time if MATLAB or a similar package is readily available, two approximate methods for calculating the batch time that can be implemented on a scientific calculator are described here. The first is a rigorous mathematical approximation to the batch integral, and the second is the purely empirical one due to Cheryan [4]. In order to derive the first approximation, the 'x' variable is used again. Thus, we let

$$x = \frac{c_0}{c} = \frac{V}{V_0}.$$ (5.46)

This is often known as the 'volume reduction factor' by UF practitioners. Thus Eq. 5.36 becomes

$$t = \frac{V_0}{kA} \int_{x_{\text{f}}}^{1} \frac{dx}{B + \ln x}.$$ (5.47)

The parameter, B, has been introduced for convenience and is defined by

$$B = \ln \frac{c_{\text{lim}}}{c_0}.$$ (5.48)

It can also be noted that $x_{\text{f}} = c_0/c_{\text{f}} = V_{\text{f}}/V_0$.

Equation 5.47 is useful because since $x < 1$, this version of the batch time integral has potential to be approximated in a mathematically sound manner. Now, using a Taylor series approximation for the natural logarithm and neglecting *third* order and higher terms, Eq. 5.47 can be written as

$$t = \frac{V_0}{kA} \int_{x_{\text{f}}}^{1} \frac{dx}{B + (x - 1) - \frac{1}{2}(x - 1)^2}.$$ (5.49)

This is an awkward integral that is best done with the MATLAB symbolic integrator, as shown in the command sequence below:

```
syms B x
f=1/(B+(x-1)-1/2*(x-1)^2);
int(f,x)
ans =
- (atan(((2*x - 4)*i)/(2*(2*B + 1)^(1/2)))*2*i)/(2*B + 1)^(1/2)
```

Using the above expression for the indefinite integral, the batch time is given as

$$t = -\frac{V_0}{kA} \left. \frac{2i \tan^{-1} \dfrac{(2x-4)i}{2\sqrt{2B+1}}}{\sqrt{2B+1}} \right|_{x_f}^{1}. \tag{5.50}$$

While this expression will compute to give a real number if coded in MATLAB (but not in spreadsheet software), there is no advantage to doing so since the exact integral is actually easier to code. However, with the aid of the Mathematica™ integrator, it can be shown that the batch time can also be written

$$t = \frac{V_0}{kA} \frac{2}{\sqrt{2B+1}} \left[\tanh^{-1}\left(\frac{-1}{\sqrt{2B+1}} \right) - \tanh^{-1}\left(\frac{x_f - 2}{\sqrt{2B+1}} \right) \right]. \tag{5.51}$$

This is obviously more amenable to calculation with a calculator or spreadsheet software, although it is still a little cumbersome for an equation that is only an approximation. In Problem 5.12, you are asked to derive an even simpler equation by neglecting second order terms in the Taylor expansion of $\ln x$.

The second approximation that is often used is the one due to Cheryan [4]. He proposed that the average flux, J_{av}, during batch ultrafiltration is given by an expression of the form

$$J_{av} = \theta J_0 + (1 - \theta) J_f, \tag{5.52}$$

where J_0 is the initial flux and J_f is the final flux, where these fluxes are given by

$$J_0 = k \ln \frac{c_{lim}}{c_0}, \tag{5.53}$$

and

$$J_f = k \ln \frac{c_{lim}}{c_f}. \tag{5.54}$$

Therefore, the batch time is given by

$$t = \frac{V_0 - V_f}{A J_{av}}. \tag{5.55}$$

In terms of the dimensionless concentration variable, this is easily shown (Problem 5.13) to be equivalent to

$$t = \frac{V_0}{kA} \frac{1 - x_f}{B + (1-\theta) \ln x_f}. \tag{5.56}$$

In the first edition of his handbook, Cheryan proposed $\theta = \frac{1}{3}$ but subsequently changed this to $\theta = \frac{1}{2}$ in the second edition.

Before moving on to the next example, it is worth pausing briefly to comment on Eq. 5.53, which seems somewhat innocuous at first glance. This equation contains an important assumption, namely that the concentration polarisation layer is established instantaneously. One might have expected that the initial flux would be determined by the membrane properties, as it is in microfiltration where cake formation takes place much more slowly. However, UF is characterised by much more rapid dynamics in which the timescale for establishing the concentration profile adjacent to the membrane is very small. This makes modelling of ultrafiltration dynamics somewhat simpler.

Example 5.13 Approximate calculation of batch time

Calculate the batch time using Eqs. 5.51 and 5.56 for the conditions described in Example 5.11.

Solution. Using the numerical values provided in Example 5.11, one gets

$$B = \ln \frac{c_{\text{lim}}}{c_0} = \ln \frac{250}{14} = 2.882,$$

$$x_f = \frac{c_0}{c_f} = \frac{14}{56} = 0.25,$$

$$\sqrt{2B + 1} = \sqrt{2 \times 2.882 + 1} = 2.601.$$

Therefore Eq. 5.51 becomes

$$t = \frac{60 \times 10^{-3}}{3 \times 10^{-6} \times 1.3} \frac{2}{2.601} \left[\tanh^{-1} \left(-\frac{1}{2.601} \right) - \tanh^{-1} \left(\frac{0.25 - 2}{2.601} \right) \right] = 4857 \, \text{s}.$$

This value *underestimates* the true time (5072 s) by 4.2%.
Now, Eq. 5.56 with $\theta = 0.5$, becomes

$$t = \frac{60 \times 10^{-3}}{3 \times 10^{-6} \times 1.3} \frac{1 - 0.25}{2.882 + (1 - 0.5) \ln(0.25)} = 5271 \, \text{s}.$$

Therefore, this equation *overestimates* the time by about 3.9%. As a matter of interest, using Cheryan's original value of $\theta = \frac{1}{3}$ gives a time of 5894 s, which represents an error of just over 16%.

Having discussed a number of approaches to computing the batch time integral, it is important to remind ourselves that in real systems, Eq. 5.36 is itself an approximation. This is because of the viscosity dependence of the mass transfer coefficient and, perhaps more importantly, because of membrane fouling. In the case of the former, the batch time is easily solved using the original ODE. Quantifying the effect of fouling is more difficult, as is seen in Chapter 10.

5.4 Fed-batch operation

The basic layout of a fed-batch configuration is shown in Fig. 5.6.

As mentioned previously in the context of CFMF, ultrafiltration equipment often comes from the manufacturer as a skid incorporating membranes, pumps and tanks. If the capacity of the retentate tank is less than the volume of feed to be processed, a fed-batch method of operating can be used. This configuration is also used sometimes for bench-scale studies employing low volume stirred cell units that are fed from a separate

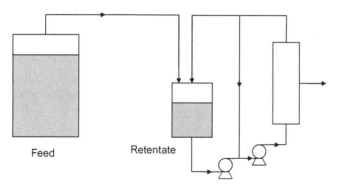

Figure 5.6 Fed-batch operation.

reservoir, as explained in Chapter 1. In this section, the key equations for this process are formulated and the time taken to process a certain amount of feed is computed.

It is assumed that there is a *total* initial volume, V_0, of feed with a solute concentration, c_0. Of this, a volume, V_f, is placed in the retentate tank and the rest is fed from an external tank. It is assumed also that V_f is the final desired volume. The system is now operated at constant retentate volume, i.e., V_f is kept constant, by balancing the permeate and feed flowrates, throughout the operation. The solute balance for the system, assuming complete rejection, can be written

$$V_f \frac{dc}{dt} = JAc_0,$$ (5.57)

where J is the flux at any time, t. Assuming that operation is at the limiting flux, this becomes

$$\frac{dc}{dt} = \frac{kA}{V_f} c_0 \ln \frac{c_{\lim}}{c}.$$ (5.58)

The final concentration is given by the familiar expression

$$c_f = \frac{c_0 V_0}{V_f}.$$ (5.59)

Thus, the time taken to perform this operation is given by

$$t = \frac{V_0}{kAc_f} \int_{c_0}^{c_f} \frac{dc}{\ln \frac{c_{\lim}}{c}}.$$ (5.60)

Now defining $s = c_{\lim}/c$ as before, this becomes

$$t = \frac{V_0}{kA} \frac{c_{\lim}}{c_f} \int_{c_{\lim}/c_f}^{c_{\lim}/c_0} \frac{ds}{s^2 \ln s}.$$ (5.61)

To solve this integral using the symbolic integrator, the following commands are entered into MATLAB:

```
syms s
int(1/s^2/log(s),s)
ans =
Ei(-log(s))
```

Thus, the fed-batch time is given by

$$ t = \frac{V_0}{kA} \frac{c_{\text{lim}}}{c_{\text{f}}} \left[Ei \left(- \ln \frac{c_{\text{lim}}}{c_0} \right) - Ei \left(- \ln \frac{c_{\text{lim}}}{c_{\text{f}}} \right) \right]. \tag{5.62} $$

Example 5.14 illustrates the use of this equation.

Example 5.14 Calculation of fed-batch time and comparison with batch time

A volume of 180 l of a protein solution is to be ultrafiltered by a fed-batch process in which 40 l is placed in the retentate tank and the remainder is fed from a feed tank at a rate equal to the permeate flowrate. The feed concentration is 10 g/l and the limiting concentration is assumed to be 250 g/l. The membrane area is 1.3 m^2 and the mass transfer coefficient is 3.0×10^{-6} m/s. Calculate the time required to empty the feed tank and compare that value with the time required to reduce 180 l of the same solution down to 40 l by a batch process.

Solution. We first compute the final concentration, i.e.,

$$ c_{\text{f}} = \frac{c_0 V_0}{V_{\text{f}}} = \frac{10 \times 180}{40} = 45 \, \text{g/l}. $$

A MATLAB script for computing the time using Eq. 5.62 can be written as follows:

```
t=0.18/(3e-6*1.3)*250/45*(-expint(log(250/10))
+expint(log(250/45)))
```

giving $t = 16\,202$ s.

For a conventional batch process where the volume is reduced from V_0 to V_{f}, we can use Eq.5.45 which can be written in MATLAB script as

```
t=0.18/(3e-6*1.3)*10/250*(-expint(-log(250/10))
+expint(-log(250/45)))
```

which gives $t = 13\,928$ s.

Thus, the fed-batch process takes longer than the batch process, which is precisely what was found in Chapter 3 on CFMF.

The above analysis can obviously be extended for the case of a viscosity dependent mass transfer coefficient, although computation of the process time requires solution of the relevant ODE. For laminar flow, Eq. 5.58 becomes

$$ \frac{dc}{dt} = \frac{k_0 A}{V_{\text{f}}} c_0 e^{\gamma z (c - c_{\text{lim}})} \ln \frac{c_{\text{lim}}}{c}, \tag{5.63} $$

while for turbulent flow, we have

$$\frac{dc}{dt} = \frac{k_0 A}{V_f} c_0 e^{-0.5\gamma c} e^{\gamma z (c - c_{lim})} \ln \frac{c_{lim}}{c}.$$ (5.64)

Use of Eq. 5.64 is explored in Problem 5.14.

5.5   Single pass operation

As the final example of a process configuration, we consider a single-stage, single pass process as illustrated in Fig. 5.7.

The key complicating feature of the single pass system is that there is a significant axial variation in tangential flowrate. This means that there must be a significant variation in both the concentration and the mass transfer coefficient. In Section 5.5.1, only the variation in mass transfer coefficient due to the variation in tangential flowrate is included in the analysis. This is the standard analysis of this problem although the solution presented here is new in that the exponential integral is found to appear again. In Section 5.5.2, the analysis is extended to account for a viscosity dependent mass transfer coefficient.

5.5.1   Single pass analysis with viscosity independent mass transfer coefficient

The volume balance for a single system can be written

$$\frac{dQ}{dA} = -J,$$ (5.65)

where dQ is the change in tangential flowrate over a differential area, dA, of membrane. A solute balance, for complete rejection, gives

$$c = \frac{c_0 Q_0}{Q},$$ (5.66)

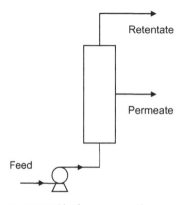

Figure 5.7 Single pass operation.

where c and Q are the concentration and tangential flowrate at an arbitrary position in the membrane and c_0 and Q_0 are the inlet values of these parameters. Combining Eqs. 5.65 and 5.66 and assuming that the flux is at the limiting values at all points within the module gives

$$\frac{dc}{dA} = \frac{kc^2}{c_0 Q_0} \ln \frac{c_{\text{lim}}}{c}. \tag{5.67}$$

The similarity to Eq. 5.33 should be noted. This similarity between the single pass system and the batch system will be familiar to chemical engineers who have studied batch and plug flow chemical reactors. It is now assumed that the flow in the module is turbulent. In laminar flows, the mass transfer coefficient is technically a function of position even for constant tangential flowrate [5]. Assuming k is viscosity independent, the variation in tangential flowrate is the sole cause of the variation in mass transfer coefficient within the flow channel. This variation can be expressed with the following equation

$$k = k_0 \left(\frac{Q}{Q_0} \right)^n, \tag{5.68}$$

where k is the mass transfer coefficient at a point in the module where the flowrate is Q, k_0 is the mass transfer coefficient evaluated at the inlet flowrate, Q_0, and n is an empirical constant which can reasonably be taken to be 0.8 for turbulent flow as discussed in Chapter 4. Substituting into Eq. 5.67 and noting Eq 5.66 gives

$$\frac{dc}{dA} = k_0 \frac{c^2}{c_0 Q_0} \left(\frac{c_0}{c} \right)^n \ln \frac{c_{\text{lim}}}{c}. \tag{5.69}$$

Use of this equation to tackle the process analysis problem, i.e., to find the exit concentration for a fixed area, is illustrated in the next example.

Example 5.15 Single pass analysis
A 10 g/l protein solution is to be concentrated in a continuous, single pass ultrafiltration process where the flow in the module is turbulent and $n = 0.8$. The mass transfer coefficient at the inlet conditions is 2.0×10^{-5} m/s and the limiting concentration is 280 g/l. The feed flowrate is 10 l/min. Calculate the exit concentration if the membrane area is 8 m^2.

Solution. This problem can be readily solved by numerical solution of Eq. 5.69. A simple MATLAB code to do this using **ode45** is shown below.

```
function example515
[A,c]=ode45(@singlepassUF,[0 8],10)
function dcdA=singlepassUF(A,c)
clim=280; c0=10; k0=2e-5; Q0=10e-3/60; n=0.8;
dcdA=k0*c^2/c0/Q0*(c0/c)^n*log(clim/c);
```

Running the code gives an exit concentration of 109.8 g/l.

5.5.2   Single pass analysis with viscosity dependent mass transfer coefficient

To account for a viscosity dependent mass transfer coefficient, one only has to change the mass transfer coefficient term. Allowing for the wall correction factor and assuming an exponential dependence of viscosity on concentration, the governing differential equation becomes

$$\frac{dc}{dA} = k_{c0} e^{0.5\gamma(c_0-c)} e^{\gamma z(c-c_{\lim})} \frac{c^2}{c_0 Q_0} \left(\frac{c_0}{c}\right)^n \ln \frac{c_{\lim}}{c}, \tag{5.70}$$

where k_{c0} represents the un-corrected mass transfer coefficient at the inlet conditions. Example 5.16 illustrates the use of this equation.

Example 5.16   Single pass analysis with viscosity dependent mass transfer coefficient

Repeat Example 5.15 with $\gamma = 0.01$ L/g and $z = 0.14$.

Solution. MATLAB script is shown below:

```
function example516
[A,c]=ode45(@singlepassUF,[0 8],10)
function dcdA=singlepassUF(A,c)
clim=280; c0=10; kc0=2e-5; Q0=10e-3/60; n=0.8;
g=0.01; z=0.14;
kc=kc0*exp(0.5*g*(c0-c))*exp(g*z*(c-clim));
dcdA=kc*c^2/c0/Q0*(c0/c)^n*log(clim/c);
```

Running the complete code gives an exit concentration of 57.5 g/l, a substantial reduction from the previous example, as expected.

5.5.3   Single pass design with constant mass transfer coefficient

Here the goal is to calculate the area required to achieve a specified exit concentration. Again, we express the problem as an integral as it leads to an analytical solution involving a special function. Rearranging Eq. 5.69 gives

$$A = \frac{c_0^{1-n} Q_0}{k_0} \int_{c_0}^{c_f} \frac{c^{n-2} dc}{\ln \frac{c_{\lim}}{c}}. \tag{5.71}$$

Letting $s = c_{\lim}/c$ and differentiating gives

$$dc = -\frac{c_{\lim}}{s^2} ds. \tag{5.72}$$

Therefore

$$c^{n-2} dc = -\left(\frac{c_{\lim}}{s}\right)^{n-2} \frac{c_{\lim}}{s^2} ds = -\frac{c_{\lim}^{n-1}}{s^n} ds. \tag{5.73}$$

Equation 5.71 thus transforms to the neat form

$$A = \frac{Q_0}{k_0} \left(\frac{c_0}{c_{\lim}} \right)^{1-n} \int_{c_{\lim}/c_f}^{c_{\lim}/c_0} \frac{ds}{s^n \ln s}. \tag{5.74}$$

For $n = 0$, this reduces to an expression essentially identical in form to Eq. 5.39 for a batch system, as one would expect. To solve the indefinite integral with the MATLAB symbolic integrator, one enters the following at the command window:

```
syms s n
int(1/(s^n*log(1/s)), s)
ans =
piecewise([n = 1, log(log(s))], [n <> 1, Ei(-log(s)*(n - 1))])
```

Thus for the relevant situation of $n \neq 1$, the design equation takes the concise form:

$$A = \frac{Q_0}{k_0} \left(\frac{c_0}{c_{\lim}} \right)^{1-n} \left[Ei \left(-(n-1) \ln \frac{c_{\lim}}{c_0} \right) - Ei \left(-(n-1) \ln \frac{c_{\lim}}{c_f} \right) \right]. \tag{5.75}$$

Example 5.17 illustrates the use of this equation.

Example 5.17 Calculation of the area of a single pass module
A protein solution is to be concentrated from 10 g/l to 25 g/l in a continuous single-pass ultrafiltration process where the flow in the module is turbulent with $n = 0.8$. The mass transfer coefficient at the inlet conditions is 2.0×10^{-5} m/s and the limiting concentration is 280 g/l. The feed flowrate is 50 l/min. Calculate the required membrane area.

Solution. The key equation here is Eq. 5.75. MATLAB script is shown below.

```
c0=10; cf=25; clim=280; Q0=50e-3/60; k0=2e-5; n=0.8
A=Q0/k0*(c0/clim)^(1-n)*(-expint((n-
1)*log(clim/c0))+expint((n-1)*log(clim/cf)))
```

Running the script gives $A = 12.2$ m^2.

When the mass transfer coefficient is viscosity-dependent, the single pass design problem is best done by solving the governing ODE and using an event function to stop the calculations at the appropriate end point. This is illustrated in Problem 5.15.

5.6 Membrane fouling and limiting flux operation

When describing membrane filtration processes, it is commonplace to neglect membrane fouling as was done in this chapter. Recall that in this book membrane fouling is defined as the increase in *membrane resistance* that arises as a result of membrane pore clogging.

It should not have escaped the reader's notice that membrane resistance is not mentioned anywhere in this chapter. So, an obvious question is: if the equations do not include the membrane resistance, how is membrane fouling accounted for in the calculations? The short answer is that it is not accounted for and it cannot be incorporated into the limiting flux model. By definition, a system operating under limiting flux conditions is unaffected by membrane properties.

When membrane systems are operated under limiting flux conditions, they are typically operated at a pressure that is just high enough to ensure that the flux is at the limiting value. There is no point, from an energy utilisation point of view, in going to higher pressures. But this choice of pressure is only acceptable if membrane fouling is likely to be insignificant. If the pressure is too low, the flux may be at its limiting value initially but membrane fouling may cause it to drop out of the limiting region. To truly understand how and why this happens requires us to integrate the knowledge gained in Chapter 8, which covers application of the osmotic pressure model, and Chapter 10, which covers membrane fouling. Essentially, a number of concepts need to be combined, including concentration polarisation, osmotic pressure and membrane fouling.

5.7   Conclusions

In this chapter, a wide variety of computational problems in modelling of continuous, batch and fed-batch UF systems has been explored. None of these problems posed any real computational difficulties. The key point, however, is that these problems are easy because modern mathematical and computational techniques, aided by computer software, are available. In the past, many of these problems would have been solved with laborious trial and error methods. Such strategies are clearly no longer needed.

One of the philosophies that has driven this chapter has been the idea that there are many ways to solve process problems, not just in membrane engineering but in all fields of engineering. As long as methods are equally accurate, there is no real objective reason for saying one is 'better' than another. None of the problems solved here is computationally demanding, so there is usually very little time to be gained by choosing one method over another. To a large degree, therefore, the choice of method used for any given problem is determined by personal preference. The next chapter is similar in approach, but examines analysis and design of diafiltration processes.

References

1. Rautenbach, R. and Albrecht, R. (1989). *Membrane Processes*. Wiley-Blackwell, Hoboken, New Jersey, USA.
2. Abramowitz, M. and Stegun, I.A. (Eds.) (1972). *Handbook of Mathematical Functions*. Dover, New York, USA.
3. Foley, G. (2011). Some classic ultrafiltration problems solved with the Exponential Integral. *Education for Chemical Engineers*, **6**, e90–e96.

4. Cheryan, M. (1998). *Ultrafiltration and Microfiltration Handbook*. CRC Press, Florida, USA.
5. Cussler, E.L. (1997). *Diffusion: Mass Transfer in Fluid Systems*, 2nd Edition. Cambridge University Press, UK.

Problems

Problem 5.1   Governing equation for feed-and-bleed UF
Derive Eq. 5.4 from Eqs. 5.1 to 5.3.

Problem 5.2   General governing equation for the *i*th stage
Derive Eq. 5.6 and generalise it for a system with unequal areas. Denote the area of each stage by A_i.

Problem 5.3   Three-stage analysis for unequal membrane areas
Repeat the problem solved in Examples 5.2 and 5.3, except for making the areas of each stage as follows: $A_1 = 1.2$ m^2, $A_2 = 0.9$ m^2 and $A_3 = 0.6$ m^2. Use a sequential or simultaneous approach, whichever you prefer.

Problem 5.4   ODE formulation of a CSTR operating at steady state
This problem should be familiar to chemical engineers. Consider a continuous stirred tank reactor (CSTR) in which a reaction with non-integer order, s, is occurring. The governing equation for this reactor can be written

$$Q\,(c_0 - c_1) - V k c_1^s = 0,$$

where Q is the volumetric flowrate through the reactor, V is the reactor volume, c_0 is the inlet concentration of reactant and c_1 is the exit concentration of that reactant. Formulate this equation as an ODE and use it to compute c_1 when $c_0 = 1.0$, $Q = 0.5$, $V = 5$, $s = 0.7$ and $k = 0.4$, where all parameters are in arbitrary but consistent units.

Problem 5.5   Graphical feed-and-bleed analysis with unequal areas
Repeat Example 5.4 but assume that there are two stages, the first of area 1.5 m^2 and the second of area 1.2 m^2.

Problem 5.6   Derivation of Eq. 5.12
Derive Eq. 5.12, starting from Eq. 4.14.

Problem 5.7   Rautenbach and Albrecht approximation for a three-stage system
Show that the Rautenbach and Albrecht approximation (Eq. 5.22) implies that in a three-stage system

$$x_1 = x_3^{1/3} \text{ and } x_2 = x_3^{2/3}.$$

Problem 5.8   Three-stage feed-and-bleed system with laminar flow and viscosity effects
Re-examine Example 5.8 assuming laminar tangential flow.

Problem 5.9   Batch UF with turbulent flow and viscosity effects
Re-examine Example 5.10 but assume turbulent flow.

Problem 5.10   Batch UF calculations using numerical integration

Re-examine the problem solved in Example 5.11 by recasting it as an integral. Solve the integral numerically using the **trapz** function in MATLAB.

Problem 5.11   Effect of c_{lim} on batch time

In this problem, your task is to investigate the effect of the limiting concentration on the batch time. For the conditions described in Example 5.12, use the exponential integral method to calculate the batch time for values of c_{lim} in the range 200–350 g/l and generate a plot of batch time versus c_{lim}.

Problem 5.12   Approximate method for computing batch time

Show that if second order terms are neglected in the Taylor expansion for $\ln x$, the batch time can be computed from the expression

$$t = \frac{V_0}{kA} \ln \frac{B}{(B-1) + x_f}.$$

Hence calculate the batch time for the conditions of Example 5.11.

Problem 5.13   Cheryan equation written in terms of x variable

Derive Eq. 5.56 assuming Cheryan's approximation for the average flux.

Problem 5.14   Fed-batch time for turbulent flow with viscosity effects

Using the same parameter values as in Example 5.14, compute the fed-batch time for the situation where the flow is turbulent and the mass transfer coefficient is viscosity dependent. Take $z = 0.14$ and $\gamma = 0.01$ l/g.

Problem 5.15   Single pass area with viscosity effects

For the same conditions as Example 5.17, compute the membrane area assuming a viscosity dependent mass transfer coefficient with $\gamma = 0.01$ l/g and $z = 0.14$.

Further problems

Problem 5.16   Feed-and-bleed design using the Rautenbach–Albrecht approximation

In this problem we will build on Problem 5.7. Consider a three-stage, continuous feed-and-bleed system again. Show that when the Rautenbach–Albrecht approximation is used, the total membrane area is given by

$$A = \frac{Q_0}{k}(1 - x_3^{1/3})$$

$$\times \left[\frac{1}{\ln(c_{lim}/c_0) + \dfrac{1}{3}\ln x_3} + \frac{x_3^{1/3}}{\ln(c_{lim}/c_0) + \dfrac{2}{3}\ln x_3} + \frac{x_3^{2/3}}{\ln(c_{lim}/c_0) + \ln x_3} \right].$$

Hence deduce that the area of the ith stage of an N-stage system is given by

$$A_i = \frac{Q_0}{k} \frac{x_N^{(i-1)/N}(1 - x_N^{1/N})}{\ln(c_{lim}/c_0) + \dfrac{i}{N}\ln x_N}.$$

Problem 5.17   Effect of retentate volume in fed-batch processes

When modelling the fed-batch process, we assumed that the volume of solution in the retentate tank was equal to the required final volume. Let us now consider a situation where the volume of solution placed in the retentate tank is greater than the required volume. This means that the fed-batch process must be followed by a batch process. Now, for the conditions specified in Example 5.14, calculate the total process time if the total feed is still 180 l, the required volume is still 40 l but 80 l of solution is placed in the tank for the fed-batch step.

Problem 5.18   Is there a better value of θ in Cheryan's equation

Here we revisit Example 5.13. Cheryan has chosen $\alpha = 0.5$ in his empirical equation. What value of α would make Cheryan's expression exact, at least in this case.

Problem 5.19   ODE formulation of multi-stage system with viscosity effects

In Section 5.2.3 it was shown how multi-stage problems can be formulated as a system of ODEs. Generalise the system of ODEs that is Eq. 5.7 for the case of a concentration-dependent mass transfer coefficient.

6 Diafiltration at the limiting flux

6.1 Introduction

Diafiltration (DF) is the removal of low molecular weight solutes, such as salts, from a solution of a high molecular weight solute, such as a protein. The basis of the separation is that the high molecular weight solute ('macrosolute'), or particle, does not pass through the membrane but the low molecular weight 'microsolute' does. It is somewhat similar to dialysis but, in the latter, a dialysing fluid flows counter-currently on the permeate side, providing a mass transfer driving force that enhances removal of the microsolute. Dialysis is discussed briefly towards the end of this chapter.

There are many ways to perform DF and in this chapter a number of different approaches are examined. Throughout the analyses presented, the solution is assumed to be composed of two solutes; a macrosolute whose concentration is denoted by c_A and a microsolute whose concentration is denoted by c_B. Complete rejection of the macrosolute is assumed throughout and the microsolute is assumed to have a rejection coefficient of zero. An additional and important assumption is that the flux is determined by the macrosolute concentration only. Furthermore, the flux is assumed to be at the limiting value as given by the limiting flux model. The one exception to this is in Section 6.4.3, where a more complex, essentially empirical, model for the flux is used. The most notable feature of that model is that the flux depends on the concentrations of both the macrosolute and the microsolute.

The two key aspects of DF that are examined throughout are water ('diluent') consumption and process time. In the next section discontinuous diafiltration (DDF) is examined from the perspective of these two characteristics.

6.2 Discontinuous diafiltration

Discontinuous diafiltration is essentially a combination of ultrafiltration and dilution and it can be done in two ways. The first approach is a volume reduction method, the second is a dilution method, and each approach is examined in the next two sections.

6.2.1 Volume reduction method

The steps involved in a discontinuous diafiltration by volume reduction are outlined in Table 6.1.

Table 6.1 Steps in DDF by the volume reduction method.

Step 1 Reduce the solution volume from V_0 to V_f.
  Macrosolute concentration increases, microsolute concentration remains the same.
Step 2 Replace lost permeate with diluent (pure water or buffer).
  Macrosolute concentration returns to initial concentration, microsolute concentration reduced.
Step 3 Repeat as required.

Let us consider a single 'stage', i.e., steps 1 and 2 in Table 6.1. Clearly, the net effect on the macrosolute is zero and its concentration remains unchanged. A mass balance on the microsolute, however, gives the following expression for its concentration at the end of the two-step process:

$$c_{Bf} = c_{B0}\frac{V_f}{V_0}. \tag{6.1}$$

For example, if the intial volume is 100 l and UF is used to bring the volume down to 10 l before adding water to bring it back to 100 l, the microsolute concentration is reduced by a factor of 10. If this process is repeated n times, the reduction factor for the microsolute, β, is given by

$$\beta = \frac{c_{B0}}{c_{Bf}} = \left(\frac{V_0}{V_f}\right)^n, \tag{6.2}$$

and therefore

$$\frac{V_0}{V_f} = \beta^{1/n}. \tag{6.3}$$

Thus, if one wants to reduce the microsolute concentration by a factor of 10 in a single-stage process, the volume must be reduced by a factor of 10. However, if three stages are used, the volume must be reduced in each stage by a factor of $10^{1/3} = 2.15$.

The total amount of water used in DDF by volume reduction is then given by

$$V_w = n(V_0 - V_f) = nV_0\left(1 - \frac{V_f}{V_0}\right) = nV_0(1 - \beta^{-1/n}). \tag{6.4}$$

Figure 6.1 is a plot of total dimensionless water consumption, V_w/V_0 versus n for $\beta = 10$, i.e., a ten-fold reduction in microsolute concentration. Clearly, the more stages involved, the greater the amount of water used and in Problem 6.1 you are asked to show that the dimensionless water consumption approaches $\ln\beta$ as $n \to \infty$.

So, would it ever be a good idea to adopt a multi-stage process? To answer this question, the total process time must be considered. From Eq. 5.45 in Chapter 5, the total time required to carry out the UF steps is given by

$$t = \frac{nV_0}{kA}\frac{c_{A0}}{c_{lim}}\left[Ei\left(\ln\frac{c_{lim}}{c_{A0}}\right) - Ei\left(\ln\frac{c_{lim}}{c_{Af}}\right)\right], \tag{6.5}$$

where c_{Af} is the macrosolute concentration at the end of each UF step and the mass transfer coefficient is assumed to be viscosity independent. Note that c_{lim} represents the

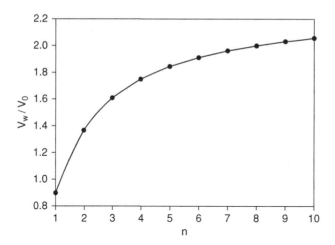

Figure 6.1 Water consumption versus the number of DDF stages for a ten-fold microsolute reduction.

limiting concentration of the macrosolute, but for neatness we dispense with the 'A' subscript for this parameter. Given that

$$\frac{c_{Af}}{c_{A0}} = \frac{V_0}{V_f} = \beta^{1/n},$$ (6.6)

Eq. 6.5 becomes

$$t = \frac{n V_0}{kA} \frac{c_{A0}}{c_{\lim}} \left[Ei \left(\ln \frac{c_{\lim}}{c_{A0}} \right) - Ei \left(\ln \frac{c_{\lim}}{c_{A0}\beta^{1/n}} \right) \right].$$ (6.7)

Use of this equation is illustrated in the example below.

Example 6.1 Effect of number of stages on total DDF time

A solution with an initial macrosolute of 10 g/l and a limiting concentration of 250 g/l is to be diafiltered using DDF by the volume reduction method. The microsolute concentration is to be reduced by a factor of 10. Generate a plot of dimensionless process time, tkA/V_0, versus the number of stages (1 – 10 inclusive), see Fig. 6.2, assuming that the time required to add water after each volume reduction is negligible.

Solution. A simple MATLAB script for this problem is shown below.

```
%Example 61
beta=10; cA0=10; clim=250;
for n=1:10
cAf=cA0*beta^(1/n);
time(n)=n*cA0/clim*(-expint(-log(clim/cA0))+expint(-
log(clim/cAf)));
stages(n)=n;
end
```

```
plot(stages,time,'s')
xlabel('Number of Stages')
ylabel('Time (-)')
```

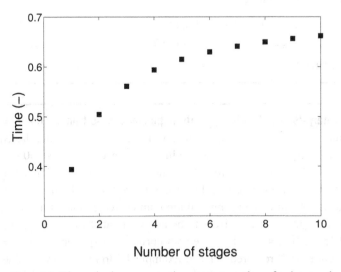

Figure 6.2 Dimensionless process time versus number of volume reduction stages for a ten-fold reduction in microsolute concentration.

As an additional exercise (Problem 6.2), you should run the above code and show that for very large n, the dimensionless time approaches $\ln\beta\,/\ln(c_{\lim}/c_{A0})$. Thus the conclusion from both the water consumption calculations and the process time calculations is that DDF with volume reduction is best performed in a single stage rather than multiple stages. However, with the single-stage approach, the reduction in volume may be sufficiently large to makes viscosity effects important. The next two examples illustrate calculation of water consumption and process time for viscosity independent and viscosity dependent mass transfer coefficients.

Example 6.2 Analysis of DDF with constant mass transfer coefficient

A certain solution containing a protein and a salt impurity is to be diafiltered by DDF by volume reduction. The protein concentration is 10 g/l and the limiting concentration is 300 g/l. The salt is to be reduced by a factor of 8. Given $V_0 = 60$ l, $A = 1.3$ m^2 and $k = 2 \times 10^{-5}$ m/s, compute the process time and the water consumption for a two-stage process.

Solution. Using Eq. 6.4 we get

$$V_w = 2 \times 0.06 \times (1 - 8^{-1/2}) = 77.6 \text{ l}.$$

Now Eq. 6.7 becomes

$$t = 2 \times \frac{0.06}{1.3 \times 2 \times 10^{-5}} \frac{10}{300} \left[Ei \left(\ln \frac{300}{10} \right) - Ei \left(\ln \frac{300}{10 \times 8^{1/2}} \right) \right].$$

In MATLAB script, this can be written

```
t=2*0.06/1.3/2e-5*10/300*(-expint(-
log(300/10))+expint(-log(300/(10*8^.5)))))
```

giving a time of 1015 s.

Example 6.3 Analysis of DDF with viscosity-dependent mass transfer coefficient

A certain solution containing a protein and a salt impurity is to be diafiltered by DDF. The protein concentration is 40 g/l and the limiting concentration is 300 g/l. The feed vessel contains 100 l and the microsolute concentration is to be reduced by a factor of 5. Compute the process time to achieve this reduction for both a one-stage and a two-stage process. Assume the infinite dilution, un-corrected mass transfer coefficient, k_0, to be 2×10^{-5} m/s. Let the viscosity be an exponential function of concentration with $\gamma = 0.015$ l/g (Eq. 4.21). Take the wall correction exponent to be $z = 0.14$ and assume turbulent flow. If required, refer to Example 4.5 to recall how the mass transfer coefficient varies with concentration in this case.

Solution. Solution of the time problem now requires numerical solution of the governing differential equations for batch ultrafiltration as described in Chapter 5. The relevant code for this is shown below.

```
function example63
global n cA0
cA0=40;
options=odeset('events', @eventfcn);
for n=1:2
[t,cA]=ode45(@DDF,[0 1e4], [cA0],options);
time=n*tend/60
end
function dcAdt=DDF(t,cA)
global beta cA0
k0=2e-5; g=0.015; z=0.14; clim=300; A=1.3;
beta=5;V0=0.1;
k=k0*exp(g*z*(cA-clim))*exp(-0.5*g*cA);
dcAdt=k*A/V0*cA^2/cA0*log(clim/cA);%Eq.5.33
function [value, isterminal, direction]=eventfcn(t,cA)
global n beta cA0
value = cA-cA0*beta^(1/n);% final concentration from Eq 6.6
isterminal = 1;
direction = 0;
```

Running the code gives a time of 137.9 minutes for the single-stage method and 113.2 minutes for the two-stage method, illustrating how the increased solution viscosities encountered in the single-stage approach can lead to inferior performance.

As well as raising the important issue of fluid viscosity when making large reductions in solution volume, the above example has raised an equally important and practical aspect of this operation. This is the fact that in reality most UF systems require a minimum operating volume. Thus, if one starts with a certain volume in the feed (retentate) tank, one does not have the ability to reduce the volume to arbitrary low levels, as might be required by the DDF specification. Thus, multi-stage operation, where less severe volume reductions are required, might be necessary for purely practical reasons.

6.2.2 Dilution method

In the previous section a process in which the solution was first concentrated and then returned to its initial volume by the addition of diluent was described. In this section, a different approach is examined in which the water addition and UF steps are reversed, i.e., the solution is first diluted and then ultrafiltered back to its original volume. Our instinct should suggest to us that this is going to require more water. However, the fact that the macrosolute concentration is lower on average than in the volume reduction method means that it is possible that the dilution method may require less time. Table 6.2 summarises the steps involved in the dilution method.

Table 6.2 Steps in DDF by dilution.

Step 1 Add water to increase solution volume from V_0 to V_f.
  Macrosolute and microsolute concentrations decrease.
Step 2 Ultrafilter solution from V_f back to V_0.
  Macrosolute concentration returns to initial value, microsolute concentration remains unchanged.
Step 3 Repeat as required.

A microsolute balance on the first step gives

$$\beta = \frac{c_{B0}}{c_{Bf}} = \frac{V_f}{V_0}, \tag{6.8}$$

where it should be emphasised that in this case $V_f > V_0$. For the macrosolute, one gets

$$\frac{c_{Af}}{c_{A0}} = \frac{V_0}{V_f} = \frac{1}{\beta}. \tag{6.9}$$

Extending this analysis to n stages gives

$$\frac{V_f}{V_0} = \beta^{1/n} \tag{6.10}$$

and

$$\frac{c_{Af}}{c_{A0}} = \beta^{-1/n}. \tag{6.11}$$

The water consumption for this process is thus given by

$$V_w = n(V_f - V_0) = nV_0 \left(\frac{V_f}{V_0} - 1 \right) = nV_0(\beta^{1/n} - 1). \tag{6.12}$$

Figure 6.3 is a plot of total dimensionless water consumption versus n for $\beta = 10$, i.e., a ten-fold reduction in microsolute concentration.

From the above, it is seen that increasing the number of stages reduces the amount of water required. Problem 6.3 involves a simple numerical exercise to show that in the limit $n \to \infty$, the dimensionless volume approaches $\ln \beta$. Comparing these results with Figure 6.1, however, shows that the water consumption in a dilution process is considerably higher than in a volume reduction process.

The other aspect of diafiltration is, of course, the process time, which in this case is given by

$$t = \frac{nV_f}{kA} \frac{c_{Af}}{c_{lim}} \left[Ei \left(\ln \frac{c_{lim}}{c_{Af}} \right) - Ei \left(\ln \frac{c_{lim}}{c_{A0}} \right) \right], \tag{6.13}$$

where it should be recalled that in this case, the coordinate (c_{Af}, V_f) is the starting point of the UF step. Using Eq. 6.11, and since the protein is completely retained $(c_{A0}V_0 = c_{Af}V_f)$, this becomes

$$t = \frac{nV_0}{kA} \frac{c_{A0}}{c_{lim}} \left[Ei \left(\ln \left(\frac{c_{lim}\beta^{1/n}}{c_{A0}} \right) \right) - Ei \left(\ln \frac{c_{lim}}{c_{A0}} \right) \right]. \tag{6.14}$$

As before, if one assumes that the water addition step is rapid in comparison with the ultrafiltration time, Eq. 6.14 represents the total DDF time. This equation is examined in the next example, which is analogous to Example 6.1.

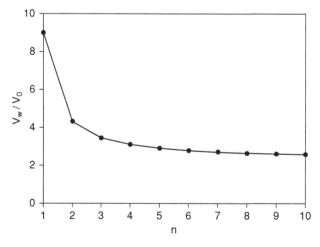

Figure 6.3 Dimensionless water volume versus number of stages for DDF by dilution for a ten-fold microsolute reduction.

Example 6.4 Effect of number of stages on total DDF time – dilution method

A solution with an initial macrosolute concentration of 10 g/l and a limiting concentration of 250 g/l is to be diafiltered using DDF by the dilution method. The microsolute concentration is to be reduced by a factor of 10. Generate a plot of dimensionless process time tkA/V_0 versus the number of stages (1 – 10 inclusive), see Fig. 6.4, assuming that the time required to add water after each volume reduction is negligible.

Solution. The script from Example 6.1 is modified as shown below

```
%Example64
beta=10;cA0=10;clim=250;
for n=1:10
cAf=cA0*beta^(-1/n)
time(n)=n*cA0/clim*(-expint(-log(clim/cAf))+expint(-
log(clim/cA0)));
stages(n)=n;
end
plot(stages,time,'s')
xlabel('Number of Stages'); ylabel('Time(-)')
```

In Problem 6.4, you are asked to show that in the limit as n beomes very large the dimensionless time converges to $\ln\beta/\ln(c_{\lim}/c_{A0})$, meaning that the volume reduction and dilution methods converge to the same value, which is not entirely unexpected.

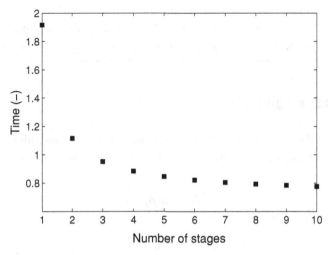

Figure 6.4 Dimensionless process time versus number of stages for a ten-fold reduction in microsolute concentration.

Given that in the dilution method, the macrosolute is decreased by dilution, it is likely that viscosity effects are less important than in the volume reduction method. The next example finishes this section by performing a calculation in which the volume reduction and dilution methods are compared.

Example 6.5 Comparison between volume reduction and dilution methods

A microsolute is to be removed from a 12 g/l protein solution and its concentration is to be reduced by a factor of 5. The process is performed in a UF system with area 1.1 m² and the mass transfer coefficient is 1×10^{-5} m/s. The limiting protein concentration is 280 g/l and the solution volume is 100 l. Compare the water consumption and the process times for *single-stage* volume reduction and dilution methods. The viscosity dependence of the mass transfer coefficient should be included, and take $z = 0.14$ and $\gamma = 0.015$ l/g and assume turbulent flow in the module.

Solution

(i) Water Consumption. For the volume reduction method, the water consumption is given by Eq. 6.4 as

$$V_{\mathrm{w}} = 100 \, (1 - 5^{-1}) = 80 \text{ l}.$$

For the dilution method, the water consumption is given by Eq. 6.12 as

$$V_{\mathrm{w}} = 100 \, (5 - 1) = 400 \text{ l}.$$

(ii) Process time. Solution of this part of the problem requires the use of the code in Example 6.3.

Running the code for the volume reduction method with $n = 1$ only, $c_{A0} = 12$ g/l, $V_0 = 0.1$ m³ and other parameters taking the appropriate values gives a process time of 102.3 min.

Running the code for the dilution method with $c_{A0} = 2.4$ g/l $(= 12/5)$ and $V_0 = 0.5$ m³ $(= 0.1 \times 5)$ gives a time of 166.7 min. Therefore, even accounting for viscosity effects, the dilution method takes considerably longer, at least in this particular example.

6.3 Constant volume diafiltration

It was shown in the analysis of DDF that the volume reduction and dilution methods converge to essentially the same process when $n \to \infty$. In particular, it was found that for large n

$$\frac{V_{\mathrm{w}}}{V_0} \to \ln \beta, \tag{6.15}$$

while

$$\frac{t k A}{V_0} \to \frac{\ln \beta}{\ln (c_{\mathrm{lim}}/c_{A0})}. \tag{6.16}$$

This special case of $n \to \infty$ is referred to as constant volume diafiltration, denoted CVD. Somewhat confusingly, perhaps, this is also referred to as continuous diafiltration, although that terminology is avoided in this work. CVD is commonly used in industry and, in terms of efficiency, it lies somewhere between DDF by dilution and DDF by volume reduction. However, CVD is a safe approach because viscosity effects are likely to be insignificant and problems with the need for very low working volumes to achieve a large reduction in microsolute concentration do not arise. Furthermore, CVD is easily adapted to become what is known as a generalised variable volume diafiltration (VVD) process. Such processes can be designed using optimal control theory to achieve any desired process objective [1].

In the next section, the basic equations for a CVD process are formulated, the key assumptions being that the macrosolute exhibits complete rejection and the microsolute has a rejection coefficient of zero.

6.3.1   Water consumption in CVD

Let us consider a solution containing a macrosolute at a concentration c_{A0} and a microsolute at a concentration c_{B0}. The goal is to reduce the concentration of microsolute by a factor, β, where β is defined as before, i.e., c_{B0}/c_{Bf}. The solution volume is constant at V_0. The basic setup for a CVD process is shown in Fig. 6.5. The key point is that the flowrate of water exactly balances the permeate flowrate.

The fact that the microsolute has a rejection coefficient of zero means that its concentration in the retentate is identical to its concentration in the permeate at all times. Therefore, the microsolute balance can be written

$$V_0 \frac{dc_B}{dt} = -JAc_B.$$ (6.17)

The water consumption can be expressed by the following:

$$\frac{dV_w}{dt} = JA.$$ (6.18)

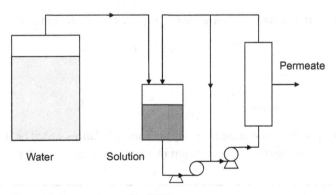

Figure 6.5 Process layout for CVD.

Now dividing Eq. 6.17 by Eq. 6.18 gives

$$V_0 \frac{dc_B}{dV_w} = -c_B. \tag{6.19}$$

Integrating gives

$$c_B = c_{B0}e^{-V_w/V_0}. \tag{6.20}$$

Thus, to achieve the desired final concentration of microsolute, one gets the following expression for the amount of water:

$$\frac{V_w}{V_0} = \ln \frac{c_{B0}}{c_{Bf}} = \ln \beta. \tag{6.21}$$

This is the familiar expression mentioned earlier as a limiting case of DDF. For example, if one wanted to reduce the macrosolute by a factor of 10 from a 100 l solution, the water consumption would be about 230 l. In contrast, a single-stage DDF by dilution process would require 900 l while a DDF by volume reduction would require only 90 l.

Example 6.6 Mass balances in CVD

A 25 l volume of a protein solution is to undergo CVD to reduce the microsolute concentration, which is at 1.5 g/l. 100 l of diafiltration water is used for this purpose. The permeate from this operation is retained in a separate permeate tank. Calculate the macrosolute concentration in both the retentate and permeate tanks at the end of the operation.

Solution. The microsolute concentration at the end of the CVD process is given by Eq. 6.20 as

$$c_{Bf} = 1.5e^{-100/25} = 0.027 \, \text{g/l}.$$

Now, given that the volume in the permeate tank is the same as the volume used for diafiltration, the microsolute balance becomes

$$c_{B0}V_0 = c_{Bf}V_0 + c_{Bp}V_w,$$

where c_{Bp} is the microsolute concentration in the permeate. Putting in numbers gives

$$1.5 \times 25 = 0.027 \times 25 + c_{Bp} \times 100.$$

Therefore

$$c_{Bp} = \frac{(1.5 - 0.027) \times 25}{100} = 0.37 \, \text{g/l}.$$

A useful exercise in formulating unsteady state mass balances would be to derive an expression for the microsolute concentration in the permeate as a function of time (see Problem 6.5).

6.3.2 Time taken for CVD

When the macrosolute is completely rejected and it is the sole determinant of the flux, the flux during CVD is constant. Thus, the time for CVD takes the simple form

$$t_{\text{CVD}} = \frac{V_w}{kA \ln(c_{\text{lim}}/c_{A0})}, \tag{6.22}$$

i.e.,

$$t_{\text{CVD}} = \frac{V_0}{kA} \frac{\ln \beta}{\ln(c_{\text{lim}}/c_{A0})}, \tag{6.23}$$

which is the familiar expression first presented as a limiting case of DDF, Eq. 6.16. Thus computing CVD times is simply a matter of 'plugging in numbers'. Of far more interest is ultrafiltration with constant volume diafiltration, i.e., a process in which the objective is to both concentrate a macrosolute by batch UF and remove a microsolute by CVD. This is the subject of the next section where it is shown that an interesting, and well known, optimisation arises.

6.3.3 Ultrafiltration with constant volume diafiltration (UFCVD)

Consider a process in which the goal is to concentrate a macrosolute from c_{A0} to c_{Af} and diafilter a microsolute from c_{B0} to c_{Bf}. The key question is this: is there a quickest way to perform this operation? For example, if the UF step is done first, it will mean that the CVD step will require a lot of water (V_0 is large in Eq. 6.21) but the flux will be higher (c_{A0} is small in Eq. 6.23). Conversely, if the UF step is done second, very little water will be required (V_0 small in Eq. 6.21) but the flux will be low (c_{A0} will be large in Eq. 6.23). As one might expect, the answer lies somewhere in between, i.e., it is best, from the point of view of time reduction, to do the CVD step at some intermediate concentration, c_A. The total process time in that case can be written

$$t = \frac{V_0 c_{A0}}{kA c_{\text{lim}}} \left[Ei\left(\ln \frac{c_{\text{lim}}}{c_{A0}}\right) - Ei\left(\ln \frac{c_{\text{lim}}}{c_A}\right) \right] + \frac{V \ln \beta}{kA \ln(c_{\text{lim}}/c_A)}$$

$$+ \frac{V c_A}{kA c_{\text{lim}}} \left[Ei\left(\ln \frac{c_{\text{lim}}}{c_A}\right) - Ei\left(\ln \frac{c_{\text{lim}}}{c_{Af}}\right) \right], \tag{6.24}$$

where V is the solution volume corresponding to the intermediate concentration, c_A. The first term is the time for the first UF step, the second term accounts for CVD and the third accounts for the second UF step. Noting the usual macrosolute balance, $c_{A0} V_0 = c_A V$, Eq. 6.24 simplifies to

$$t = \frac{V_0 c_{A0}}{kA c_{\text{lim}}} \left[Ei\left(\ln \frac{c_{\text{lim}}}{c_{A0}}\right) - Ei\left(\ln \frac{c_{\text{lim}}}{c_{Af}}\right) \right] + \frac{V \ln \beta}{kA \ln(c_{\text{lim}}/c_A)}. \tag{6.25}$$

This implies that the time taken for the UF steps is fixed and thus the total process time is minimised when the diafiltration time is minimised. Therefore, the criterion for minimising the total process time becomes

$$\frac{d}{dc_A}\left[\frac{\ln\beta}{kA}\frac{V}{\ln(c_{\lim}/c_A)}\right] = 0. \tag{6.26}$$

But since the macrosolute is completely rejected, a mass balance means that we get

$$\frac{d}{dc_A}\left[\frac{V_0c_{A0}\ln\beta}{kA}\frac{1}{c_A\ln(c_{\lim}/c_A)}\right] = 0. \tag{6.27}$$

Mathematically, this implies

$$\frac{d}{dc_A}[c_A\ln(c_{\lim}/c_A)] = 0. \tag{6.28}$$

Differentiating and denoting the optimum as $c_{A\mathrm{opt}}$, one gets

$$c_{A\mathrm{opt}}\left(-\frac{1}{c_{A\mathrm{opt}}}\right) + \ln\frac{c_{\lim}}{c_{A\mathrm{opt}}} = 0. \tag{6.29}$$

Therefore

$$\ln\frac{c_{\lim}}{c_{A\mathrm{opt}}} = 1, \tag{6.30}$$

or

$$c_{A\mathrm{opt}} = \frac{c_{\lim}}{e} \approx \frac{c_{\lim}}{2.72}, \tag{6.31}$$

where e is the base of the natural logarithm. This is a remarkably simple result first mentioned in the literature by Ng and co-workers [2]. But is this an important result in practice? It is only important if (i) the optimum is 'deep', i.e., the minimum time is significantly different from times under non-optimum conditions, and (ii) the diafiltration time is a significant fraction of the total processing time. The next example illustrates these points.

Example 6.7 Optimum concentration for CVD in UFCVD

A 10 g/l protein solution is to undergo UFCVD in which the protein concentration is to be increased to 160 g/l and the microsolute concentration is to be reduced by a factor of 10. The limiting concentration is 250 g/l, the initial volume is 200 l, the membrane area is 1.5 m^2 and the mass transfer coefficient can be assumed to be constant and equal to 1×10^{-5} m/s. Calculate and plot the CVD time as a function of the macrosolute concentration at which CVD is performed.

Solution. MATLAB code to solve the problem, based on the relevant term in Eq. 6.25, is given below.

```
%Example67
cA0=12; cAf=160; clim=250; V0=0.2; k=1e-5; A=1.5;
beta=10;
```

```
for n=1:50;
cA(n)=40+(160--40)*(n-1)/50;
V=cA0*V0/cA(n);
tcvd(n)=V/k/A*log(beta)/log(clim/cA(n))/60;
end
plot(cA,tcvd,'-k');
axis([40 160 60 90])
xlabel('c_A (g/l)'); ylabel('Time (mins)')
```

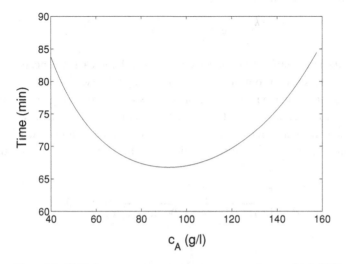

Figure 6.6 CVD time versus macrosolute concentration at which CVD is performed.

Figure 6.6 illustrates the optimum and, at first glance, it would seem that it is a deep optimum. However, if one looks closely at the data, it is seen that one has to be a long way from the optimum to make a significant difference to the process time. A quick look at the numbers suggests that a 25% departure from the optimum concentration, either above or below, only produces about a 4% increase in CVD time. Therefore, as long as one is in the general region of c_{lim}/e, a significant reduction in process time will be achieved.

6.3.4 UFCVD time optimisation with viscosity dependent mass transfer coefficient

In the previous section it was assumed throughout that the mass transfer coefficient was constant. However, the optimum concentration for diafiltration, c_{lim}/e, is rather high. For example, for a solution with an exponential dependence of the viscosity on concentration and $\gamma = 0.01$ l/g (see Eq. 4.21), and $c_{lim} = 250$ g/l, the bulk solution viscosity at c_{lim}/e is more than 2.5 times that of the solvent. Consider now a solution with an initial concentration, c_{A0}, of 20 g/l. In turbulent flow, the mass transfer coefficient at $c_{\lim}/e$, is related to the mass transfer coefficient at 20 g/l by the expression (see Example 4.2)

$$\frac{k_c}{k_{c0}} = e^{-0.5\gamma(c_{\lim}/e - c_{A0})} e^{0.14\gamma(c_{\lim}/e - c_{A0})}. \tag{6.32}$$

For $c_{\lim} = 250$ g/l and $\gamma = 0.01$ l/g, this ratio works out as

$$\frac{k_c}{k_{c0}} = 0.772. \tag{6.33}$$

For laminar flow, the calculation is

$$\frac{k_c}{k_{c0}} = e^{0.14\gamma(c_{\lim}/e - c_{A0})} = 1.106. \tag{6.34}$$

The fact that the mass transfer coefficient decreases with increasing concentration in turbulent flow suggests that the optimum concentration for CVD will shift to concentrations below $c_{\lim}/e$ in that case. Conversely, the optimum should shift to higher concentrations in laminar flow. Let us prove now that these two predictions are true. As was the case with a constant mass transfer coefficient, the process time is minimised if the CVD time is minimised. Therefore, in turbulent flow, the optimum is found by solution of the equation

$$\frac{d}{dc_A}\left[\frac{1}{c_A e^{-0.5\gamma c_A} e^{\gamma z(c_A - c_{\lim})} \ln(c_{\lim}/c_A)} \right] = 0. \tag{6.35}$$

Therefore

$$\frac{d}{dc_A}\left[c_A e^{\gamma c_A(z - 0.5)} \ln(c_{\lim}/c_A) \right] = 0. \tag{6.36}$$

Performing the differentiation and using the same notation as before gives

$$\ln\frac{c_{\lim}}{c_{A\mathrm{opt}}}\left[1 + \gamma(z - 0.5)c_{A\mathrm{opt}} \right] = 1. \tag{6.37}$$

This is a non-linear algebraic equation that must be solved numerically to get the optimum concentration. The analogous expression for laminar flow is obviously

$$\ln\frac{c_{\lim}}{c_{A\mathrm{opt}}}\left[1 + \gamma z c_{A\mathrm{opt}} \right] = 1. \tag{6.38}$$

Both expressions reduce to the classic result when $\gamma = 0$ as expected. In the next example, the predictions of these expressions are examined.

Example 6.8 Optimum concentration for UFCVD with viscosity dependent mass transfer coefficient

A protein solution has a limiting concentration of 300 g/l and a γ value of 0.012 l/g. Compute the optimum concentration for diafiltration for laminar and turbulent flows.

Solution. Numerical solution of the relevant equations can be performed with the MATLAB function **fzero** as shown in the code below:

```
function example68;
guess=100;
copt=fzero(@ufcvdopt,guess)
function f=ufcvdopt(cAopt);
clim=300; g=0.012; z=0.14; flow=0.5;
f=log(clim/cAopt)*(1+g*(z-flow)*cAopt)-1;
```

For turbulent flow, the optimum is at 70.9 g/l. For laminar flow (`flow = 0`) it is at 132.4 g/l. These values compare with a value of 110.4 g/l obtained with the classic expression c_{lim}/e. Thus, the optimum shifts towards lower concentrations in turbulent flow, as predicted, and higher concentrations in laminar flow, also as predicted earlier.

6.3.5 Economic optimisation of UFCVD

In the previous sections, the optimum concentration was derived based on the criterion that the total process time should be minimised. However, the process time and the resulting energy costs are not the only factors that could be considered when performing a process optimisation. Often in diafiltration processes the water costs can be substantial because highly pure water or buffer must be used. Therefore, in this section optimisation of the process cost, based on a time (energy) cost and a water cost, is derived. Initially, it is assumed that the mass transfer coefficient is constant. The total cost of the process is given by

$$\text{Total cost} = \text{UF time cost} + \text{CVD time cost} + \text{diafiltration water cost.} \quad (6.39)$$

The UF time cost is fixed, therefore all one needs to consider are the CVD time cost and the water cost. Thus, the function, P, must be minimised where

$$P = \kappa_1 V_w + \kappa_2 t_{CVD}, \quad (6.40)$$

where κ_1 and κ_2 represents cost coefficients, the former being a cost per unit volume and the latter being a cost per unit time. The unit time cost would obviously be the product of the power consumed and the unit energy cost. The power consumption would depend on the particular module and the configuration of the system [3], while the unit energy cost would be specific to the particular country. Thus, the function to be minimised is given by

$$P = \kappa_1 \frac{V_0 c_{A0}}{c_A} \ln \beta + \frac{\kappa_2 V_0 c_{A0} \ln \beta}{k A c_A \ln (c_{lim}/c_A)}. \quad (6.41)$$

Therefore, to minimise the total cost, one imposes the criterion

$$\frac{d}{dc_A} \left[\frac{\kappa_1}{c_A} + \frac{\kappa_2}{k A c_A \ln (c_{lim}/c_A)} \right] = 0. \quad (6.42)$$

Differentiating and using the same notation as before gives

$$\frac{-\kappa_1}{c_{A\text{opt}}^2} + \frac{-\kappa_2 \left(-1 + \ln \left(c_{\text{lim}}/c_{A\text{opt}} \right) \right)}{kA \left(c_{A\text{opt}} \ln \frac{c_{\text{lim}}}{c_{A\text{opt}}} \right)^2} = 0. \tag{6.43}$$

Therefore

$$\kappa_1 \left(\ln \frac{c_{\text{lim}}}{c_{A\text{opt}}} \right)^2 + \frac{\kappa_2}{kA} \left(\ln \frac{c_{\text{lim}}}{c_{A\text{opt}}} - 1 \right) = 0. \tag{6.44}$$

This clearly reduces to the classic result in the limit where there is no water cost, i.e., $\kappa_1 = 0$. Solving Eq. 6.44 gives

$$\ln \frac{c_{\text{lim}}}{c_{A\text{opt}}} = \left[-\frac{\kappa_2}{kA\kappa_1} + \sqrt{ \left(\frac{\kappa_2}{kA\kappa_1} \right)^2 + 4 \left(\frac{\kappa_2}{kA\kappa_1} \right) } \right] \Bigg/ 2. \tag{6.45}$$

Simplifying gives

$$\ln \frac{c_{\text{lim}}}{c_{A\text{opt}}} = \left(\frac{\kappa_2}{2kA\kappa_1} \right) \left[\sqrt{1 + 4 \left(\frac{kA\kappa_1}{\kappa_2} \right)} - 1 \right]. \tag{6.46}$$

To neaten this expression, we define a dimensionless number, N_c, representing the relative costs of time and water, giving

$$\ln \frac{c_{\text{lim}}}{c_{A\text{opt}}} = \frac{N_c}{2} \left[\sqrt{1 + 4/N_c} - 1 \right]. \tag{6.47}$$

To see more clearly the meaning of N_c, it is worth multiplying it above and below by V_0 to give

$$N_c = \frac{\kappa_2}{\kappa_1 kA} = \frac{\kappa_2 V_0 / kA}{\kappa_1 V_0}. \tag{6.48}$$

The combination of terms, V_0/kA, can be considered as a characteristic time for the process while V_0 can be considered a characteristic volume. Thus N_c represents a characteristic time cost relative to a characteristic water cost. In the limit where the water cost is negligible, i.e.,

$$N_c \to \infty, \tag{6.49}$$

one can use the approximation

$$\sqrt{1 + x} \approx 1 + \frac{x}{2}, \tag{6.50}$$

and Eq. 6.47 reduces to the classic result as expected. In Problem 6.6, you are asked to examine the situation where the water cost is dominant. Figure 6.7 illustrates the prediction of Eq. 6.48.

The above calculations and example were limited by the assumption of constant mass transfer coefficient. If the mass transfer coefficient becomes viscosity dependent, the analysis becomes a little bit more complicated, as shown below.

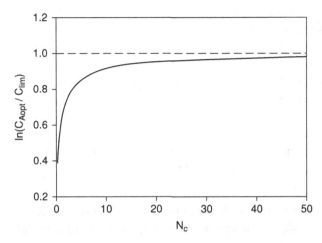

Figure 6.7 Effect of relative costs of time and water on optimum concentration for CVD.

6.3.6 Economic optimisation with viscosity dependent mass transfer coefficient

The same methodology as outlined in the previous section can be used again, the only difference being that the algebra is a bit more cumbersome. The analogous expression to Eq. 6.42 can be written for turbulent flow as

$$\frac{d}{dc_A}\left[\frac{\kappa_1}{c_A} + \frac{\kappa_2}{k_0 A e^{-\gamma z c_{\lim}} e^{\gamma c_A(z-0.5)} c_A \ln(c_{\lim}/c_A)}\right] = 0, \qquad (6.51)$$

where k_0 is the infinite-dilution, uncorrected mass transfer coefficient. Therefore

$$-\frac{1}{c_{A\text{opt}}^2} - N_c e^{\gamma z c_{\lim}} \frac{(\ln(c_{\lim}/c_{A\text{opt}}) - 1) + \gamma(z - 0.5)c_{A\text{opt}}\ln(c_{\lim}/c_{A\text{opt}})}{e^{\gamma c_{A\text{opt}}(z-0.5)}\left[c_{A\text{opt}}\ln(c_{\lim}/c_{A\text{opt}})\right]^2} = 0, \quad (6.52)$$

where N_c is defined in terms of k_0. This is clearly a complex equation and it will be kept in the above form and solved numerically as shown in the next example.

Example 6.9 Economic optimisation with viscosity dependent mass transfer coefficient
Use Eq. 6.52 to compute and plot the optimum concentration for CVD for N_c values in the range 0.5–20. Take $c_{\lim} = 300$ g/l, $\gamma = 0.01$ l/g, $z = 0.14$.

Solution. MATLAB code for solving this problem is given below.

```
function example69
global clim Nc
clim=300;
for n=1:40;
Nc=n*0.5;
Ncout(n)=n;
guess=clim/exp(1);
```

```
cA=fzero(@costopt,guess);
cAout(n)=cA/clim;
end
plot(Ncout,cAout,'-k');
xlabel('N_{\rm c}'); ylabel('c_A_o_p_t / c_l_i_m')
function f=costopt(cA)
global clim Nc
g=0.01; z=0.14;
f=-1/cA^2-Nc*exp(g*z*clim)*((log(clim/cA)-1)+g*(z-
0.5)*cA*log(clim/cA))/exp(g*cA*(z-
0.5))/(cA*log(clim/cA))^2;
```

Results from the calculation are shown in Fig. 6.8. As shown previously for the case where the mass transfer coefficient was constant, the concentration shifts to higher concentrations when the water costs become dominant (N_c small). When the energy costs are dominant (N_c large), the optimum in this case shifts to the value computed from Eq. 6.37.

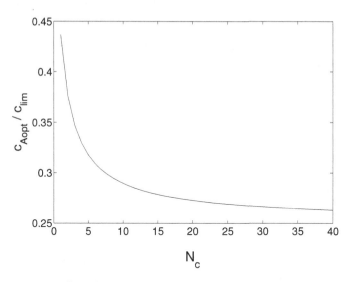

Figure 6.8 Effect of relative costs of time and water on optimum concentration for CVD with viscosity dependent mass transfer coefficient.

6.3.7   Optimisation of UFCVD when $c_{Af} < c_{Aopt}$

In the previous sections the focus has been on investigating optimum concentrations for beginning the CVD step. But what should one do if the final macrosolute concentration, as demanded by the process specification, is actually less than the optimum concentration? In that case, is it correct to say that the question of an optimum does not arise and one should simply perform the CVD step at c_{Af}? This should ensure that the process time is minimised, as a quick glance at Fig. 6.6 will confirm. However, as pointed out by Paulen *et al.* [4], it may well be a better option to 'over concentrate' the solution to $c_{A\text{opt}}$. Then, one can do the CVD step and rapidly dilute back to c_{Af}. Let us consider such an approach and derive an expression for the total process time in that case. The first step involves batch UF from c_{A0} to $c_{A\text{opt}}$, and the time taken is given by

$$t_{UF} = \frac{V_0 c_{A0}}{kA c_{\lim}} \left[Ei \left(\ln \frac{c_{\lim}}{c_{A0}} \right) - Ei \left(\ln \frac{c_{\lim}}{c_{A\text{opt}}} \right) \right]. \tag{6.53}$$

The diafiltration time is given by

$$t_{CVD} = \frac{V_w}{kA \ln \left(c_{\lim}/c_{A\text{opt}} \right)}, \tag{6.54}$$

where V_w is the amount of water used in the diafiltration step. Assigning a value to this needs a little bit of thought. The naive approach would be simply to apply Eq. 6.21, use a macrosolute balance and write

$$V_w = \frac{c_{A0} V_0}{c_{A\text{opt}}} \ln \beta, \tag{6.55}$$

where β is the desired microsolute reduction. But Eq. 6.55 ignores the fact that when the solution is diluted from $c_{A\text{opt}}$ to c_{Af}, this brings about a reduction in microsolute concentration as well. Thus, the amount of water required in the CVD step is less than that predicted by Eq. 6.55. If the microsolute concentration reaches its required value, c_{Bf}, when $c_A = c_{Af}$, then its concentration, $c_{B\text{opt}}$, at the end of the CVD step done at $c_{A\text{opt}}$, need only be as low as

$$c_{B\text{opt}} = c_{Bf} \frac{c_{A\text{opt}}}{c_{Af}}, \tag{6.56}$$

where $c_{B\text{opt}} > c_{Bf}$. The CVD step must only reduce the microsolute concentration to this value, the dilution step will do the rest. Thus, the water required for the CVD step is given by

$$V_w = \frac{c_{A0} V_0}{c_{A\text{opt}}} \ln \left(\beta \frac{c_{Af}}{c_{A\text{opt}}} \right). \tag{6.57}$$

Use of these equations is illustrated in the next example.

Example 6.10 UFCVD with a dilution step

A protein solution with limiting concentration 300 g/l is to be concentrated from 20 g/l to 80 g/l by UFCVD. The microsolute concentration is to be reduced by a factor of 10. The initial solution volume is 150 l, the membrane area is 1.5 m² and the mass transfer coefficient is constant and equal to 2×10^{-5} m/s. Calculate the time taken to perform this operation if (i) CVD is done at c_{Af}, (ii) CVD is done at $c_{\lim}/e$ and a dilution step is incorporated. You may neglect the time taken for the dilution step. What would the flowrate of water delivered by the pump during the dilution step have to be to make these two methods take the same time?

Solution. Looking at the more conventional approach first, the total time is given by

$$t = \frac{V_0 c_{A0}}{k A c_{\lim}} \left\{ Ei\left(\ln\frac{c_{\lim}}{c_{A0}}\right) - Ei\left(\ln\frac{c_{\lim}}{c_{Af}}\right) + \frac{c_{\lim}\ln\beta}{c_{Af}\ln(c_{\lim}/c_{Af})} \right\}.$$

Putting in numbers gives

$$t = \frac{0.15 \times 20}{2 \times 10^{-5} \times 1.5 \times 300} \left[Ei\left(\ln\frac{300}{20}\right) - Ei\left(\ln\frac{300}{80}\right) + \frac{300\ln(10)}{80\ln(300/80)} \right].$$

Using the **expint** function in MATLAB, as before, this works out as 3968 s.

For the 'over-concentration' approach, we have

$$t = \frac{V_0 c_{A0}}{k A c_{\lim}} \left\{ Ei\left(\ln\frac{c_{\lim}}{c_{A0}}\right) - Ei\left(\ln\frac{c_{\lim}}{c_{\lim}/e}\right) + \frac{e\ln(\beta c_{Af}/(c_{\lim}/e))}{\ln(c_{\lim}/(c_{\lim}/e))} \right\}.$$

Simplifying gives

$$t = \frac{V_0 c_{A0}}{k A c_{\lim}} \left\{ Ei\left(\ln\frac{c_{\lim}}{c_{A0}}\right) - Ei(1.0) + e\ln(\beta c_{Af}/(c_{\lim}/e)) \right\}.$$

Putting in numbers gives

$$t = \frac{0.15 \times 20}{2.0 \times 10^{-5} \times 1.5 \times 300} \left\{ Ei\left(\ln\frac{300}{20}\right) - Ei(1.0) + e\ln(10 \times 80/(300/e)) \right\}$$

which works out as 3881 s, which represents a 2% reduction in time compared to the more conventional approach.

Now the actual time difference between the two methods is 87 s, and in this time the pump must supply a dilution volume $V_f - V_{opt}$, where V_f is the retentate volume when $c_A = c_{Af}$ and V_{opt} is the retenate volume when $c_A = c_{\lim}/e$.

Therefore the pump flowrate, Q_{pump} is given by

$$Q_{pump} = \frac{V_f - V_{opt}}{\Delta t} = \frac{V_0 c_{A0}/c_{Af} - V_0 c_{A0}/(c_{\lim}/e)}{87}.$$

Putting in numbers gives

$$Q_{pump} = \frac{0.15 \times 20}{87}\left(\frac{1}{80} - \frac{e}{300}\right) = 1.186 \times 10^{-4}\,\mathrm{m^3/s} = 7.1\ \mathrm{l/min}.$$

As long as the pump can exceed this flowrate, the dilution method is (marginally) faster. In Problem 6.7, you are asked to derive a criterion for when the 'over-concentration' approach is faster than the conventional approach.

6.4   Dynamic modelling of UFDF processes

So far, discussion of diafiltration has focused on discontinuous diafiltration and constant volume diafiltration, with the emphasis on water consumption and process time. However, there is a wide range of possibilities when it comes to devising methods by which a macrosolute can be concentrated and a microsolute removed. In this section, the details of the solute dynamics during ultrafiltration–diafiltration (UFDF) processes are examined and it is shown how the basic UFCVD process can be modified and generalised, ultimately providing a framework for dynamic optimisation of any arbitrary UFDF processes. To start, the dynamics of UFCVD are modelled and, from there, the emergence of the concept of variable volume diafiltration is described.

6.4.1   Dynamic modelling of UFCVD

Let us recall the basic layout of a DF process and now introduce the α parameter (not to be confused with specific cake resistance), as indicated in Fig. 6.9. In this formulation, the flow of water into the retentate tank is set at a fraction, α, of the permeate flowrate.

The balances for the system can be written as follows:

$$\frac{dV}{dt} = (\alpha - 1)\,JA, \tag{6.58}$$

$$\frac{d\,(Vc_A)}{dt} = 0, \tag{6.59}$$

$$\frac{d\,(Vc_B)}{dt} = -JAc_B, \tag{6.60}$$

$$\frac{dV_{\mathrm{w}}}{dt} = \alpha JA, \tag{6.61}$$

where the symbols have their usual meaning. Now, for UFCVD, the α parameter takes on the functional form described in the pseudocode below:

```
IF  cA ≥ cAi  and  cB > cBf
THEN  α = 1
ELSE  α = 0
```

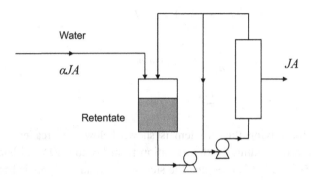

Figure 6.9 UFDF process using α formulation.

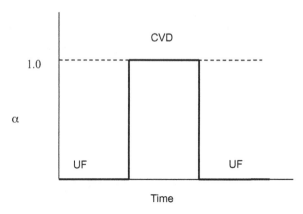

Figure 6.10 α function for UFCVD.

where c_{Ai} is the macrosolute concentration at which CVD is begun. Technically, the first criterion should be $c_A = c_{Ai}$ but for computational purposes it is better to use the inequality. Thus, for UFCVD, α is a square pulse as indicated in Fig. 6.10.

The next example illustrates use of this formulation to generate the concentration–time profiles for both the macrosolute and the microsolute.

Example 6.11 Dynamic modelling of UFCVD

A protein solution is to be concentrated from 50 g/l to 150 g/l and the microsolute is to be reduced from 1 g/l to 0.2 g/l. The limiting protein concentration, c_{lim}, is 280 g/l. The initial volume is 100 l, the mass transfer coefficient is constant and equal to 2×10^{-5} m/s and the membrane area is 0.8 m². Generate plots of c_A/c_{A0} and c_B/c_{B0} versus time if CVD is commenced at the optimum concentration $c_{\text{lim}}/e = 103$ g/l.

Solution. To start, Eqs. 6.58 to 6.61 are expanded to give the following:

$$\frac{dV}{dt} = (\alpha - 1) J A,$$

$$\frac{dc_A}{dt} = \frac{c_A}{V} (1 - \alpha) J A,$$

$$\frac{dc_B}{dt} = -\frac{c_B}{V} \alpha J A,$$

$$\frac{dV_{\text{w}}}{dt} = \alpha J A.$$

MATLAB code for solving this problem is shown below. The reader should pay particular attention to the coding for the transition from UF to CVD and back to UF. Note that the **ode15s** function is used and the step size is reduced. This is because the transition from UF to CVD leads to a few minor numerical errors when **ode45** is used. The function **ode15s** is a routine used for stiff problems and is often the first choice if **ode45** is not working as well as one might wish.

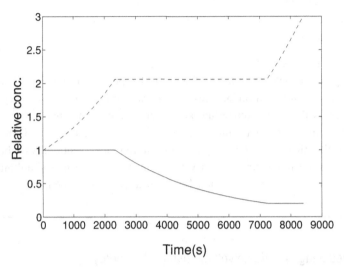

Figure 6.11 Macrosolute and microsolute profiles in UFCVD. The dashed curve denotes macrosolute and the solid curve denotes the microsolute.

```
function example611
V0=0.1; cA0=50; cB0=1; Vw0=0;
options=odeset('events', @eventfcn,'MaxStep',10);
[t,y]=ode15s(@ufcvd,[0 36000],[V0 cA0 cB0 Vw0], options)
cA=y(:,2)/cA0; cB=y(:,3)/cB0;
plot(t,cA,t,cB);
axis([0 9000 0 3]);
xlabel('Time(s)');ylabel('Relative Conc.')
function dydt=ufcvd(t,y)
A=0.8; cAi=103; cBf=0.2; k=2e-5;clim=280;
J=k*log(clim/y(2));
if and(y(2)>=cAi,y(3)>cBf);
alpha=1;
else
alpha=0;
end
%y(1) is V, y(2) is cA, y(3) is cB, y(4) is Vw
Veqn=(alpha-1)*J*A;
cAeqn=y(2)/y(1)*(1-alpha)*J*A;
cBeqn=-y(3)/y(1)*alpha*J*A;
Vweqn=alpha*J*A;
```

```
dydt=[Veqn;cAeqn;cBeqn;Vweqn];
function [value, isterminal, direction]=eventfcn(t,y)
cAf=150;
value = y(2)-cAf;
isterminal = 1;
direction = 0;
```

Results from this code are shown in Fig. 6.11.

The plot gives a nice illustration of what is happening in a UFCVD process. In a sense, the two concentration profiles mirror each other. One of the key points to note, however, is that the macrosolute and microsolute do not reach their respective targets at the same time. The microsolute reaches its target first and then remains constant for the final UF step. As is seen in the next section, variable volume diafiltration has the interesting property that both targets are met simultaneously. In Problem 6.8, this example is repeated but with a viscosity-dependent mass transfer coefficient.

6.4.2   Dynamic modelling of variable volume diafiltration (VVD)

In variable volume diafiltration (VVD), the processes of UF and DF occur at the same time. The macrosolute concentration increases and the microsolute concentration decreases steadily throughout the process. The respective targets are met simultaneously. This is a very elegant solution to the UFDF problem, but as will be seen, it is not a very efficient way of doing UFDF, at least when the flux is given by the simple limiting flux model used thus far.

Let us consider the four equations given in Example 6.11. VVD is characterised by a constant value of α where $0 \leq \alpha \leq 1$. This means that the retentate volume is declining but the addition of the water has the effect of flushing out the microsolute. Now, dividing the last equation of those four by the first gives

$$\frac{dV_w}{dV} = \frac{\alpha}{\alpha - 1}. \tag{6.62}$$

Integrating gives

$$V_w = \frac{\alpha}{1 - \alpha} (V_0 - V_f). \tag{6.63}$$

This expression relates the water consumption to the retentate volume reduction but an appropriate value for α is needed. This depends on the problem specification, i.e., both the increase in macrosolute concentration (or reduction in solution volume) *and* the microsolute reduction. Dividing the c_A equation by the c_B equation in Example 6.11 gives

$$\frac{dc_A}{dc_B} = \frac{\alpha - 1}{\alpha} \frac{c_A}{c_B}. \tag{6.64}$$

Integrating gives

$$\ln \frac{c_{Af}}{c_{A0}} = \frac{1-\alpha}{\alpha} \ln \frac{c_{B0}}{c_{Bf}}. \tag{6.65}$$

Rearranging leads to

$$\alpha = \frac{\ln (c_{B0}/c_{Bf})}{\ln (c_{Af}/c_{A0}) + \ln (c_{B0}/c_{Bf})}. \tag{6.66}$$

This is a remarkable result because it predicts that if a suitable value of α is chosen, namely a value computed using Eq. 6.66, VVD will naturally lead to both the macrosolute and microsolute reaching their respective targets at exactly the same time. This is illustrated in Fig. 6.12, which was obtained with the MATLAB code of Example 6.11, modified by simply making α constant and computed from Eq. 6.66.

The next example examines water consumption in VVD based on the equations developed above.

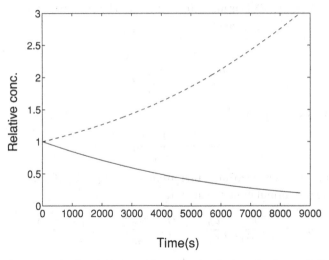

Figure 6.12 Concentration profiles in VVD based on data in Example 6.11. The dashed curve denotes macrosolute, solid curve denotes the microsolute.

Example 6.12 Water consumption in VVD

A 150 l volume of a macrosolute solution is to be increased in concentration from 14 g/l to 140 g/l and the microsolute concentration reduced by a factor of 10 using VVD. Calculate how much water is consumed in the process and compare it with the amount of water used in a UFCVD process where the CVD step is done when $c_A = 100$ g/l.

Solution. Using Eq. 6.66, the required value of α is given by

$$\alpha = \frac{\ln (10)}{\ln (10) + \ln (10)} = 0.5.$$

Now a macrosolute balance gives

$$V_f = \frac{V_0 c_{A0}}{c_{Af}} = \frac{150 \times 14}{140} = 15 \, l.$$

Therefore using Eq. 6.63, the water consumption in VVD is given by

$$V_w = \frac{0.5}{1 - 0.5}(150 - 15) = 135 \, l.$$

Now for UFCVD with CVD done at 100 g/l, the volume of water required is given by Eq. 6.21 as

$$V_w = \frac{150 \times 14}{100} \ln 10 = 48.4 \, l.$$

The VVD approach is clearly much less efficient from a water consumption point of view than VVD. Perhaps VVD is more time efficient? The exact time for a VVD process can be obtained by simply rearranging Eq. 6.58 to give

$$t_{VVD} = \frac{1}{1 - \alpha} \int_{V_f}^{V_0} \frac{dV}{JA}, \tag{6.67}$$

which can be written simply as

$$t_{VVD} = \frac{1}{1 - \alpha} t_{UF}, \tag{6.68}$$

where t_{Uf} is the time taken to ultrafilter from V_0 to V_f. Therefore, using the usual exponential integral formulation we can write

$$t_{VVD} = \frac{V_0}{(1 - \alpha) kA} \frac{c_{A0}}{c_{lim}} \left[Ei \left(\ln \frac{c_{lim}}{c_{A0}} \right) - Ei \left(\ln \frac{c_{lim}}{c_{Af}} \right) \right]. \tag{6.69}$$

In the next example, the time taken for a VVD process is compared with the time taken for a UFCVD process where the CVD step is performed at the optimum concentration. Problem 6.9 involves calculating VVD times using an approximate approach.

Example 6.13 Comparison of VVD and UFCVD times

A 250 l volume of a 10 g/l macrosolute solution is to be concentrated to 150 g/l while the macrosolute is to be reduced by a factor of 12. The limiting concentration is 250 g/l, the mass transfer coefficient is 1×10^{-5} m/s and the membrane area is 2.2 m^2. Calculate the process time assuming (i) VVD, and (ii) UFCVD with CVD done at c_{lim}/e.

Solution. Looking at the VVD process first, we have

$$\alpha = \frac{\ln(12)}{\ln(150/10) + \ln(12)} = 0.479.$$

Therefore, using Eq. 6.69, one gets

$$t_{\text{VVD}} = \frac{0.25}{(1 - 0.479) \times 10^{-5} \times 2.2} \frac{10}{250} \left[Ei \left(\ln \frac{250}{10} \right) - Ei \left(\ln \frac{250}{150} \right) \right].$$

Using the **expint** function in MATLAB gives a time of 9617 s or approximately 160 min. For the UFCVD case, the time is given by

$$t = \frac{V_0 c_{A0}}{k A c_{\text{lim}}} \left\{ Ei \left(\ln \frac{c_{\text{lim}}}{c_{A0}} \right) - Ei \left(\ln \frac{c_{\text{lim}}}{c_{Af}} \right) + e \ln \beta \right\}.$$

Putting in numbers gives

$$t = \frac{0.25 \times 10}{1 \times 10^{-5} \times 2.2 \times 250} \left\{ Ei \left(\ln \frac{250}{10} \right) - Ei \left(\ln \frac{250}{150} \right) + e \ln 12 \right\}.$$

Using the **expint** function in MATLAB gives a time of 8081 s or approximately 135 min.

The general conclusion from the previous calculations is that VVD, while elegant, is not an especially efficient method for doing ultrafiltration and diafiltration. The key problem is that by starting the diafiltration from the beginning, the water consumption is inevitably going to be very high and this also results in large process times. A significant improvement to the VVD process can be made by introducing a pre-concentration step. Essentially the VVD step is delayed until the solution volume has been reduced. The abbreviation UFVVD has been coined for this modified process and it has been shown to lead to significantly reduced water consumption and process times [5]. However, as proved by Paulen *et al.* [4], *no* UFDF process can be more time efficient than UFCVD, at least when the flux is given by the limiting flux model.

6.4.3   Generalised VVD processes and dynamic optimisation

One of the advantages of writing models of UFDF processes in terms of the α parameter is that it allows one to consider an essentially infinite number of ways of doing UFDF. This is because α can take any functional form, whether it be constant or time dependent. One can then use the powerful technique of optimal control theory, also known as dynamic optimisation, to devise highly complex α functions that will optimise any arbitrary process metric. This is possible regardless of the complexity of the flux model. For example, Paulen *et al.* [6] examined optimisation of an albumin–ethanol mixture in water where the flux was given by a rather complicated empirical expression

$$J = \left[b_1 + b_2 c_A + b_3 c_B + b_4 c_A c_B + b_5 c_A^2 + b_6 c_B^2 \right]^{-1}, \tag{6.70}$$

where the b_i are empirical constants, c_A is the albumin concentration and c_B is the ethanol concentration. Using optimal control theory they then showed that the time would be minimised if α was initially zero until a predetermined point (c_{A1}, c_{B1}) was reached. The α parameter was then shown to be given by

$$\alpha = \frac{0.5b_4 c_A c_B + b_5 c_A^2}{b_5 c_A^2 + b_4 c_A c_B + b_6 c_B^2},\qquad(6.71)$$

until another defined point, (c_{A2}, c_{B2}) was reached. Finally, $\alpha = 1$ until the required target (c_{Af}, c_{Bf}) is reached. Some simulation results, based on the Paulen *et al.* equations, are shown in Fig. 6.13, and illustrate how α varies with time to optimise the process time for a particular process specification. The details of the simulation are not important, the key point being that optimal UFDF processes may not be simple and may require complex α profiles.

While a profile such as the one described above is complicated, it can be achieved in practice with the availability of modern process instrumentation and control equipment. While the details of the underlying mathematics of optimal control theory are far beyond the scope of this book, this example illustrates how advanced mathematics, coupled with technological developments, can be employed to make improvements to real industrial processes. The 'rule of thumb' approach will gradually become obsolete as advanced concepts and tools are employed.

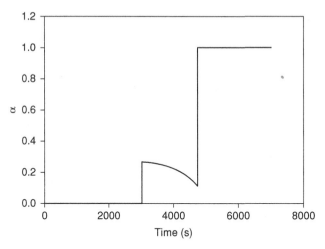

Figure 6.13 α profile for minimisation of process time during UFDF of a solution of albumin and ethanol in water.

6.4.4   Diafiltration of suspensions

Everything done so far in this chapter has focused on diafiltration done in an ultrafiltration context, i.e., diafiltration of *solutions*. But what about diafiltration in a microfiltration context, a typical example being cell suspension washing? Here, the problem is a little bit trickier, not least because the models available for microfiltration processes are much less rigorous than those available for ultrafiltration. However, even if we were to have an accurate model of crossflow microfiltration, the problem of microfiltration with diafiltration (MFCVD) is inherently more complicated. The reason is as follows: in UF, formation of the concentration polarisation layer is essentially instantaneous and, therefore, when the macrosolute concentration is changing, the flux immediately establishes itself at the flux for that concentration. This might be the flux as predicted by the limiting flux model, for example. In CFMF, cake formation is a slower process and the dynamics of cake formation must be incorporated into the analysis. To illustrate, let us examine a generalised MFCVD process as described above and use the kinetic model for cake formation outlined in Chapter 3.

The volume balance for the problem is the same as before, i.e.,

$$\frac{dV}{dt} = (\alpha - 1)JA. \tag{6.72}$$

Now, the balance for the particles is written

$$\frac{d\,(Vc_A)}{dt} = -A\frac{dM}{dt}, \tag{6.73}$$

where M is the cake mass per unit area. Recall that Eq. 6.59 is the equivalent relation for UF and in Problem 6.10 you are asked to investigate the effect of using that equation instead of Eq. 6.73 in the MFCVD model. As before, the microsolute balance is

$$\frac{d\,(Vc_B)}{dt} = -JAc_B, \tag{6.74}$$

and the water consumption balance is

$$\frac{dV_w}{dt} = \alpha JA. \tag{6.75}$$

The cake formation equation is taken from Chapter 3 (Eq. 3.19) and can be written

$$\frac{dM}{dt} = c_AJ - k_r\tau_wM, \tag{6.76}$$

where k_r is the cake removal constant and τ_w is the wall shear stress. In Chapter 3, a non-dimensionalised version of this model was explored in detail for the case of batch CFMF and the role of changing suspension viscosity was investigated. Here, we avoid getting bogged down in details and use Eq. 6.76 with constant τ_w. It is now easy to show that the complete MFCVD model can then be summarised in Eqs. 6.72, 6.75, 6.76 and the following:

$$\frac{dc_B}{dt} = -\frac{\alpha c_B JA}{V}, \tag{6.77}$$

$$\frac{dc_A}{dt} = -\frac{A}{V}\,(c_AJ\alpha - k_r\tau_wM). \tag{6.78}$$

For CFMF, the standard expression for the flux is used, i.e.,

$$J = \frac{\Delta P}{\mu\,(R_m + \alpha M)}, \tag{6.79}$$

where ΔP is the trans-membrane pressure, R_m is the membrane resistance, μ is the filtrate viscosity and α is the specific cake resistance. In the next example, the predictions of this model are explored.

Example 6.14 MFCVD of a cell suspension

A 500 ml volume of a 10 g/l yeast suspension is to undergo MFCVD to reduce the volume to 100 ml and reduce the concentration of a low molecular weight impurity from 1 g/l to 0.2 g/l. The CVD step is to be done when a volume of 200 ml remains, i.e., when the yeast concentration is 25 g/l. The yeast forms cakes with a specific resistance of 6×10^{12} m/kg, the membrane resistance is 2×10^{11} m^{-1}, the filtrate viscosity is 0.001 Ns/m^2, the membrane area is 0.008 m^2 and the product $k_r\tau_w$ can be taken to be 0.002 s^{-1}. Generate a plot of macrosolute concentration versus time.

Solution. MATLAB code for solving this problem is shown below. Again the reader should note the coding for the transitions between UF and CVD, expressed in terms of the *volume* and the microsolute concentration. The function **ode15s** is used as previously.

```
function example614
global A cA0 Vi Vf cB0 cBf SR Rm u DP krtauw
A=0.008; cA0=10; cB0=1; cBf=0.2; V0=500e-6; Vi=200e-
6; Vf=100e-6
krtauw=0.002; DP=1e5;Rm=2e11; SR=6e12;u=0.001; Vw0=0; M0=0;
options=odeset('events', @eventfcn,'MaxStep',1.0);
[t,y]=ode15s(@MFCVD,[0 36000],[V0,cA0,cB0,Vw0,M0],
options);
plot(t,y(:,2));
xlabel('Time (s)'); ylabel('c_A (g/l)');
function dydt=MFCVD(t,y)
global A Vi cBf SR Rm u DP krtauw
J=DP/u./(Rm+SR*y(5))
if and(y(1)<=Vi,y(3)>=cBf)
alpha=1;
else
alpha=0;
end
%y(1) is V, y(2) is cA, y(3) is cB, y(4) is Vw, y(5) is M
Veqn=(alpha-1)*J*A;
cAeqn=(-A*y(2)*J*alpha+A*krtauw*y(5))/y(1);
cBeqn=-y(3)/y(1)*alpha*J*A;
Vweqn=alpha*J*A;
Meqn=y(2)*J-krtauw*y(5);
```

```
dydt=[Veqn;cAeqn;cBeqn;Vweqn;Meqn];
function [value, isterminal, direction]=eventfcn(t,y)
global Vf
value = y(1)-Vf;
isterminal = 1;
direction = 0;
```

The results are presented in Fig. 6.14 and indicate that the cell concentration is not constant during the CVD step.

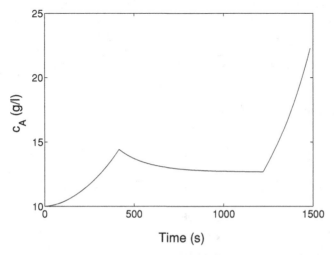

Figure 6.14 Model plot of concentration versus time during MFCVD.

The fact that the cell concentration is changing during the CVD step means that the flux is not constant during CVD. However, even if the concentration did not change, the flux would also change because during CVD the system behaves exactly like a continuous system operating at constant composition. As seen in Chapter 3, such a system is characterised by a decline in flux over measurably long time scales. Figure 6.15, which illustrates the flux behaviour predicted in Example 6.14, shows that decline in flux is smooth with no obvious sign of the CVD step save for some slight evidence of the transition from the CVD step to the second UF step towards the end of the run. Figure 6.16 shows some actual experimental flux data on MFCVD which verifies that such a smooth flux profile is observed in practice.

Unlike UFCVD, MFCVD cannot be optimised in a simple analytical way. The only way to do it is to compute the process time for various intermediate concentrations (or volumes) and locate any optimum that might exist by examination of the results. This is the subject of Problem 6.11, the answer to which suggests that the optimum occurs at the final particle concentration. Thus, it would seem to be always better to do the MF step first and then do the CVD step.

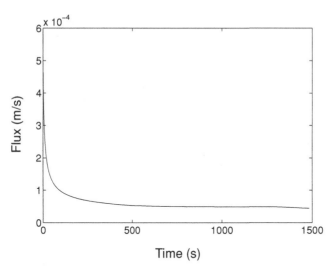

Figure 6.15 Model plot of flux decline in MFCVD.

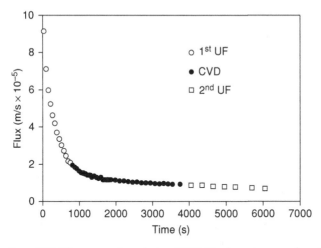

Figure 6.16 Flux versus time during MFCVD of yeast suspensions [7].

6.5   Continuous diafiltration

Although constant volume diafiltration is often referred to as continuous diafiltration, it is really better understood as a batch process. However, let us consider a continuous feed-and-bleed process in which the macrosolute is to be increased in concentration from c_{A0} to c_{Af} while the microsolute is to be reduced from c_{B0} to c_{Bf}. In this section, two simple approaches to performing this operation are described.

6.5.1 Two-stage continuous UFDF

When discussing UFCVD it was shown that the process time is minimised if a three-step process is employed involving UF to reduce the solution volume, CVD to remove the microsolute and a final UF step to increase the macrosolute concentration as required. The analogous process for a feed-and-bleed system is the two-stage, cross-current configuration shown in Fig. 6.17.

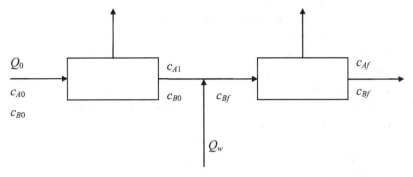

Figure 6.17 Two stage cross-current UFDF process. Recirculation loops are omitted for clarity.

Here, Q_w is the water flowrate and the other symbols have their usual meaning. The first stage is where the first concentration step occurs, while the mixing point is where the diafiltration occurs. This point is not unlike a DDF step by dilution. The second stage then acts as the second concentration step.

Let us consider the design of a two-stage, equal area, UFDF system. The logical place to start is at the mass balances. Combining macrosolute and volume balances for the *whole* system gives (see Problem 6.12):

$$\frac{Q_0}{kA}(1 - x_f) + \frac{Q_w}{kA} - (B + \ln x_1) - (B + \ln x_f) = 0, \qquad (6.80)$$

where $B = \ln(c_{lim}/c_{A0})$ and $x_1 = c_{A0}/c_{A1}$.

The usual combination of macrosolute and volume balances on the first stage gives (see Eq. 5.4)

$$\frac{Q_0}{kA}(1 - x_1) - (B + \ln x_1) = 0. \qquad (6.81)$$

Noting that the microsolute concentration remains unchanged in the first stage, a simple microsolute balance on the diluent addition point gives

$$Q_w = Q_0 x_1 (\beta - 1), \qquad (6.82)$$

where β is c_{B0}/c_{Bf} as before. Use of these equations is illustrated in the next example.

Example 6.15 Two stage continuous cross-current diafiltration

A two stage ultrafiltration rig is to be operated in cross-current diafiltration mode. The feed flowrate is 2.5 l/min, the microsolute is to be reduced by a factor of 5 and the macrosolute concentration is to be doubled. Take $B = 4.0$ while the mass transfer coefficient in each stage is estimated to be 1.0×10^{-5} m/s. Compute the water consumption, stage area and the permeate flowrate if the diafiltration water is added after the first stage.

Solution. MATLAB code for solving this problem is shown below.

```
function example615
x=fsolve(@crossufdf,[0.5,1])
function f=crossufdf(x)
Q0=2.5e-3/60; k=1.0e-5; B=4.0; beta=5; xf=0.5;
Qw=Q0*x(1)*(beta-1);
Qwout=Qw*1000*60% water consumption in L/min
permeate=(Q0*(1-xf)+Qw)*1000*60%overall volume balance
eq1=Q0/k/x(2)*(1-xf)+Qw/k/x(2)-(B+log(x(1)))-
(B+log(xf));%Eq. 6.80
eq2=Q0/k/x(2)*(1-x(1))-(B+log(x(1)));% Eq. 6.81
f=[eq1,eq2];
```

Running the code gives a stage area of $A = 1.11$ m^2, $Q_w = 2.76$ l/min and the permeate flowrate is 4.01 l/min.

In Problem 6.13, you are asked to investigate continuous diafiltration in a single-stage system with the dilution step being done either before or after the ultrafiltration step.

6.5.2   Counter-current diafiltration

In counter-current diafiltration, permeate is recycled to act as diluent. This is illustrated in the multi-stage system in Fig. 6.18.

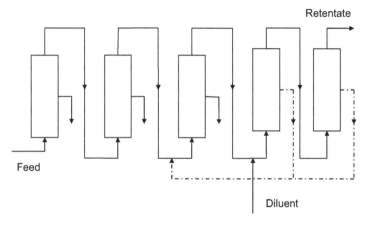

Figure 6.18 Counter-current diafiltration (recirculation loops omitted for clarity).

The analysis of such a system is inherently more complicated than the simple cross-current configuration described earlier. This is because the macrosolute balances for each stage are coupled. However, the general view is that although more membrane area is required, less diluent must be added and less permeate is produced. However, details of such calculations are not available in the literature and it is probably a subject that deserves some detailed modelling in the future.

6.6   Dialysis

It has been seen so far that removal of a low molecular weight solute from a solution using ultrafiltration or microfiltration technology inevitably requires water addition, i.e., diafiltration. But suppose we wanted to reduce the volume of a solution, while removing a low molecular weight solute, without adding any water to the system. This is something that is done routinely during haemodialysis. People with chronic kidney failure typically suffer from two major problems: first; an undesirable accumulation in the blood of metabolites such as potassium, creatinine and urea; and second, the retention of large amounts of fluid which can cause serious problems if the fluid accumulates in the lungs. When a person goes for a haemodialysis session, the blood is essentially ultrafiltered to reduce the overall fluid volume in the body. At the same time, the concentration of waste metabolites is reduced to more acceptable levels. This could of course be done by any of the UFDF techniques described above but such a process would take inordinately long and lead to unacceptably large losses in essential blood components. The answer is to employ mass transfer principles to selectively remove the undesired metabolites without adding fluid to the patient. The basic set-up for dialysis operation is shown in Fig. 6.19.

The dialysate, which is a complex solution of electrolytes, acts as a mass transfer sink for selective removal of microsolutes such as creatine, urea and potassium from

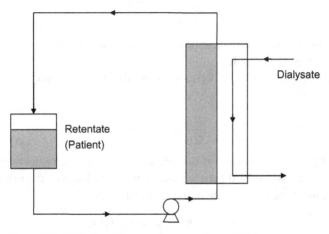

Figure 6.19 Dialysis with counter-current flow of dialysate.

the person. There will, however, be some undesirable losses, especially of vitamins. At any point in the membrane, the flux, J_B, of a microsolute through the membrane can be described, in a simplistic way, by an expression of the form

$$J_B = J c_B + K_B (c_B - c_D),\qquad(6.83)$$

where c_B is the microsolute concentration in the retentate, c_D is its concentration in the dialysate and K_B is an overall mass transfer coefficient.

Let us now consider the batch dialysis of a solution and formulate the governing set of differential equations. A key point to note is that haemodialysis is typically operated at a low trans-membrane pressure (<1 bar) and the flux is controlled by the equipment at a constant value that is well below the limiting value. A typical clinical scenario would be the removal of 1.5 l over a three-hour period. At a constant flux, the volume at any time, t, is given by

$$V = V_0 - JAt.\qquad(6.84)$$

There is no need for a macrosolute balance in this case as it is decoupled from the flux and of no interest anyway. The balance for the microsolute can be written

$$\frac{d\,(Vc_B)}{dt} = -JAc_B - K_B A\,(c_B - c_D)_{\mathrm{LM}},\qquad(6.85)$$

where LM denotes the log-mean. Students with a background in chemical engineering should be familiar with this quantity. It quantifies, precisely, the mean concentration (or temperature) difference for situations where the driving force for the particular transfer process is position dependent.

Using the product rule of differentiation and combining with Eq. 6.84 gives

$$\frac{dc_B}{dt} = -\frac{K_B A\,(c_B - c_D)_{\mathrm{LM}}}{V_0 - JAt}.\qquad(6.86)$$

In this case, the log-mean is defined by

$$(c_B - c_D)_{\mathrm{LM}} = \frac{(c_{Be} - 0) - (c_B - c_{De})}{\ln\left[(c_{Be} - 0)/(c_B - c_{De})\right]},\qquad(6.87)$$

where the 'e' subscript denotes concentrations leaving the membrane. The dialysate is solute-free as it enters the module.

The key to making progress is to derive an expression for c_{Be}. A microsolute balance gives

$$c_{Dy} = \frac{Q_B}{Q_D}\left(c_{By} - c_{Be}\right),\qquad(6.88)$$

where Q_B and Q_D are the flowrates of the retentate and dialysate respectively. The 'y' notation denotes the position in the membrane, with $y = 0$ denoting the retentate inlet. The axial variation of the microsolute in the retentate can be written

$$-Q_B dc_{By} = K_B\left(c_{By} - c_{Dy}\right)dA.\qquad(6.89)$$

Combining Eqs. 6.88 and 6.89 and rearranging gives

$$\int_{c_B}^{c_{Be}} \frac{dc_{By}}{(1 - Q_B/Q_D)c_{By} + Q_B c_{Be}/Q_D} = -\int_0^A \frac{K_B}{Q_B} dA. \tag{6.90}$$

Integrating and simplifying leads to

$$c_{Be} = \frac{(1 - Q_B/Q_D)\,e^{-\lambda}}{1 - (Q_B/Q_D)\,e^{-\lambda}} c_{Be}, \tag{6.91}$$

where

$$\lambda = K_B A (1/Q_D - 1/Q_B). \tag{6.92}$$

In summary, solving for c_B as a function of time requires one to solve Eq. 6.86. The log-mean temperature difference depends on c_{Be}, which is calculated from Eq. 6.91, and c_{De}, which can be computed from a special case of Eq. 6.88, i.e.,

$$c_{De} = \frac{Q_B}{Q_D} (c_B - c_{Be}). \tag{6.93}$$

The use of this model is outlined in the next example.

Example 6.16 Modelling of solute clearance in batch dialysis

A person is to undergo a 3-hour dialysis session during which their fluid volume is to be reduced from 10 l to 8 l. The concentration of a certain metabolite is to be reduced from its initial concentration of 0.75 g/l. The dialyser area is 1.0 m^2, the retentate flowrate is 300 ml/min, the dialysate flowrate is 500 ml/min and the overall mass transfer coefficient is 5×10^{-7} m/s. Generate a plot of metabolite concentration versus time (Fig. 6.20) and note the final concentration after the three hour session.

Solution. MATLAB code for the problem is shown below

```
function example617
[t,cB]=ode45(@dialysis, [0 10800], 0.75)
plot(t,cB);
ylabel('c_B (g/l)'); xlabel('Time (s)');
function dcBdt=dialysis(t,cB)
A=1.0; KB=5e-7;QD=500e-6/60;QB=350e-6/60;
J=2e-3/3/3600/A;
V=0.01-J*A*t;
lambda=KB*A/(1/QD-1/QB);
cBe=cB*(1-QB/QD)*exp(-lambda)/(1-QB/QD*exp(-lambda))
cDe=QB/QD*(cB-cBe);
DeltaLM=((cBe-0)-(cB-cDe))/(log((cBe-0)/(cB-cDe)));
dcBdt=-KB*A/V*DeltaLM;
```

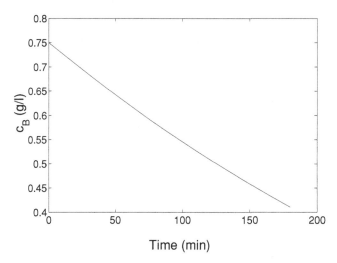

Figure 6.20 Microsolute concentration versus time during batch dialysis.

The final concentration works out as 0.411 g/l. It is worth noting that when one examines the detailed output from this program, there is actually negligible difference between c_{Be} and c_B. This is not unexpected as the change in retentate concentration in a single pass through the module is likely to be very small indeed.

6.7   Conclusions

One thing we can conclude from this chapter is that the underlying principles of UFDF are relatively simple, but applying the basic theory to existing and potential process configurations leads to a rich supply of computational problems. Indeed, as is seen in Chapter 7, the extension of this work, and the work of Chapter 5, to systems with arbitrary rejection coefficients uncovers an even deeper vein of computational problems in the whole area of UFDF.

References

1. Paulen, R., Fikar, M., Kovacs, Z. and Czermak, P. (2011). Process optimization with time-dependent water adding for albumin production. *Chemical Engineering and Processing*, **50**, 815–821.
2. Ng, P., Lundblad, J. and Mitra, G. (1976). Solute separation by diafiltration. *Separation Science*, **11**, 499–502.
3. Cheryan, M. (1998). *Ultrafiltration and Microfiltration Handbook*, 2nd Edition. CRC Press, Boca Raton, Florida, USA.
4. Paulen, R., Foley, G., Fikar, M., Kovacs, Z. and Czermak, P. (2011). Minimizing the process time for ultrafiltration/diafiltration under gel polarization conditions. *Journal of Membrane Science*, **380**, 148–154.

5. Foley, G. (2006). Ultrafiltration with variable volume diafiltration: a novel approach to water saving in diafiltration processes. *Desalination*, **199**, 220–221.
6. Paulen, R., Fikar, M., Foley, G., Kovacs, Z. and Czermak, P. (2012). Optimal feeding strategy of diafiltration buffer in batch membrane processes. *Journal of Membrane Science*, **411**, 160–172.
7. These data were obtained by one of my students, Colin Barr, as part of his undergraduate research project. The work was done using Baker's yeast suspensions and a tubular ceramic membrane.

Additional reading

Barba, D. Beolchini, F. and Veglio, F. (1998). Water saving in a two stage diafiltration for the production of whey protein concentrates. *Desalination*, **199**, 187–188.
Lipnizki, F., Boelsmand, J. and Madsen, R.F. (2002). Concepts of industrial scale diafiltration systems. *Desalination*, **144**, Special Issue, 179–184.
Madsen, R.F. (2001). Design of sanitary and sterile UF and diafiltration plants. *Separation and Purification Technology*, **22**, Special Issue, 79–87.
Takaci, A., Zikic-Dosenovic, T. and Zavargo, Z. (2009). Mathematical model of variable volume diafiltration with time dependent water adding. *Engineering Computations*, **26**, 857–867.

Problems

Problem 6.1 Limiting water consumption in DDF by volume reduction

Prove that in DDF by the volume reduction method, the dimensionless water consumption approaches $\ln \beta$ as $n \to \infty$, where n is the number of stages.

Problem 6.2 Limiting time for DDF by volume reduction

Show that for $n \to \infty$, the dimensionless time for DDF by the volume reduction method approaches $\ln \beta / \ln(c_{\lim} / c_{A0})$.

Problem 6.3 Limiting water consumption in DDF by dilution

Prove that in DDF by the dilution method, the dimensionless water consumption approaches $\ln \beta$ as $n \to \infty$, where n is the number of stages.

Problem 6.4 Limiting time for DDF by dilution

Show that for $n \to \infty$, the dimensionless time for DDF by the dilution method approaches the limit $\ln \beta / \ln(c_{\lim} / c_{A0})$.

Problem 6.5 Dynamics of microsolute accumulation in permeate

Show that during CVD, the concentration of the microsolute in the permeate, c_{Bp}, can be described by the following ordinary differential equation:

$$\frac{dc_{Bp}}{dt} = \frac{JA \left(c_B - c_{Bp} \right)}{V_w}.$$

Hence, or otherwise, show that

$$c_{Bp} = \frac{V_0 c_{B0}}{J A t} \left(1 - e^{-J A t / V_0}\right).$$

All symbols are as defined in Section 6.3.

Problem 6.6   CVD optimisation when water costs are dominant
Using Eq. 6.47, derive an expression for the optimum diafiltration concentration in the limit where the water cost is much greater than the time cost.

Problem 6.7   Criterion for superiority of 'over-concentration' method
Considering the problem of UFCVD with $c_{Af} < c_{\lim}/e$, and using Example 6.10 as a guide, develop a general criterion for when the 'over-concentration' method for UFCVD is faster than simply doing the CVD step at c_{Af}. Is it possible to say if it is always faster?

Problem 6.8   UFCVD with viscosity dependent mass transfer coefficient
Repeat Example 6.11 but account for a viscosity-dependent mass transfer coefficient. Use $z = 0.14$ and $\gamma = 0.015$ l/g and assume laminar flow.

Problem 6.9   Approximate time for VVD
Using Cheryan's approximate equation for the batch ultrafiltration time, Eq. 5.56, derive an analytical expression for the VVD time. Estimate the VVD time for the conditions of Example 6.13.

Problem 6.10   Simplified analysis of MFCVD
Modify the MATLAB code in Example 6.14 for a simplified model where Eq. 6.73 is written as

$$\frac{d(V c_A)}{dt} = 0.$$

Run the program and comment on any obvious differences between your results and the results presented in Figure 6.14.

Problem 6.11   Optimisation of MFCVD
In Example 6.14, the initial volume of the suspension was 500 ml and CVD was done when the volume had been reduced to 200 ml. Use the MATLAB code in that example to determine if there is an optimum suspension volume for commencing the CVD step.

Problem 6.12   Derivation of Eq. 6.80
Derive Eq. 6.80. A key point to remember for formulating balances in this kind of system is:

$$Q_0 c_{A0} = Q_1 c_{A1} = Q_f c_{Af}.$$

Problem 6.13   Alternative approaches to cross-currrent diafiltration
For the process specification of Example 6.15, calculate the membrane area, water consumption and permeate flowrate if (i) water is added before the concentration step, and (ii) water is added after the concentration step. Compare your answer with the answer to Example 6.15.

Further problems

Problem 6.14 Ultrafiltration with variable volume diafiltration (UFVVD)

A solution is to undergo UFDF to increase its macrosolute concentration and reduce its microsolute concentration. Variable volume diafiltration is to be employed but it is to be preceded by a concentration (UF) step which increases the macrosolute concentration to some intermediate concentration that is less than the final, desired concentration. Derive expressions for the value of α required in the VVD step and the volume of water consumed.

Problem 6.15   Continuous UFDF with permeate recycle

Consider the continuous diafiltration process shown in Fig. 6.21. It is a simplification of Fig. 6.18 and is essentially a form of counter-current operation. Repeat Example 6.15 with this configuration.

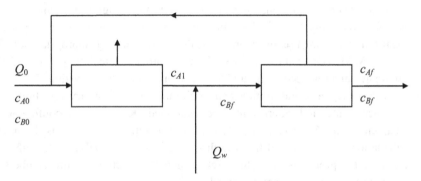

Figure 6.21 Continuous UFDF with permeate recycle.

Problem 6.16   Simplified model of haemodialysis

Assuming the concentration of the microsolute in the dialysate is essentially zero everywhere, derive an analytical expression for the microsolute concentration as a function of time. Assume constant flux operation.

7 Ultrafiltration and diafiltration with incomplete rejection

7.1 Introduction

Chapters 5 and 6 focused on ultrafiltration and diafiltration for situations where the macrosolute exhibited complete rejection and the microsolute showed no rejection. Even then, the resulting equations for process analysis and design were found to be relatively complex, requiring numerical solution in many cases. This is all a consequence of the logarithmic term in the expression for the flux. However, with appropriate software tools, such as MATLAB, trial and error solutions of process calculations can, and should, be eliminated and replaced with rapid and accurate numerical algorithms.

In this chapter, ultrafiltration and diafiltration with arbitrary rejection coefficients of macrosolute and microsolute are examined. Key process configurations from Chapters 5 and 6 are re-examined and the separation of solutes is discussed. In all calculations, it is assumed that the flux is at its limiting value. Although the formulation of the problems is a little trickier, and the algebra is more cumbersome, few computational problems are presented.

7.2 Quantifying rejection

In Chapter 1, the concept of rejection was introduced and a rejection coefficient was defined by the expression

$$\sigma = 1 - \frac{c_p}{c}, \tag{7.1}$$

where c_p is the solute concentration in the permeate and c is the solute concentration in the retentate. Clearly, a solute is completely rejected if $c_p = 0$, i.e., $\sigma = 1$. A solute shows no rejection at all if $c_p = c$ and hence $\sigma = 0$. In this chapter, more than one type of rejection coefficient is defined, and the one defined in Eq. 7.1 is referred to as an *apparent* rejection coefficient. The reason for this terminology is that the concentration that the membrane 'sees' is actually the wall concentration and it makes more physical sense to speak about rejection in terms of that concentration. Indeed, the *intrinsic* rejection coefficient, σ_i, is defined by the expression

$$\sigma_i = 1 - \frac{c_p}{c_w}, \tag{7.2}$$

where c_w is the wall concentration of the solute. This would typically be the rejection coefficient used by manufacturers to define the molecular weight cut-off of a membrane. Intrinsic rejection coefficients are typically obtained in stirred cells employing a low pressure and high agitation rate, thus eliminating concentration polarisation. In that case, the wall concentration and the bulk concentration are identical.

Equating expressions for c_p from Eqs. 7.1 and 7.2 gives

$$\sigma = 1 - \frac{c_w}{c}(1 - \sigma_i). \tag{7.3}$$

Thus, the apparent rejection coefficient of a solute is determined by its intrinsic rejection properties, but also the degree of concentration polarisation. It is worth pointing out here that c_w is technically not independent of the rejection coefficients. So one cannot put seemingly reasonable and independent values of c_w/c and σ_i into Eq. 7.3 to predict meaningless values of σ. In fact, once σ_i values begin to depart significantly from 1.0, the degree of concentration polarisation that is possible becomes constrained. This whole issue is revisited in Chapter 8, because it can only be understood within the framework of the osmotic pressure model.

The intrinsic rejection coefficient is a complex parameter and is not a simple constant. A complete understanding of this parameter requires one to consider transport mechanisms in membranes, and that is beyond the scope of this chapter where engineering calculations are emphasised. Transport in membranes is revisited briefly in Chapters 8 and 9.

However, one parameter affecting intrinsic rejection that is worth mentioning now is the membrane Peclet number. Peclet numbers are well known to chemical engineers as they arise in a variety of chemical engineering systems, especially chemical reactors. In membrane transport, the Peclet number is defined as

$$Pe_m = \frac{J\delta_m}{\varepsilon D}\frac{K_c}{K_d} = \frac{J\delta_m}{D_{eff}}K_c, \tag{7.4}$$

where J is the flux, δ_m is the membrane thickness, D is the diffusivity of the solute in free solution, ε is the membrane porosity, D_{eff} is the effective diffusivity of the solute in the membrane, while K_c and K_d are hindrance factors for convection and diffusion respectively. It is reasonable to assume that when $Pe_m \gg 1$, i.e., when the solute transport is predominantly a convective process, σ_i can be assumed to be constant and equal to a limiting value of that (flux dependent) parameter [1]. This is normally the case in UF. That said, most process analysis and design equations are formulated in terms of the apparent rejection coefficient, mainly because it is easily measured and, perhaps more importantly, this is the coefficient that is of practical interest.

Before moving on to the next section, it is worth mentioning some terminology that is often used in the context of rejection. The apparent *sieving coefficient*, S, and the intrinsic sieving coefficient, S_i are defined by the expressions

$$S = 1 - \sigma, \tag{7.5}$$

$$S_i = 1 - \sigma_i. \tag{7.6}$$

The sieving coefficient is essentially the same concept as *transmission*, discussed in Chapter 2.

7.3   Concentration polarisation

The process of concentration polarisation was analysed in Chapter 4 and it was shown (see Eq. 4.3) that the flux is related to the wall and bulk concentrations (of macrosolute) by the expression

$$J = k \ln \frac{c_w - c_p}{c - c_p}. \tag{7.7}$$

Only when the solute is completely rejected does one get the well known expression

$$J = k \ln \frac{c_w}{c}. \tag{7.8}$$

Equation 7.7 can be written in terms of the apparent rejection coefficient as

$$J = k \ln \frac{c_w - c(1 - \sigma)}{c - c(1 - \sigma)} \tag{7.9}$$

or

$$J = k \ln \left(\frac{c_w}{\sigma c} - \frac{1 - \sigma}{\sigma} \right). \tag{7.10}$$

The limiting flux is thus given by

$$J_{\text{lim}} = k \ln \left(\frac{c_{\text{lim}}}{\sigma c} - \frac{1 - \sigma}{\sigma} \right). \tag{7.11}$$

This expression is the flux equation for nearly all subsequent analysis and design calculations. As before, and simply for convenience, the 'lim' subscript on J is dropped throughout this chapter. An important characteristic of this equation is that for a given value of c, lower values of σ lead to higher values of the flux.

Returning to a point made in the last section, it is important to note that Eq. 7.11 is not necessarily valid, regardless of the value of σ. If σ is sufficiently low, the wall concentration 'drops out' of the limiting region (at a fixed trans-membrane pressure) and one has to use Eq. 7.10 in combination with the osmotic pressure model. This is examined in detail in the next chapter.

7.4   Flux dependence of the apparent rejection coefficient

While changes in the intrinsic rejection coefficient might be a good indicator of membrane fouling (i.e. changes in the membrane characteristics), changes in the apparent rejection coefficient are due to both membrane fouling and concentration polarisation effects. From the definitions of the rejection coefficients, one can write

$$J = k \ln \left(\frac{c_p/(1 - \sigma_i) - c_p}{c_p/(1 - \sigma) - c_p} \right) = k \ln \frac{\sigma_i/(1 - \sigma_i)}{\sigma/(1 - \sigma)}. \tag{7.12}$$

Therefore

$$\frac{\sigma/(1 - \sigma)}{\sigma_i/(1 - \sigma_i)} = e^{-J/k}. \tag{7.13}$$

After some algebra, this becomes

$$\sigma = \frac{1}{1 + (1/\sigma_i - 1)e^{J/k}}. \tag{7.14}$$

Alternatively, if one wishes to compute σ_i from a measurement of σ, Eq. 7.13 can be written

$$\sigma_i = \frac{1}{1 + (1/\sigma - 1)e^{-J/k}}. \tag{7.15}$$

Figure 7.1 shows the sigmoidal relationship between apparent rejection coefficient and flux for a fixed intrinsic rejection coefficient.

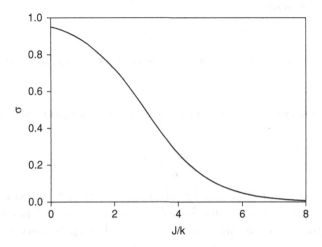

Figure 7.1 Effect of flux on apparent rejection coefficient for $\sigma_i = 0.95$.

Clearly, then, as the flux declines, as in a batch UF process for example, the apparent rejection coefficient increases. It would be an understandable mistake to make if one assumed that such a phenomenon was due to a membrane fouling effect. A better way to check for membrane fouling would be to evaluate σ_i from an experimentally determined value of σ using Eq. 7.15. An increase in σ_i, as opposed to σ, over the course of a run would be a good indicator of some sort of pore blocking or constriction phenomenon.

7.5 Continuous feed-and-bleed ultrafiltration

This is the simplest place to start an analysis of process systems exhibiting incomplete solute rejection. While the algebra is a little bit more involved than in the case of complete rejection, no new concepts arise and calculations can proceed in a very similar manner to that employed in Chapter 5.

7.5.1   Analysis and design of a single-stage system

Let us consider a single-stage system as discussed in Chapter 5 and formulate the appropriate balances. The process layout is illustrated in Fig. 5.1. The overall balance for the system can be written

$$Q_0 = Q_1 + JA, \tag{7.16}$$

where all symbols are as defined in Section 5.2. The solute balance is now a little more complicated than for the complete rejection case, and can be expressed as

$$Q_0 c_{A0} = Q_1 c_{A1} + JA c_{A1}(1 - \sigma_A). \tag{7.17}$$

The 'A' subscript is used to remind us that we are dealing with the macrosolute here. Later analyses in this chapter involve diafiltration and the microsolute is, as usual, given the 'B' subscript. Therefore

$$Q_1 = Q_0 x_{A1} - JA(1 - \sigma_A), \tag{7.18}$$

where x_{A1} is c_{A0}/c_{A1} as defined in Chapter 5. Combining with Eq. 7.16 gives

$$Q_0(1 - x_{A1}) - JA\sigma_A = 0. \tag{7.19}$$

In passing, it is of interest to note that the membrane area is computed from the simple equation

$$A = \frac{Q_0(1 - x_{A1})}{\sigma_A J}. \tag{7.20}$$

This is a generalisation of the expression obtained in Chapter 5 for the case where $\sigma_A = 1$, namely Eq. 5.11. Now, assuming that the system is operating at the limiting flux and Eq. 7.11 applies, the governing equation for the single-stage system can be written in full as

$$Q_0(1 - x_{A1}) - kA\sigma_A \ln\left(\frac{c_{\lim}}{\sigma_A c_{A0}} x_{A1} - \frac{1 - \sigma_A}{\sigma_A}\right) = 0. \tag{7.21}$$

For numerical purposes it is convenient to write this as

$$\frac{Q_0}{kA\sigma_A}(1 - x_{A1}) - \ln\left(\frac{c_{\lim}}{c_{A0}} \frac{x_{A1}}{\sigma_A} - \frac{1 - \sigma_A}{\sigma_A}\right) = 0. \tag{7.22}$$

The example below illustrates use of this equation and is essentially a repeat of Example 5.1.

Example 7.1 Analysis of a single-stage system

A 1.0 l/min flowrate of a protein solution enters a single-stage feed-and-bleed system. The feed concentration is 10 g/l and the limiting concentration is 300 g/l. The mass transfer coefficient is 3.5×10^{-6} m/s and the membrane area is 2.7 m^2. If the apparent rejection coefficient is 0.9, calculate the exit concentration. Compute the percentage product loss.

Solution. MATLAB code for solving the problem is shown below

```
function example71
guess=0.5;
x=fzero(@singlestage,guess)
function f=singlestage(x)
Q0=1.0e-3/60;s=0.9;A=2.7;k=3.5e-6;clim=300;cA0=10;
f=Q0/(k*A*s)*(1-x)-log(clim/cA0*x/s-(1-s)/s);
```

Running the code gives $x_1 = 0.159$ or $c_{A1} = 62.81$ g/l. This compares with 66.89 g/l for the analogous problem involving complete rejection, i.e., Example 5.1. The solute mass flowrate in the feed is

$$Q_0 c_{A0} = \frac{1.0 \times 10^{-3}}{60} \times 10 = 1.667 \times 10^{-4} \text{ kg/s.}$$

The solute mass flowrate in the permeate is given by

$$J A c_{A1}(1 - \sigma_A) = 3.5 \times 10^{-6} \times 2.7 \times 62.81 \times 0.1$$
$$\times \ln\left(\frac{300}{0.9 \times 10} \times 0.159 - \frac{0.1}{0.9}\right) = 9.78 \times 10^{-5} \text{ kg/s.}$$

The percentage loss is thus given by

$$\frac{9.78 \times 10^{-5}}{1.67 \times 10^{-4}} \times 100 = 58.8\%.$$

In Problem 7.1 you are asked to derive an analytical expression for the fraction of solute lost in terms of x_{A1} and σ_A only.

The design problem for a single-stage system is obviously trivial, as Eq. 7.22 can be rearranged to give

$$A = \frac{Q_0(1 - x_{A1})}{\sigma_A k \ln\left(\dfrac{c_{\lim}}{c_{A0}} \dfrac{x_{A1}}{\sigma_A} - \dfrac{1 - \sigma_A}{\sigma_A}\right)}. \tag{7.23}$$

The answer is obtained by simply 'plugging in numbers' as long as k is known.

7.5.2   Solving single-stage problems using a graphical method

In Chapter 5, it was shown that continuous feed-and-bleed analysis problems, including multi-stage ones, can be solved readily using a simple graphical method that is similar to those employed in cross-current liquid–liquid extraction. Here, it is shown that a similar method can be applied even when there is incomplete solute rejection. The method relies on equating different expressions for the flux. Let us consider the dimensionless exit concentration, x_A, as the independent variable. Then, as shown above, the flux can be written using concentration polarisation theory as

$$\frac{J}{k} = \ln\left(\frac{c_{\lim}}{\sigma_A c_{A0}} x_A - \frac{1 - \sigma_A}{\sigma_A}\right). \tag{7.24}$$

Likewise, mass balance considerations (Eq. 7.19) imply that

$$\frac{J}{k} = \frac{Q_0(1 - x_A)}{\sigma_A k A}. \tag{7.25}$$

Equation 7.24 can be used to generate a plot of J/k versus x_A while Eq. 7.25 is the equation of a straight line in x_A. The intersection of these two is the solution to the problem, i.e., x_{A1}. The next example is a repeat of Example 7.1 and uses the graphical method.

Example 7.2 Graphical method for single-stage analysis

Repeat Example 7.1 but use the graphical method rather than a numerical one.

Solution. Equation 7.25 represents a straight line whose x-intercept is 1.0 and whose J/k-intercept is $Q_0/\sigma_A k A$. For the numbers supplied in the problem statement, this intercept is

$$\frac{Q_0}{\sigma_A k A} = \frac{1 \times 10^{-3}/60}{0.9 \times 3.5 \times 10^{-6} \times 2.7} = 1.957.$$

Thus the 'operating line' joins the two coordinates $(0, 1.960)$ and $(1.0, 0)$. The flux curve is

$$\frac{J}{k} = \ln\left(\frac{300}{9}x_A - \frac{1}{9}\right).$$

The required construction is shown in Fig. 7.2.

A quick inspection of the graph shows a point of intersection (x_{A1}) of 0.16, which is in close agreement with the numerical approach.

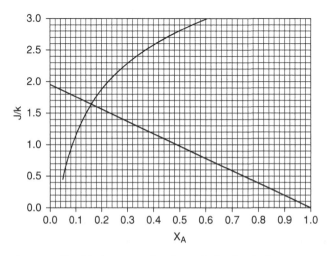

Figure 7.2 Graphical construction for analysis of a single-stage system.

The graphical method described in the above example is not so easily extended to multi-stage systems, as shown in the next section.

7.5.3   Analysis and design of two-stage systems

Let us consider the second stage of a two-stage, equal area, system. First, Eq. 7.19 can be applied to the first stage to give

$$Q_0(1 - x_{A1}) - J_1 A \sigma_A = 0, \tag{7.26}$$

where J_1 is the flux in the first stage stage. An overall balance on the second stage gives

$$Q_1 - Q_2 - J_2 A = 0, \tag{7.27}$$

where J_2 is the flux in the second stage. The solute balance for the second stage is

$$Q_2 c_{A2} = Q_1 c_{A1} - J_2 A c_{A2}(1 - \sigma_A). \tag{7.28}$$

Combining the two balances give

$$Q_1 \left(1 - \frac{x_{A2}}{x_{A1}}\right) - J_2 A \sigma_A = 0. \tag{7.29}$$

Using the expression for Q_1 in Eq. 7.18, this becomes

$$(Q_0 x_{A1} - J_1 A(1 - \sigma_A)) \left(1 - \frac{x_{A2}}{x_{A1}}\right) - J_2 A \sigma_A = 0. \tag{7.30}$$

Equations 7.26 and 7.30 constitute the model of the two-stage system and form the basis of two-stage process analysis and design. Use of them is illustrated in the next example. In Problem 7.2, you are asked to develop an expression for the fractional solute loss in a two-stage system. In Problem 7.3, a graphical solution to Example 7.3 is required, while in Problem 7.4 the task is to develop the required third equation for the analysis of a three-stage system.

Example 7.3   Analysis of a two-stage system
A 1.0 l/min, 10 g/l solution is fed to a two-stage continuous feed-and-bleed system. The mass transfer coefficient is 3.5×10^{-6} m/s, the limiting concentration is 300 g/l, the area of each stage is 1.35 m^2 and the solute has a rejection coefficient of 0.9. Calculate the composition of the fluid exiting the second stage and the fractional product loss.

Solution. The equations to be solved are Eqs. 7.26 and 7.30, and these are first written as follows:

$$\frac{Q_0}{\sigma_A k A}(1 - x_{A1}) - \frac{J_1}{k} = 0,$$

$$\frac{Q_0}{k A \sigma_A}\left(x_{A1} - \frac{J_1 A}{Q_0}(1 - \sigma_A)\right)\left(1 - \frac{x_{A2}}{x_{A1}}\right) - \frac{J_2}{k} = 0.$$

In each case the flux is given by the limiting flux model, i.e.,

$$\frac{J_1}{k} = \ln\left(\frac{c_{\lim}}{\sigma_A c_{A0}} x_{A1} - \frac{1 - \sigma_A}{\sigma_A}\right)$$

and

$$\frac{J_2}{k} = \ln\left(\frac{c_{\lim}}{\sigma_A c_{A0}} x_{A2} - \frac{1 - \sigma_A}{\sigma_A}\right).$$

MATLAB code for solving these equations, using **fsolve**, is given below.

```
function example73
guess=[0.4,0.1];
x=fsolve(@twostage,guess)
function f=twostage(x)
Q0=1.0e-3/60;s=0.9;A=1.35;k=3.5e-6;clim=300;cA0=10;
J1=k*log(clim/cA0/s*x(1)-(1-s)/s)
J2=k*log(clim/cA0/s*x(2)-(1-s)/s)
stage1=Q0/(k*A*s)*(1-x(1))-J1/k;
stage2=Q0/(k*A*s)*(x(1)-J1*A/Q0*(1-s))*(1-x(2)/x(1))-J2/k;
loss=(J1*A*cA0/x(1)*(1-s)+J2*A*cA0/x(2)*(1-s))/(Q0*cA0)
f=[stage1;stage2];
```

Running the code gives $x_{A1} = 0.365$ and $x_{A2} = 0.078$ and a fractional loss of 0.525. The final concentration leaving the second stage ($10/0.078 = 128$ g/l) is a considerable improvement on the 62.8 g/l found in Example 7.1 for the equivalent single-stage system.

The above analysis can be easily applied to the design problem, i.e., a problem where x_{A2} is specified and A must be computed. This is the focus of Problem 7.5. Furthermore, the general approach used here can be extended to allow for a viscosity dependent mass transfer coefficient, just as in Chapter 5. This is the subject of Problem 7.6.

7.6   Batch ultrafiltration

In Chapter 5 it was seen that in the case of constant mass transfer coefficient, the batch UF problem could be solved in a variety of ways. These included numerical solution of the governing ODEs, the exponential integral method and a number of approximate methods. Of course the problem could also be solved by using numerical integration with Simpson's rule or similar but this approach was left for the problems in that chapter. It is, however, used on more than one occasion in this chapter.

It was seen at the start of this chapter that one of the complicating factors associated with use of an apparent rejection coefficient is that technically, it is a function of flux. Thus, one can really tackle ultrafiltration problems in two ways; assuming constant apparent rejection or assuming constant intrinsic rejection. In the last section, the flux dependence of the rejection coefficient was ignored but in this section, both the constant σ and the constant σ_i scenarios are considered. First, the constant σ case is examined.

7.6.1   Batch UF with constant apparent rejection

The process layout for batch UF has previously been described in Fig. 1.3 and Fig. 5.4. The starting point for any batch UF analysis is formulation of the overall balance for the system. This is written

$$\frac{dV}{dt} = -AJ, \tag{7.31}$$

where V is the retentate volume at time, t. This is true regardless of the rejection coefficient of the macrosolute. The macrosolute balance can be written

$$\frac{d(Vc_A)}{dt} = -JAc_A(1 - \sigma_A). \tag{7.32}$$

Expanding gives

$$V\frac{dc_A}{dt} + c_A\frac{dV}{dt} = -JAc_A(1 - \sigma_A). \tag{7.33}$$

Combining with Eq. 7.31 gives

$$\frac{dc_A}{dt} = \frac{JAc_A\sigma_A}{V}. \tag{7.34}$$

A simple relation between c_A and V at any time is easily obtained by dividing Eq. 7.34 by Eq. 7.31, giving

$$\frac{dc_A}{dV} = -\frac{c_A}{V}\sigma_A. \tag{7.35}$$

Separating the variables gives

$$\int_{c_{A0}}^{c_A} \frac{dc_A}{c_A} = -\sigma_A \int_{V_0}^{V} \frac{dV}{V}. \tag{7.36}$$

Integrating gives

$$\ln\frac{c_A}{c_{A0}} = -\sigma_A \ln\frac{V}{V_0}, \tag{7.37}$$

i.e.,

$$\frac{c_A}{c_{A0}} = \left(\frac{V_0}{V}\right)^{\sigma_A}. \tag{7.38}$$

Obviously in the case $\sigma_A = 1$, this reduces to the expected result;

$$c_{A0}V_0 = c_AV. \tag{7.39}$$

Now, the batch time can be computed, using Eq. 7.31 and assuming constant mass transfer coefficient, from the integral

$$t = \frac{1}{kA} \int_{V_f}^{V_0} \frac{dV}{\ln\left[\dfrac{c_{\lim}}{\sigma_A c_A} - \dfrac{1-\sigma_A}{\sigma_A}\right]}. \tag{7.40}$$

Using Eq. 7.38, this becomes

$$t = \frac{1}{kA} \int_{V_f}^{V_0} \frac{dV}{\ln\left[\dfrac{c_{\lim}}{\sigma_A c_{A0}} \left(\dfrac{V}{V_0}\right)^{\sigma_A} - \dfrac{1-\sigma_A}{\sigma_A}\right]}. \tag{7.41}$$

There appears to be no analytic solution to this integral, even in terms of special functions. Thus to compute the batch time, one either has to go back to the original ODEs or use a numerical integration technique, such as that employed in the next example.

Example 7.4 Computation of batch time using Simpson's rule

A 60 l volume of a protein solution is to undergo a four-fold increase in concentration from 14 g/l to 56 g/l. The initial concentration is 14 g/l and limiting concentration is 250 g/l. The mass transfer coefficient is 3.0×10^{-6} m/s and the membrane area is 1.3 m^2. Calculate the time required to carry out this operation. The apparent rejection coefficient is 0.9 and constant.

Solution. MATLAB code for solving this problem is given below. It uses the **quad** function which employs an adaptive Simpson's rule algorithm. The code below is based on Eq. 7.41

```
function example74
global V0 sA cA0
V0=60e-3; k=3e-6; A=1.3;sA=0.9;cA0=14;cAf=56;
Vf=60e-3*(cA0/cAf)^(1/sA); %eq.7.38
a=Vf; b=V0
Vint=quad(@intfun,a,b);
time=1/k/A*Vint
function Vint=intfun(V)
global V0 sA cA0;
clim=250;
Vint=1./log(clim/sA/cA0*(V/V0).^sA-(1-sA)/sA);
```

Running the code gives a time of 5060 s, which is slightly lower than the 5072 s required when $\sigma_A = 1$ (see Example 5.12 where the $\sigma_A = 1$ case is analysed with the exponential integral).

7.6.2   Batch UF with constant intrinsic rejection

In this section, the situation where the intrinsic rejection coefficient is fixed is studied. The simplest way to tackle this problem is to recognise that if the system is operating at the limiting flux ($c_w = c_{lim}$) and the intrinsic rejection coefficient is constant, the permeate concentration is also constant. This means that the apparent rejection coefficient is given by

$$c_p = c_A(1 - \sigma_A) = c_{A0}(1 - \sigma_{A0}), \tag{7.42}$$

where σ_{A0} is the apparent rejection coefficient at time zero. Therefore

$$\sigma_A = 1 - \frac{c_{A0}}{c_A}(1 - \sigma_{A0}). \tag{7.43}$$

Therefore Eq. 7.35 becomes

$$\frac{dc_A}{dV} = -\frac{c_A - c_{A0}(1 - \sigma_{A0})}{V}. \tag{7.44}$$

Separating the variables gives

$$\int_{c_{A0}}^{c_A} \frac{dc_A}{c_A - c_{A0}(1 - \sigma_{A0})} = \int_{V}^{V_0} \frac{dV}{V}. \tag{7.45}$$

Integrating leads to

$$\frac{c_A}{c_{A0}\sigma_{A0}} - \frac{(1 - \sigma_{A0})}{\sigma_{A0}} = \frac{V_0}{V}, \tag{7.46}$$

which is the analogous expression to Eq. 7.38 which is valid for constant apparent rejection. In Problem 7.7, the task is to compare Eqs. 7.46 and 7.38 when applied to a system with constant intrinsic rejection.

For the purposes of calculating the batch time, it is wise to take the algebra no further and simply solve the problem by numerical solution of the two differential equations, Eqs. 7.31 and 7.35, and using Eq. 7.46 for the variation of the rejection coefficient. This is done in the next example.

Example 7.5 Computation of batch time with constant intrinsic rejection

A 60 l volume of a protein solution is to undergo a four-fold increase in concentration from 14 to 56 g/l. The limiting concentration is 250 g/l. The mass transfer coefficient is 3.0×10^{-6} m/s and the membrane area is 1.3 m^2. The initial apparent rejection coefficient is 0.9 and the intrinsic rejection coefficient is constant. Calculate the time required to carry out this operation.

Solution. The solution is based on numerical solutions of Eqs. 7.31 and 7.35, along with Eq. 7.43. MATLAB code is shown below.

```
function example75
options=odeset('events', @eventfcn);
```

```
[t,x]=ode45(@varrej,[0 24000],[0.06 14],options)
function f=varrej(t,x)
%x(1) is V, x(2) is cA
A=1.3; k=3.0e-6;cA0=14;sA0=0.9;clim=250;
sA=1-cA0/x(2)*(1-sA0)
J=k*log(clim/x(2)/sA-(1-sA)/sA);
Veqn=-A*J;
ceqn=A*J*x(2)/x(1)*sA;
f=[Veqn; ceqn];
function [value, isterminal, direction]=eventfcn(t,x)
value = x(2)-56 ;
isterminal = 1;
direction = 0;
```

Running the code gives a time of 5043 s. This compares with a time of 5060 s obtained in Example 7.4 where the rejection coefficient was assumed constant. The increased time is due to the fact that the apparent rejection increases during the run from 0.9 to $(1-0.25 \times (1-0.9)) = 0.975$. As mentioned in Example 7.4, the batch time for complete rejection is 5072 s so the time in this example is approaching that value. The differences involved, however, are small indeed.

7.7   Fed-batch ultrafiltration

The basic layout of a fed-batch system was shown in Fig. 5.6. The initial total volume is V_0, of which a volume V_f is placed in the retentate tank while the remainder is placed in the feed reservoir. The system is operated at constant retentate volume until the feed reservoir is emptied. The balance for solute A is given by

$$V_f \frac{dc_A}{dt} = JAc_{A0} - JAc_A(1 - \sigma_A),$$ (7.47)

where c_{A0} is the composition of the feed and the initial composition of the retentate. Rearranging gives

$$t = \frac{V_f}{kA} \int_{c_{A0}}^{c_{Af}} \frac{dc_A}{(c_{A0} - c_A(1 - \sigma_A)) \ln \left[\frac{c_{lim}}{\sigma_A c_A} - \frac{1 - \sigma_A}{\sigma_A} \right]},$$ (7.48)

where we assume that the apparent rejection coefficient is constant and hence

$$c_{Af} = c_{A0} \left(\frac{V_0}{V_f} \right)^{\sigma_A}.$$ (7.49)

Use of these equations in illustrated in the next example, which is analogous to Example 5.14.

Example 7.6 Fed-batch ultrafiltration with incomplete rejection

A 180 l volume of a protein solution is to be ultrafiltered in a fed-batch process in which 40 l is placed in the retentate and the remainder is fed from the feed reservoir. The feed concentration is 10 g/l and the limiting concentration is 250 g/l. The membrane area is 1.3 m² and the mass transfer coefficient is 3.0×10^{-6} m/s. The solute has an apparent rejection coefficient of 0.9 which is assumed constant. Calculate the time required to empty the feed tank.

Solution. MATLAB code to solve Eq. 7.48 by numerical integration is shown below.

```
function example76
global cA0 sA;
V0=180e-3; Vf=40e-3; k=3e-6; A=1.3; sA=0.9;cA0=10;
cAf=cA0*(V0/Vf)^sA
a=cA0; b=cAf;
cint=quad(@intfun,a,b);
time=Vf*sA/k/A*cint
function cint=intfun(cA)
global cA0 sA;
clim=250;
cint=1./(cA0-cA.*(1-sA))./log(clim/sA./cA-(1-sA)/sA);
```

Running the code gives a time of 14 802 s, which is less than the value of 16 202 s obtained when $\sigma_A = 1$, as can be easily shown using the above code or using the exponential integral method of Example 5.14.

7.8 Single pass ultrafiltration

Single pass UF was analysed for the case of complete rejection in section 5.5. It is similar to continuous feed-and-bleed ultrafiltration but the absence of a recirculation loop means that the system is not well mixed. Hence the variation of flowrate, concentration and mass transfer coefficient with position must be considered. The volume balance for the system can be written

$$\frac{dQ}{dA} = -J, \tag{7.50}$$

where dQ is the change in tangential flowrate over a differential area, dA, of membrane. The solute balance can be written

$$\frac{d(c_A Q)}{dA} = -Jc_A(1 - \sigma_A). \tag{7.51}$$

Combining these two equations gives

$$\frac{dc_A}{dA} = \frac{Jc_A\sigma_A}{Q}. \tag{7.52}$$

Taking into account the flowrate dependence of the mass transfer coefficient, the mass transfer coefficient is given by

$$k_c = k_{c0}\left(\frac{Q}{Q_0}\right)^n, \tag{7.53}$$

where k_{c0} is the un-corrected mass transfer coefficient evaluated at the inlet flowrate, Q_0.

One could of course solve this problem by simply solving the two ODES, Eq. 7.50 and 7.52. However, the numerical integration approach is used here. Combining these two equations and integrating precisely, as in the batch case, one gets

$$c_A = c_{A0}\left(\frac{Q_0}{Q}\right)^{\sigma_A}. \tag{7.54}$$

Therefore, Eq. 7.52 written in full becomes

$$\frac{dc_A}{dA} = \frac{k_{c0}}{Q_0}\left(\frac{c_A}{c_{A0}}\right)^{1/\sigma_A}\left(\frac{c_{A0}}{c_A}\right)^{n/\sigma_A} c_A\sigma_A \ln\left[\frac{c_{\lim}}{\sigma_A c_A} - \frac{1-\sigma_A}{\sigma_A}\right]. \tag{7.55}$$

For the process analysis problem, i.e, calculation of the exit concentration for fixed A, this equation can be solved numerically as required in Problem 7.8. For the design problem, i.e., calculation of A for a given exit concentration, Eq. 7.55 can be rearranged to give

$$A = \frac{Q_0 c_{A0}^{(1-n)/\sigma_A}}{\sigma_A k_{c0}} \int_{c_{A0}}^{c_{Af}} \frac{c_A^{(n-1-\sigma_A)/\sigma_A}}{\ln\left(\dfrac{c_{\lim}}{\sigma_A c_A} - \dfrac{1-\sigma_A}{\sigma_A}\right)} dc_A. \tag{7.56}$$

Reassuringly, this reduces to Eq. 5.74 when $\sigma_A = 1$. Use of this equation is illustrated in the next example, which is analogous to Example 5.17.

Example 7.7 Single pass design with incomplete rejection

A 10 g/l protein solution is to be concentrated to 25 g/l in a single pass UF process where the flow in the module is turbulent and $n = 0.8$. The mass transfer coefficient at the inlet conditions is 2×10^{-5} m/s and the limiting concentration is 280 g/l. The feed flowrate is 50 l/min. Calculate the membrane area if the rejection coefficient of the solute is 0.9.

Solution. MATLAB code for solving this problem, based on Eq. 7.56, is shown below.

```
function example77
global n sA;
Q0=0.05/60; cA0=10;cAf=25;kc0=2e-5;sA=0.9;n=0.8;
a=cA0; b=cAf;
```

```
cint=quad(@intfun,a,b);
A=Q0/kc0*(cA0^((1-n)/sA))*cint/sA
function cint=intfun(cA)
global n sA;
clim=280;
cint=(cA.^((n-1-sA)/sA))./log(clim./cA/sA-(1-sA)/sA);
```

Running the code gives $A = 12.94$ m^2 which is slightly higher than the 12.18 m^2 required when there is complete rejection, as one would expect.

7.9   Constant volume diafiltration

The basic process layout for a CVD process was shown in Fig. 6.5. The first step in analysis of this process is to formulate the solute balances. Let us consider an arbitrary solute with an apparent rejection coefficient of σ which is assumed constant. Since the solution volume is constant (V_0) in CVD, the balance for any solute can be written

$$\frac{dc}{dt} = -\frac{AJ}{V_0}c(1-\sigma).$$ (7.57)

The consumption of diafiltration water is given by

$$\frac{dV_w}{dt} = AJ.$$ (7.58)

Dividing the first of these equations by the second gives

$$\frac{dc}{dV_w} = -\frac{c(1-\sigma)}{V_0}.$$ (7.59)

Separating the variables, this becomes

$$\int_{c_0}^{c} \frac{dc}{c} = -\frac{1-\sigma}{V_0} \int_{0}^{V_w} dV_w.$$ (7.60)

Therefore

$$c = c_0 e^{-(1-\sigma)V_w/V_0}.$$ (7.61)

Thus the solute concentration declines exponentially with the volume of water added, and the magnitude of this decline depends on the rejection coefficient. For a solute showing complete rejection, $c = c_0$, i.e., there is no change in concentration. For $\sigma = 0$, the simpler form of Eq. 7.61, which was derived in Chapter 6, is attained. This can be written

$$c = c_0 e^{-V_w/V_0}.$$ (7.62)

Equation 7.61 can be rearranged to give the water requirement of a given CVD process as follows:

$$V_w = \frac{V_0}{1 - \sigma} \ln \frac{c_0}{c_f}, \tag{7.63}$$

where c_f is the required final concentration of a solute, usually the microsolute of a macrosolute–microsolute mixture. Equation 7.63 can thus be written in terms of the microsolute as

$$\frac{V_w}{V_0} = \frac{\ln \beta}{1 - \sigma_B}, \tag{7.64}$$

where, as before,

$$\beta = \frac{c_{B0}}{c_{Bf}}. \tag{7.65}$$

Use of this equation is illustrated in the next example.

Example 7.8 Product loss in CVD

A certain low molecular weight solute is to be removed from a protein solution by diafiltration. The rejection coefficient of the protein is 0.9 while that of the low molecular weight solute is 0.1. A CVD process is used to reduce the microsolute concentration by a factor of 10. What percentage of the protein is lost in this process?

Solution. To achieve the microsolute reduction target, we must have

$$\frac{V_w}{V_0} = \frac{\ln 10}{1 - 0.1}.$$

Therefore, for the macrosolute

$$\frac{c_{Af}}{c_{A0}} = \exp\left(-(1 - 0.9)\frac{\ln 10}{1 - 0.1}\right) = 0.77.$$

Therefore the percentage loss is 23%. An alternative approach is to use a little bit of algebra to get

$$\frac{c_{Af}}{c_{A0}} = \left(\frac{c_{B0}}{c_{Bf}}\right)^{-(1 - \sigma_A)/(1 - \sigma_B)},$$

which, for this problem, becomes

$$\frac{c_{Af}}{c_{A0}} = 10^{-0.1/0.9}.$$

This gives the same answer as the method used above.

The above example raises the possibility of solute fractionation, i.e., the separation of a complex mixture of solutes into separate solutions containing high purities of each solute. When a solution undergoes CVD, it becomes richer in the solute with the higher rejection coefficient. Similarly, the permeate becomes richer in the solute with the lower rejection coefficient. This is illustrated in the next example.

Example 7.9 Fractionation of a solution containing two solutes

A protein solution contains two solutes whose rejection coefficients for the particular membrane being used are $\sigma_A = 0.9$ and $\sigma_B = 0.7$. The solution is to be diafiltered to produce equal volumes of retentate and permeate. If the solution initially contains equal concentrations of each solute, calculate the final (relative) concentrations of A and B in the retentate and permeate.

Solution. Let c_{A0} and c_{B0} be the initial concentrations of each solute. Then the concentrations of each solute in the retentate after CVD are given by

$$\frac{c_{Af}}{c_{A0}} = e^{-(1-0.9)} = 0.905,$$

$$\frac{c_{Bf}}{c_{B0}} = e^{-(1-0.7)} = 0.741,$$

where it should be noted that $V_w / V_0 = 1$ in this case. The concentrations in the permeate are given by mass balances as

$$\frac{c_{Ap}}{c_{A0}} = \frac{V_0(1 - c_{Af}/c_{A0})}{V_w} = 1 - 0.905 = 0.095,$$

$$\frac{c_{Bp}}{c_{B0}} = \frac{V_0(1 - c_{Bf}/c_{B0})}{V_w} = 1 - 0.741 = 0.259.$$

The above example shows that some degree of separation, i.e., fractionation, of the solutes is possible. Of course, better separation is obtained if more diafiltration water is used but increased purity in the retentate is attained at the expense of greater loss of A into the permeate, i.e., a lower *yield* of A. Furthermore, increased purity of A in the retentate is accompanied by reduced purity of B in the permeate. This is the subject of Problem 7.9. As a general rule, therefore, simple CVD is not a good way of separating two solutes with similar rejection coefficients. Usually one has to use multi-stage methods in which different membranes, with different rejection coefficients, are used in each stage [2]. Often, one has to include one or more UF steps due to the extra water added to the system in the CVD step(s). One advantage of this, however, is the fact that UF itself has a fractionation aspect as shown by Eq. 7.38.

7.9.1 Computation of diafiltration time

One of the great simplifying factors in the analysis of CVD in Chapter 6 was the fact that the flux was constant during the diafiltration. This stemmed from the assumption that the flux was entirely dependent on the macrosolute concentration and there was complete rejection of the macrosolute. In this section, the former assumption is retained but the fact that the macrosolute concentration changes during diafiltration, as a result of incomplete rejection, means that the flux is *not* constant. Thus, computation of diafiltration time requires an integration rather than simply putting numbers into an algebraic equation. Let us consider a solution of volume, V_0, in which the concentration of a microsolute is to be reduced in concentration from c_{B0} to c_{Bf}. The required volume of diluent is given by Eq. 7.64. Rearranging Eq. 7.58 gives

$$t_{CVD} = \int_0^{V_0[\ln\beta/(1-\sigma_B)]} \frac{dV_w}{JA}. \tag{7.66}$$

Combining Eqs. 7.11 and 7.61, this becomes

$$t_{CVD} = \frac{1}{kA} \int_0^{V_0 \ln\beta/(1-\sigma_B)} \frac{dV_w}{\ln\left[\dfrac{c_{lim}}{\sigma_A c_{A0} e^{-(1-\sigma_A)V_w/V_0}} - \dfrac{1-\sigma_A}{\sigma_A}\right]}. \tag{7.67}$$

Use of this equation is illustrated in the next example.

Example 7.10 Computation of CVD time

A 100 l volume of a 25 g/l protein solution is to be diafiltered to reduce the concentration of an impurity from 5 to 1 g/l. The rejection coefficient of the protein is 0.95 while that of the impurity is 0.15. The system is operating with a mass transfer coefficient of 6 × 10^{-6} m/s and the membrane area is 1.5 m^2. The limiting protein concentration is 275 g/l. Calculate the time required to perform this operation.

Solution. MATLAB code for computing the integral is shown below.

```
function example710
k=6e-6; A=1.5;V0=0.1;beta=5;sB=0.15
a=0; b=log(beta)*V0/(1-sB);
xint=quad(@intfun,a,b);
time=xint/(k*A)
function xint=intfun(Vw)
V0=0.1; cA0=25;sA=0.95;clim=275;
xint=1./log(clim./(sA*cA0*exp(-(1-sA)*Vw/V0))-(1-sA)/sA)
```

Running the code gives a time of 8443 s. Note that running the same code with $\sigma_A = 1$ yields a slightly longer time (8774 s), which is expected because the average flux in that case is lower.

7.9.2   UFCVD with arbitrary rejection coefficients

Consider, now, a process in which a volume, V_0, of a macrosolute is to be increased in concentration from c_{A0} to c_{Af} while the microsolute concentration is to be reduced in concentration from c_{B0} to c_{Bf}. The CVD step is to be done when the macrosolute concentration has been increased to c_{Ai}. At this point the microsolute concentration is c_{Bi}. After CVD, the macrosolute concentration is further increased to c_{Af} while the microsolute concentration increases to c_{Bf}. The whole process is summarised in Table 7.1.

Table 7.1 Volume and concentration changes in UFCVD.

UF-1	CVD	UF-2
$V_0 \rightarrow V_i$	$V_i \rightarrow V_i$	$V_i \rightarrow V_f$
$c_{A0} \rightarrow c_{Ai}$	$c_{Ai} \rightarrow c_{Aif}$	$c_{Aif} \rightarrow c_{Af}$
$c_{B0} \rightarrow c_{Bi}$	$c_{Bi} \rightarrow c_{Bif}$	$c_{Bif} \rightarrow c_{Bf}$

Now, assuming constant apparent rejection, one has

$$\frac{c_{Ai}}{c_{A0}} = \left(\frac{V_0}{V_i}\right)^{\sigma_A}, \tag{7.68}$$

$$\frac{c_{Aif}}{c_{Ai}} = e^{-(1-\sigma_A)V_w/V_i}, \tag{7.69}$$

$$\frac{c_{Af}}{c_{Aif}} = \left(\frac{V_i}{V_f}\right)^{\sigma_A}. \tag{7.70}$$

Combining these three equations gives

$$\frac{c_{Af}}{c_{A0}} = \left(\frac{V_0}{V_f}\right)^{\sigma_A} e^{-(1-\sigma_A)V_w/V_i}. \tag{7.71}$$

Likewise

$$\frac{c_{Bf}}{c_{B0}} = \left(\frac{V_0}{V_f}\right)^{\sigma_B} e^{-(1-\sigma_B)V_w/V_i}. \tag{7.72}$$

Combining these last two equations, gives the following expression for the volume of diafiltration water required:

$$V_w = V_i \frac{\sigma_A \ln(c_{B0}/c_{Bf}) - \sigma_B \ln(c_{A0}/c_{Af})}{\sigma_A - \sigma_B}, \tag{7.73}$$

Use of this equation is illustrated in the next example.

Example 7.11 Water consumption in UFCVD

A 100 l volume of a solution undergoes UFCVD. The final volume of the solution is to be 30 l and CVD is done when the solution volume has been reduced to 50 l. The concentration of an inpurity, B, is to be reduced by a factor of 5. The rejection coefficients are $\sigma_A = 0.95$ and $\sigma_B = 0.15$. Calculate the amount of diafiltration water required and compare this value with the amount required when $\sigma_A = 1.0$ and $\sigma_B = 0$.

Solution. This is a simple application of Eqs. 7.38 and 7.73. First, Eq. 7.38 is applied to component A giving

$$\ln\left(\frac{c_{A0}}{c_{Af}}\right) = \sigma_A \ln\left(\frac{V_f}{V_0}\right).$$

Therefore

$$V_w = V_i \frac{\sigma_A \ln(c_{B0}/c_{Bf}) - \sigma_B \sigma_A \ln(V_f/V_0)}{\sigma_A - \sigma_B}.$$

Putting in numbers gives

$$V_w = 50 \frac{0.95 \ln 5 - 0.15 \times 0.95 \ln(30/100)}{0.95 - 0.15} = 106.2 \, \text{l}.$$

For the case where $\sigma_A = 1.0$ and $\sigma_B = 0$, we have

$$V_w = 50 \ln 5 = 80.5 \, \text{l}.$$

7.9.3 Dynamic modelling of UFCVD

As done in Chapter 6, the problem of UFCVD is now formulated as a generalised variable volume diafiltration process. Let us formulate the relevant balances as follows:

$$\frac{dV}{dt} = (\alpha - 1)AJ, \tag{7.74}$$

$$\frac{dV_w}{dt} = \alpha AJ, \tag{7.75}$$

$$\frac{dc_A}{dt} = \frac{AJc_A}{V}(\sigma_A - \alpha), \tag{7.76}$$

$$\frac{dc_B}{dt} = \frac{AJc_B}{V}(\sigma_B - \alpha). \tag{7.77}$$

The last two are the subject of Problem 7.10. For UFCVD, the function, α, is best described (in pseudocode) as follows:

```
IF V <= VI and VW < Eq. 7.73 VALUE
ALPHA=1
ELSE ALPHA=0
```

As before, the first line should technically include the condition V = Vi, but for programming purposes it is better to include the inequality in the IF statement.

This is different from the way α was defined in Chapter 6, where it was defined in terms of concentrations. Such an approach is not obviously achievable when there are arbitrary rejection coefficients. Use of this system of equations is illustrated in the next example.

Example 7.12 Simulation of UFCVD with arbitrary rejection coefficient

A protein solution is to be concentrated from 10 g/l to 30 g/l and the microsolute is to be reduced from 1 to 0.2 g/l. The limiting protein concentration, c_{lim}, is 280 g/l. The initial volume is 100 l, the mass transfer coefficient is constant and equal to 5×10^{-6} m/s and the membrane area is 1.2 m^2. The rejection coefficients are $\sigma_A = 0.9$ and $\sigma_B = 0.15$. Generate a plot of relative solute concentrations versus time if CVD is commenced at a volume of 50 l.

Solution. MATLAB code for solving this problem is shown below.

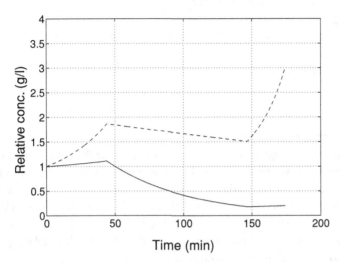

Figure 7.3 Concentration profiles in UFCVD with arbitrary rejection coefficients. Dashed curve denotes macrosolute, solid curve denotes microsolute.

```
function example712
global cA0 cB0
cA0=10; cB0=1.0;
options=odeset('events', @eventfcn, 'Maxstep',1);
[t,y]=ode15s(@ufcvd,[0 36000],[0.1 0 10 1], options);
cA=y(:,3)./cA0; cB=y(:,4)./cB0;
time=t/60;
timeout=t(end)/60
plot(time,cA,time,cB);
xlabel('Time (mins)');ylabel('Relative Conc. (g/L)')
function dydt=ufcvd(t,y)
global cA0 cB0 cAf;
%y(1) is V, y(2) is Vw, y(3) is cA, y(4) is cB
A=1.2; Vi=0.05; cBf=0.2; cAf=30;
k=5e-6;clim=280; sA=0.9;sB=0.15;V0=0.1;
J=k*log(clim/sA/y(3)-(1-sA)/sA);
Vw=Vi*(sA*log(cB0/cBf)-sB*log(cA0/cAf))/(sA-sB);
if and(y(1)<=Vi,y(2)<Vw);
alpha=1;
else
alpha=0;
end
Veqn=(alpha-1)*J*A;
Vweqn=alpha*J*A;
```

```
cAeqn=y(3)/y(1)*J*A*(sA-alpha);
cBeqn=y(4)/y(1)*J*A*(sB-alpha);
dydt=[Veqn;Vweqn;cAeqn;cBeqn];
function [value, isterminal, direction]=eventfcn(t,y)
global cAf;
value = y(3)-cAf;
isterminal = 1;
direction = 0;
```

Output from the program is shown in Fig. 7.3.

The time taken to complete the process is obtained from the code output and in this case is 174.2 min. It is worth noting that the above plots illustrate well the fact that when the rejection coefficients have arbitrary values, the targets for c_A and c_B are reached simultaneously at the end of the second UF step. This is unlike the special case of $\sigma_A = 1$ and $\sigma_B = 0$, where the target of B is reached at the end of the CVD step. The decline in macrosolute concentration during the CVD phase is what makes optimisation of UFCVD more difficult in the general case.

7.9.4   Optimisation of UFCVD with arbitrary rejection coefficients

In Chapter 6, the problem of optimising a process in which both the macrosolute and the microsolute must reach defined targets was examined in detail. A crucial aspect of that analysis was the fact that macrosolute was completely rejected and, consequently, the ultrafiltration time was fixed, regardless of when the diafiltration was performed. This meant that minimisation of the process time simply involved minimisation of the diafiltration time. This minimisation was rendered simple by the fact that the flux-determining macrosolute was constant during the diafiltration step, meaning that the optimum could be determined by simple differentiation (with respect to c_A) of the expression for the diafiltration time.

However, the case of incomplete rejection is fundamentally different from a computational point of view because the flux is not constant during the diafiltration step. Indeed, early research in this area seemed to forget that important point [3, 4]. Therefore, a cruder approach is required [5]. Of course, one could use a code like that shown in Example 7.12 and run it for a range of values of c_{Ai}. Here, however, the problem can also be expressed in terms of three integrals as follows:

$$
tkA = \int_{V_i}^{V_0} \frac{dV}{\ln\left(\dfrac{c_{\lim}}{\sigma_A c_A} - \dfrac{1-\sigma_A}{\sigma_A}\right)} + \int_0^{V_{wf}} \frac{dV_w}{\ln\left(\dfrac{c_{\lim}}{\sigma_A c_{Ai} e^{-(1-\sigma_A)V_w/V_i}} - \dfrac{1-\sigma_A}{\sigma_A}\right)}
$$

$$
+ \int_{V_f}^{V_i} \frac{dV}{\ln\left(\dfrac{c_{\lim}}{\sigma_A c_A} - \dfrac{1-\sigma_A}{\sigma_A}\right)}. \tag{7.78}
$$

At first glance it might seem that the two UF integrals should simply add to get a fixed UF time. However, it is important to note that the relationships between concentration and volume are different in each case. For the first UF integral, c_A and V are related by Eq. 7.68, while in the second UF integral, they are related by Eq. 7.70.

Getting these integrals into a format suitable for numerical integration requires a little bit of effort. For the first UF integral:

$$c_A = c_{A0} \left(\frac{V_0}{V} \right)^{\sigma_A}, \tag{7.79}$$

$$V_i = V_0 \left(\frac{c_{A0}}{c_{Ai}} \right)^{1/\sigma_A}. \tag{7.80}$$

For the CVD integral

$$\frac{V_{wf}}{V_i} = \frac{\sigma_A \ln(c_{B0}/c_{Bf}) - \sigma_B \ln(c_{A0}/c_{Af})}{\sigma_A - \sigma_B}. \tag{7.81}$$

For the second UF integral

$$c_{Aif} = c_{Ai} e^{-(1-\sigma_A)V_{wf}/V_i}, \tag{7.82}$$

$$c_A = c_{Aif} \left(\frac{V_i}{V} \right)^{\sigma_A}, \tag{7.83}$$

$$V_f = V_0 \left[\frac{c_{Aif}}{c_{Af}} \right]^{1/\sigma_A}. \tag{7.84}$$

The use of these equations is illustrated in the next example.

Example 7.13 Repeat the time calculation of Example 7.12 but use the integral formulation

Solution. MATLAB code for this solving this problem is shown below.

```
function example713
global V0 Vi clim cA0 cAif sA cAi
k=5e-6; A=1.2;V0=0.1;sA=0.9;sB=0.15;
cA0=10;cAf=30;cB0=1;cBf=0.2;clim=280;
Vi=0.05;
cAi=cA0*(V0/Vi)^sA
VwfVi=(sA*log(cB0/cBf)-sB*log(cA0/cAf))/(sA-sB);
cAif=cAi*exp(-(1-sA)*VwfVi);
Vf=Vi*(cAif/cAf)^(1/sA);
a1=Vi; b1=V0
a2=0; b2=Vi*VwfVi;
a3=Vf; b3=Vi;
t1=quad(@uf1,a1,b1)/k/A
t2=quad(@cvd,a2,b2)/k/A
```

```
t3=quad(@uf2,a3,b3)/k/A
time=(t1+t2+t3)/60
function int1=uf1(x)
global V0 clim  sA cA0
cA=cA0*(V0./x).^sA;
int1=1./log(clim/sA./cA-(1-sA)/sA);
function int2=cvd(x)
global Vi clim cAi sA
cA=cAi*exp(-(1-sA)*x/Vi);
int2=1./log(clim/sA./cA-(1-sA)/sA);
function int3=uf2(x)
global Vi clim cAif sA
cA=cAif*(Vi./x).^sA;
int3=1./log(clim/sA./cA-(1-sA)/sA);
```

Running the code gives exactly the same answer as Example 7.12 to the nearest one tenth of a minute.

Finding the optimum value of c_{Ai} in any given situation can be done by simply repeating calculations like that done above for a range of c_{Ai} values and finding the optimum by inspection.

7.10   Variable volume diafiltration

As discussed in Chapter 6, variable volume diafiltration (VVD) involves the simultaneous ultrafiltration and diafiltration of a solution. The diafiltration water is fed to the retentate tank at a constant fraction, α, of the permeate flowrate. Despite its elegance, VVD uses more water and takes longer to perform than an optimised UFCVD.

In this section, the use of VVD to separate two solutes is investigated. It was shown earlier that separation of solutes using CVD is not very efficient. So, is better separation possible with VVD? To answer that question, the appropriate mass balances must be formulated. Dividing Eqs. 7.76 and 7.77 by Eq. 7.74 gives

$$\frac{c_{Af}}{c_{A0}} = \left(\frac{V_0}{V_f}\right)^{(\sigma_A - \alpha)/(1-\alpha)}, \tag{7.85}$$

$$\frac{c_{Bf}}{c_{B0}} = \left(\frac{V_0}{V_f}\right)^{(\sigma_B - \alpha)/(1-\alpha)}. \tag{7.86}$$

In the example below these equations are used to examine how VVD can be used to separate two solutes.

Example 7.14 Fractionation by VVD

A solution contains two solutes whose rejection coefficients for the particular membrane being used are $\sigma_A = 0.9$ and $\sigma_B = 0.7$. The initial concentrations of both solutes are

1 g/l. The solution is to be processed by VVD using $\alpha = \sigma_A$ and $V_f = V_0/2$. Calculate the final concentrations of A and B in each of the retentate and permeate tanks at the end of the process.

Solution. From Eq. 7.85 we get

$$c_{Af} = c_{A0} = 1 \text{ g/l}.$$

This is an interesting quirk of using $\alpha = \sigma_A$. Now, from Eq. 7.86,

$$c_{Bf} = 2^{(0.7-0.9)/(1-0.9)} = 2^{-2} = 0.25 \text{ g/l}.$$

Now, the volume of permeate is given by dividing Eq. 7.75 by Eq. 7.74 and noting that the permeate volume is the sum of the diafiltration water used and the reduction in the retentate volume, i.e,

$$\frac{V_p}{V_0} = \frac{V_w + (V_0 - V_f)}{V_0} = \frac{\alpha}{1-\alpha}\left(1 - \frac{V_f}{V_0}\right) + \left(1 - \frac{V_f}{V_0}\right) = \frac{1}{1-\alpha}\left(1 - \frac{V_f}{V_0}\right).$$

Therefore

$$\frac{V_p}{V_0} = \frac{1}{1-0.9}\left(1 - \frac{1}{2}\right) = 5.0.$$

The permeate concentrations are given by the solute balances as follows:

$$c_{A0}V_0 = c_{Af}\frac{V_0}{2} + 5V_0 c_{Ap},$$

$$c_{B0}V_0 = c_{Bf}\frac{V_0}{2} + 5V_0 c_{Bp},$$

i.e.

$$c_{Ap} = \frac{0.5}{5.0} = 0.1 \text{ g/l},$$

$$c_{Bp} = \frac{1 - 0.25/2}{5.0} = 0.175 \text{ g/l}.$$

In Problem 7.11, the task is to examine the purities obtained with a UFCVD process using the same volume of diluent.

7.11   Conclusions

In this chapter, many of the tools developed so far in this book have been applied to ultrafiltration and diafiltration with arbitrary rejection coefficients. The various analyses have relied very strongly on the formulation of the solute balances and this illustrates the importance of this basic skill for membrane engineers. Of course, all of our analyses so far, from Chapter 5 to Chapter 7, have relied on the assumption that the system is operating at the limiting flux at all times. But this assumes that any changes in flux during operation do not cause the flux to 'drop out' of the limiting zone. It is important, however, to be able to model UF and DF processes in which the flux is always below the limiting value, or indeed to model processes in a way that can deal with fluxes that move from limiting to non-limiting. In other words, a way of doing process design and analysis under conditions of arbitrary and fixed trans-membrane pressure is required. This, of course, requires one to use the osmotic pressure model. As seen in Chapter 4, this adds an extra layer of complexity because the wall concentration is essentially unknown. Once again, however, it is seen that while this adds to the algebra, no insurmountable difficulties are presented. Applying the osmotic pressure model to process analysis and design of UF and DF systems is the subject of the next chapter.

References

1. Zeman, L.J. and Zydney, A.L. (1996). *Microfiltration and Ultrafiltration. Principles and Applications.* Marcel Dekker, New York, USA.
2. Cheang, B.L. and Zydney, A.L. (2004). A two-stage ultrafiltration process of whey protein isolate. *Journal of Membrane Science*, **231**, 159–167.
3. Ng, P. (1976). Optimization of solute separation by diafiltration. *Separation Science*, **11**, 499–502.
4. Cooper, A.R. and Booth, R.G. (1978). Ultrafiltraion of synthetic polymers Part I. Optimisation of solution flux by diafiltration. *Separation Science and Technology*, **13**, 735–744.
5. Foley, G. (1999). Minimisation of process time in ultrafiltration and continuous diafiltration: the effect of incomplete macrosolute rejection. *Journal of Membrane Science*, **163**, 349–355.

Additional reading

Ghosh, R. (2003). *Protein Fractionation Using Ultrafiltration*, Second Edition. Imperial College Press, London, UK.

Problems

Problem 7.1   Solute loss in continuous feed-and-bleed UF

Show that the fraction, f, of solute lost in a single-stage, continuous feed-and-bleed UF system is given by

$$f = \frac{(1 - x_{A1})}{x_{A1}} \frac{(1 - \sigma_A)}{\sigma_A}.$$

Is this consistent with the answer to Example 7.1?

Problem 7.2   Solute loss in a two-stage feed-and-bleed system

Show that the fractional loss of solute after two stages of a continuous feed-and-bleed system is given by

$$f_2 = f_1 \left[1 + \left(\frac{x_{A1}}{x_{A2}} - 1 \right) \frac{1 - \sigma_A}{\sigma_A} \frac{1 - f_1}{f_1} \right],$$

where f_1 is the fractional loss after the first stage as given in Problem 7.1. Check this expression by using it to compute the fractional loss for the conditions of Example 7.3.

Problem 7.3   Graphical solution of a two-stage system

Use a graphical method to solve the problem in Example 7.3. Note that in this case the operating line of the second stage is not parallel to that of the first stage, unlike the situation in systems showing complete rejection. Equation 7.30 is key here.

Problem 7.4   Operating equations for a three-stage system

Develop the operating equation for the third stage of a three-stage continuous feed-and-bleed UF system which could be used in conjunction with Eqs. 7.26 and 7.30 to form a complete performance model.

Problem 7.5   Design of a two-stage system

A 1.0 l/min, 10 g/l solution is fed to a two-stage continuous feed-and-bleed system where its concentration is to be increased to 128 g/l. The mass transfer coefficient is 3.5×10^{-6} m/s, the limiting concentration is 300 g/l and the solute has a rejection coefficient of 0.9. Calculate the area of each stage assuming equal areas.

Problem 7.6   Feed-and-bleed analysis with viscosity dependent mass transfer coefficient

Repeat Example 7.3 but assume that the mass transfer coefficient is related to composition by the expression

$$k = k_0 e^{\gamma z(c-c_{\lim})} e^{-\gamma c/2},$$

where $\gamma = 0.01$ l/g, $z = 0.14$ and $k_0 = 3.5 \times 10^{-6}$ m/s.

Problem 7.7   Accuracy of assuming constant apparent rejection

A solution is to be ultrafiltered in a system in which the initial apparent rejection coefficient is 0.9. Investigate the predictions of Eqs. 7.38 and 7.46 for volume reductions in the range 1–10.

Problem 7.8   Single pass analysis

A 10 g/l protein solution is to be concentrated in a single pass UF process where flow in the module is turbulent and $n = 0.8$. The mass transfer coefficient at the inlet conditions is 2×10^{-5} m/s and the limiting concentration is 280 g/l. The feed flowrate is 50 l/min. Calculate the exit concentration if the membrane area is 13 m^2 and the rejection coefficient of the solute is 0.9.

Problem 7.9   Fractionation with CVD

A macrosolute has a constant apparent rejection coefficient of 0.9 while a microsolute has a rejection coefficient of 0.1. Investigate the effect of diafiltration water volume on the purity of A in the retentate, the yield of A in the retentate and the purity of B in the permeate. Assume the concentrations of the solutes are initially the same.

Problem 7.10   Solute balances for generalised variable volume diafiltration

Derive Eqs. 7.76 and 7.77.

Problem 7.11   Fractionation by UFCVD

Using the same conditions as Example 7.14, compute the purity of A in the retenate and B in the permeate at the end of an equivalent UFCVD process employing the same amount of water as the VVD process. Assume that the CVD step is done at V_f. Compare the results with the results from Example 7.14.

Further problems

Problem 7.12   Approximate calculation of CVD time

Revisit Example 7.10 and make the approximation that $\sigma_A = 1$ in the code. Now, recalculate the CVD time using the methods of Chapter 6.

Problem 7.13   Optimum concentration for UFCVD

Using Example 7.13 as a starting point, investigate the effect of the intermediate volume, V_i, on the total process time. Is there an optimum?

Problem 7.14   Discontinuous diafiltration by volume reduction

Recall the volume reduction method of continuous diafiltration described in Chapter 6. This method is to be used to reduce the concentration of a certain impurity, B, from 1 to 0.1 g/l. If the solution volume is 100 l, calculate the volume of diafiltration water required and the loss of product, A. Assume that the microsolute has a rejection coefficient of zero and the macrosolute has a rejection coefficient of 0.9.

Problem 7.15   Discontinuous diafiltration by dilution

Repeat Problem 7.14 but assume that the dilution method of discontinuous diafiltration is used.

8 The osmotic pressure model applied to ultrafiltration and diafiltration

8.1 Introduction

The osmotic pressure (OP) model of ultrafiltration was described in Chapter 4 and provides a way to compute the wall concentration (i.e. the solute concentration at the membrane) a priori. Furthermore, it allows the wall concentration to vary with process conditions and provides a logical route to incorporating membrane fouling into the analysis and design of UF and DF systems. The basic idea behind the OP model is that the flux predicted by osmotic pressure theory must be the same as the flux predicted by concentration polarisation theory. The general equation for the OP model can be written

$$k(c, c_{\mathrm{w}}) \ln \left[\frac{c_{\mathrm{w}}}{\sigma c} - \frac{1 - \sigma}{\sigma} \right] - \frac{\Delta P - \sigma_0 \Delta \pi (c_{\mathrm{w}}, c_{\mathrm{p}})}{\mu R_{\mathrm{m}}} = 0, \tag{8.1}$$

where it is understood that all concentrations refer to the macrosolute. The wall concentration of this solute is c_{w}, μ is the permeate viscosity, σ is the apparent rejection coefficient, R_{m} is the membrane resistance and ΔP is the trans-membrane pressure. The above notation for k means that mass transfer coefficient is a function of both the retentate concentration, c, and the wall concentration, c_{w}, while the osmotic pressure difference across the membrane depends on the wall and permeate concentrations. The parameter, σ_0, is another type of rejection coefficient, known as the *osmotic reflection coefficient*. This is a rather complex parameter which deserves some discussion, and this is postponed until later in this chapter when systems with incomplete rejection are discussed. In the meantime, complete rejection is assumed and Eq. 8.1 reduces to

$$k(c, c_{\mathrm{w}}) \ln \left(\frac{c_{\mathrm{w}}}{c} \right) - \frac{\Delta P - \pi(c_{\mathrm{w}})}{\mu R_{\mathrm{m}}} = 0. \tag{8.2}$$

As discussed in Chapter 4, the osmotic pressure, π, is normally assumed to be a third order polynomial in c_{w} and the mass transfer coefficient is given in turbulent flow by

$$k = k_0 e^{-\gamma c/2} e^{\gamma z(c - c_{\mathrm{w}})}, \tag{8.3}$$

and in laminar flow by

$$k = k_0 e^{\gamma z(c - c_{\mathrm{w}})}. \tag{8.4}$$

The parameter, γ, is an experimental parameter characterising the concentration dependence of the solution viscosity, z is the exponent in the viscosity wall correction factor and is given the value 0.14, while k_0 is the un-corrected, infinite dilution mass transfer coefficient.

The key computational difficulty in using the OP model is the fact that where previously the wall concentration was a constant parameter, c_{lim}, now it is a variable parameter, c_w, whose value is only known by solving a non-linear algebraic equation. In practice, however, this is no major issue as while the algebra gets a little more complicated, there are no real computational challenges once the problems are formulated in a suitable manner.

Thus, there is no barrier to simply moving on and getting into the details of applying the OP model to batch and continuous UF and DF systems. As before, we start with continuous feed-and-bleed systems.

8.2 Analysis and design of continuous feed-and-bleed UF

8.2.1 Analysis and design of single-stage systems

Since this chapter deals with UF and DF, all concentrations are appropriately subscripted to make sure it is clear at all times which solute is in question. As previously, A represents the flux-determining macrosolute and B represents the microsolute. The basic process layout for a single-stage system was shown in Fig. 5.1.

The combination of overall and solute balances gives the equation derived previously (see Section 5.2):

$$\frac{Q_0}{kA}\left(1 - \frac{c_{A0}}{c_{A1}}\right) - \ln\frac{c_{Aw1}}{c_{A1}} = 0. \tag{8.5}$$

Now, for the OP model, one can write

$$\ln\frac{c_{Aw1}}{c_{A1}} - \frac{\Delta P - \pi}{k\mu R_m} = 0. \tag{8.6}$$

Now defining $x_{A1} = c_{A0}/c_{A1}$ and $y_1 = c_{A0}/c_{w1}$, these two equations become

$$\frac{Q_0}{kA}(1 - x_{A1}) - \ln\left(\frac{x_{A1}}{y_{A1}}\right) = 0, \tag{8.7}$$

$$\ln\frac{x_{A1}}{y_{A1}} - \frac{\Delta P - \pi(y_{A1})}{k\mu R_m} = 0. \tag{8.8}$$

Now, assuming that the osmotic pressure is a polynomial function as outlined in Chapter 4, one can write

$$\pi(y_{A1}) = a_1 c_{A0} y_{A1}^{-1} + a_2 c_{A0}^2 y_{A1}^{-2} + a_3 c_{A0}^3 y_{A1}^{-3}, \tag{8.9}$$

where a_1, a_2 and a_3 are empirical constants. Equations 8.7 and 8.8, in conjunction with Eq. 8.9, constitute the mathematical model of a single-stage feed-and-bleed UF system. Application of these equations is illustrated in the next example.

Example 8.1 Analysis of a single-stage feed-and-bleed system

A 1 l/min feed of a whey solution enters a single-stage feed-and-bleed UF system. The feed enters at 10 g/l, the mass transfer coefficient is constant and equal to 3.5×10^{-6} m/s and the membrane area is 2.7 m^2. Calculate the exit concentration. Use osmotic pressure data for whey from Table 4.5. The membrane resistance is 1×10^{12} m^{-1}, the permeate viscosity is 0.001 Ns/m^2 and the trans-membrane pressure is 2 bar.

Solution. This problem requires use of the MATLAB function **fsolve** as illustrated below.

```
function example81
x=fsolve(@singlestageop, [0.5,0.1])
function f=singlestageop(x)
A=2.7;Q0=1e-3/60;k=3.5e-6;u=0.001;Rm=1e12;DP=2e5;
cA0=10; a1=454.3;a2=-0.176;a3=0.0082;
%x(1) is xA1, x(2) is yA1
osm=a1*cA0/x(2)+a2*cA0^2/x(2)^2+a3*cA0^3/x(2)^3;
xA1eqn=Q0/k/A*(1-x(1))-log(x(1)/x(2));
yA1eqn=log(x(1)/x(2))-(DP-osm)/(k*u*Rm);
f=[xA1eqn;yA1eqn];
```

Running the code gives $x_{A1} = 0.184$ and $y_{A1} = 0.044$. Thus, $c_{A1} = 54.3$ g/l and $c_{Aw1} = 227.3$ g/l. It is worth pointing out that an alternative computational approach would have been to obtain an expression for x_{A1} in terms of y_{A1} using Eq. 8.8 and substituting that into Eq. 8.7 to get a single equation in y_{A1}. The reader could try using this approach as a simple exercise. In Problem 8.1 you are asked to tackle this example again but using the method employed in Example 5.1 and taking $c_{\lim} = 227.3$ g/l.

In the above example, the process analysis problem was tackled. Unlike the situation with limiting flux operation, the design calculation in this case is not a simple case of plugging in numbers. The wall concentration must first be determined by solving Eq. 8.8 before A can be computed directly from Eq. 8.7. Problem 8.2 focuses on the design problem.

8.2.2 Multi-stage systems

There is no major complication with analysis and design of multi-stage systems. There are simply more algebraic equations to be solved. Let us consider a two-stage system. The balance for the second stage can be expressed as (see Eq. 5.6)

$$\frac{Q_0}{kA}(x_{A1} - x_{A2}) - \ln\left(\frac{x_{A2}}{y_{A2}}\right) = 0. \tag{8.10}$$

Likewise the OP model equation for the second stage can be written

$$\ln\frac{x_{A2}}{y_{A2}} - \frac{\Delta P - \pi(y_{A2})}{k\mu R_m} = 0. \tag{8.11}$$

The next example illustrates a two-stage design calculation which consists of the solution to a system of *four* non-linear algebraic equations.

Example 8.2 Analysis of a two stage system

Using the same data as Example 8.1, but assuming there are two stages, each of area 1.85 m^2, calculate the exit concentration.

Solution. The MATLAB code for this problem is shown below:

```
function example82
x=fsolve(@twostageop,[0.2,0.04,0.01,0.01])
function f=twostageop(x)
A=1.35;Q0=1e-3/60;k=3.5e-6;u=0.001;Rm=1e12;DP=2e5
cA0=10; a1=454.3;a2=-0.176;a3=0.0082;
% x(1) is xA1, x(2) is yA1, x(3) is xA2 and x(4) is yA2
osm1=a1*cA0/x(2)+a2*cA0^2/x(2)^2+a3*cA0^3/x(2)^3
osm2=a1*cA0/x(4)+a2*cA0^2/x(4)^2+a3*cA0^3/x(4)^3
xA1eqn=Q0/k/A*(1-x(1))-log(x(1)/x(2));
yA1eqn=log(x(1)/x(2))-(DP-osm1)/(k*u*Rm);
xA2eqn=Q0/k/A*(x(1)-x(3))-log(x(3)/x(4));
yA2eqn=log(x(3)/x(4))-(DP-osm2)/(k*u*Rm);
f=[xA1eqn;yA1eqn;xA2eqn;yA2eqn]
```

Running the code gives $x_{A1} = 0.384$, $x_{A2} = 0.113$, $y_{A1} = 0.044$ and $y_{A2} = 0.043$. Thus $c_{A2} = 88.5$ g/l which compares with the 54.4 g/l obtained with the single-stage system of the same total area in Example 8.1. Problem 8.3 involves extending the above analysis to the design of a three-stage system.

8.3   Dynamic modelling of batch UF

Analysis and design of batch systems is probably the only place where one has to give serious consideration as to how the OP model can be incorporated into process calculations. The key difficulty is that, in principle, the wall concentration varies with time through its dependence on the time-varying macrosolute concentration. In this section, the equations for batch UF are formulated.

The process configuration is as outlined in Fig. 5.4. The retentate volume balance is

$$\frac{dV}{dt} = -AJ. \tag{8.12}$$

The macrosolute balance is

$$\frac{d(Vc_A)}{dt} = 0. \tag{8.13}$$

By combining with Eq. 8.12, this becomes

$$\frac{dc_A}{dt} = \frac{c_A^2}{c_{A0}V_0}AJ,$$ (8.14)

where the '0' subscripts denote initial values. The flux can be written as

$$J = k\ln\frac{c_{Aw}}{c_A}.$$ (8.15)

Now, the question is: what do we do with the c_{Aw} term? Of course, Eq. 8.2 must be involved in some way. There are probably a number of computational approaches one can take at this point but the one favoured here is to transform Eq. 8.2 into an ODE by implicit differentiation, as shown in the next two sections.

8.3.1 Batch UF with constant mass transfer coefficient

Let us assume for now that the mass transfer coefficient is constant. Then, implicit differentiation of Eq. 8.2 gives

$$\frac{k}{c_{Aw}}dc_{Aw} - \frac{k}{c_A}dc_A + \frac{1}{\mu R_m}d\pi = 0,$$ (8.16)

where constant pressure operation is assumed. But this can be written

$$\frac{k}{c_{Aw}}dc_{Aw} - \frac{k}{c_A}dc_A + \frac{d\pi/dc_{Aw}}{\mu R_m}dc_{Aw} = 0.$$ (8.17)

Rearranging gives

$$\frac{k}{c_{Aw}}dc_{Aw} + \frac{d\pi/dc_{Aw}}{\mu R_m}dc_{Aw} = \frac{k}{c_A}dc_A.$$ (8.18)

Therefore

$$\frac{dc_{Aw}}{dt} = \frac{k/c_A}{\dfrac{k}{c_{Aw}} + \dfrac{d\pi/dc_{Aw}}{\mu R_m}}\frac{dc_A}{dt}.$$ (8.19)

Thus, Eq. 8.14 and Eq. 8.19 can be used as the basis of the calculations. Of course, in some cases it might be more convenient to work in terms of volumes and in that case we can simply replace dc_A/dt with

$$\frac{dc_A}{dt} = -\frac{c_{A0}V_0}{V^2}\frac{dV}{dt}.$$ (8.20)

The only further complication that needs to be addressed concerns the initial condition for c_{Aw0}. As mentioned in Chapter 5, it is assumed that the concentration polarisation layer is established essentially instantaneously. Therefore, c_{Aw0} is given by solution of the equation

$$k \ln \frac{c_{Aw0}}{c_{A0}} - \frac{\Delta P - \pi(c_{Aw0})}{\mu R_{m}} = 0. \tag{8.21}$$

Application of this model is illustrated in the next example.

Example 8.3 Calculation of batch time with constant mass transfer coefficient

A whey solution is to be concentrated from 14 to 56 g/l. The membrane resistance is 5×10^{11} m^{-1}, the trans-membrane pressure is 4 bar, the membrane area is 1.3 m^2, the mass transfer coefficient is 3×10^{-6} m/s and the initial solution volume is 60 l. Calculate the time required.

Solution. MATLAB code which employs **fsolve** (to evaluate c_{Aw0}) and **ode45** is shown below. Osmotic pressure data are obtained from Table 4.5.

```
function example83
global cA0 cAw0 k DP u Rm V0 A a1 a2 a3
cA0=14;k=3e-6;DP=4e5;u=0.001;Rm=5e11;V0=0.06;A=1.3;
a1=454.3;a2=-0.176;a3=0.0082;
cAw0=fsolve(@initialop,50)
options=odeset('events', @eventfcn);
[t,y]=ode45(@batchufop,[0 10000],[cA0,cAw0],options)
function f=initialop(x)
global cA0 k DP u Rm  a1 a2 a3
osm0=a1*x+a2*x^2+a3*x^3;
f=log(x/cA0)-(DP-osm0)/(k*u*Rm);
function dydt=batchufop(t,y)
global cA0 k u Rm V0 A a1 a2 a3
%y(1) is cA; y(2) is cAw
J=k*log(y(2)/y(1));
dosm=a1+2*a2*y(2)+3*a3*y(2)^2;
ceqn=y(1)^2/cA0/V0*A*J;
cweqn=k/y(1)/(k/y(2)+dosm/(u*Rm))*ceqn;
dydt=[ceqn;cweqn];
function [value,isterminal,direction]=eventfcn(t,y)
value=y(1)-56;
isterminal=1;
direction=0;
```

Running the code gives a batch time of 4561 s. A quick perusal of the code output shows that the wall concentration remains essentially constant at about 320 g/l throughout the run, and therefore the constant wall concentration that is often ascribed to 'gel polarisation' emerges naturally from the non-linear algebraic equations of the osmotic pressure model. No 'gel' formation needs to be hypothesised at all. In Problem 8.4, you are asked to solve this problem again using the methods developed for limiting flux operation.

8.3.2   Batch UF with viscosity dependent mass transfer coefficient

In this, more complicated case, there is a dk term in the implicit differentiation of Eq. 8.2. Doing the differentiation gives

$$\frac{k}{c_{Aw}}dc_{Aw} - \frac{k}{c_A}dc_A + \left(\ln\frac{c_{Aw}}{c_A}\right)dk + \frac{1}{\mu R_m}\frac{d\pi}{dc_{Aw}}dc_{Aw} = 0. \qquad (8.22)$$

Now since

$$dk = \left(\frac{\partial k}{\partial c_A}\right)_{c_{Aw}}dc_A + \left(\frac{\partial k}{\partial c_{Aw}}\right)_{c_A}dc_{Aw}, \qquad (8.23)$$

one gets

$$\frac{dc_{Aw}}{dt} = \frac{\dfrac{k}{c_A} - \left(\dfrac{\partial k}{\partial c_A}\right)_{c_{Aw}}\ln\dfrac{c_{Aw}}{c_A}}{\dfrac{k}{c_{Aw}} + \left(\dfrac{\partial k}{\partial c_{Aw}}\right)_{c_A}\ln\dfrac{c_{Aw}}{c_A} + \dfrac{1}{\mu R_m}\dfrac{d\pi}{dc_{wA}}}\frac{dc_A}{dt}. \qquad (8.24)$$

Assuming an exponential dependence of viscosity on concentration (Eq. 4.21) and incorporating the usual wall correction factor (Eq. 4.11), this becomes

$$\frac{dc_{Aw}}{dt} = \frac{\dfrac{k_c}{c_A} - \left(k_c\gamma z + \dfrac{dk_c}{dc_A}\right)\ln\dfrac{c_{Aw}}{c_A}}{\dfrac{k_c}{c_{Aw}} - k_c\gamma z\ln\dfrac{c_{Aw}}{c_A} + \dfrac{e^{\gamma z(c_{Aw}-c_A)}}{\mu R_m}\dfrac{d\pi}{dc_{Aw}}}\frac{dc_A}{dt}, \qquad (8.25)$$

where k_c is the un-corrected mass transfer coefficient as explained in Section 4.3.

For laminar flow and considering viscosity effects only one gets

$$dk_c/dc_A = 0, \qquad (8.26)$$

and

$$k_c = k_0, \qquad (8.27)$$

where k_0 is the infinite dilution value of the uncorrected mass transfer coefficient. Therefore

$$\frac{dc_{Aw}}{dt} = \frac{\dfrac{1}{c_A} - \gamma z\ln\dfrac{c_{Aw}}{c_A}}{\dfrac{1}{c_{Aw}} - \gamma z\ln\dfrac{c_{Aw}}{c_A} + \dfrac{e^{\gamma z(c_{Aw}-c_A)}}{k_0\mu R_m}\dfrac{d\pi}{dc_{Aw}}}\frac{dc_A}{dt} \qquad (8.28)$$

Use of this model is illustrated in the next example.

Example 8.4 Dynamic simulation incorporating viscosity effects

Repeat Example 8.3 but assume a viscosity dependent mass transfer coefficient with $z = 0.14$ and $\gamma = 0.01$ l/g. Assume $k_0 = 3 \times 10^{-6}$ m/s and laminar tangential flow.

Solution. MATLAB code is shown below. The only real difference between this and the previous example is the use of Eq. 8.28 for the wall concentration and the inclusion of the wall correction factor in the expression for the mass transfer coefficient.

```
function example84
global cA0 cAw0 k0 DP u Rm V0 A a1 a2 a3 g z
cA0=14;k0=3e-6;DP=4e5;u=0.001;Rm=5e11;V0=0.06;A=1.3;
a1=454.3;a2=-0.176;a3=0.0082;g=0.01;z=0.14;
cAw0=fsolve(@initialop,500)
options=odeset('events', @eventfcn);
[t,y]=ode45(@batchufop,[0 10000],[cA0,cAw0],options)
function f=initialop(x)
global cA0 k0 DP u Rm  a1 a2 a3 g z
osm0=a1*x+a2*x^2+a3*x^3;
f=log(x./cA0)-(DP-osm0)/(k0*exp(g*z*(cA0-x))*u*Rm);
function dydt=batchufop(t,y)
global cA0 k0 u Rm V0 A a1 a2 a3 g z
%y(1) is cA; y(2) is cAw
k=k0*exp(g*z*(y(1)-y(2)))
J=k*log(y(2)/y(1));
dosm=a1+2*a2*y(2)+3*a3*y(2)^2;
ceqn=y(1)^2/cA0/V0*A*J;
top=1/y(1)-g*z*log(y(2)/y(1));
bot=1/y(2)-g*z*log(y(2)/y(1))+exp(g*z*(y(2)-
y(1)))/(k0*u*Rm)*dosm;
cweqn=top/bot*ceqn;
dydt=[ceqn;cweqn];
function [value,isterminal,direction]=eventfcn(t,y)
value=y(1)-56;
isterminal=1;
direction=0;
```

Running the code gives a time of 6871 s. The longer time than predicted by Example 8.3 is due to the reduced mass transfer coefficient. In Problem 8.5, you are asked to do a rough calculation to check if the answer to this example is consistent with the answer to Example 8.3.

In the case of turbulent flow, one can write

$$k_c = k_0 e^{-\gamma c_A/2}, \tag{8.29}$$

which implies

$$\frac{dk_c}{dc} = -\frac{\gamma}{2} k_0 e^{-\gamma c/2}. \tag{8.30}$$

Thus, Eq. 8.24 becomes, after a little rearranging,

$$\frac{dc_{Aw}}{dt} = \frac{\dfrac{1}{c_A} - (\gamma z - \gamma/2) \ln \dfrac{c_{Aw}}{c_A}}{\dfrac{1}{c_{Aw}} - \gamma z \ln \dfrac{c_{Aw}}{c_A} + \dfrac{e^{\gamma z(c_{Aw}-c_A)} e^{\gamma c_A/2}}{k_0 \mu R_m} \dfrac{d\pi}{dc_{Aw}}} \frac{dc_A}{dt}. \tag{8.31}$$

This raises no new issues beyond making the algebra and MATLAB coding a little bit more cumbersome, as seen in Problem 8.6.

8.4 Fed-batch ultrafiltration

The fed-batch mode of operation was studied in detail in Chapters 5 and 7 and the basic layout is shown in Fig. 5.6. The initial solution volume is V_0, of which a volume, V_f, is placed in the retentate tank while the remainder is placed in a feed reservoir. The system is operated at constant retentate volume until the feed reservoir is emptied. The balance for the concentration of solute A in the retentate tank can be written

$$V_f \frac{dc_A}{dt} = J A c_{A0}, \tag{8.32}$$

where c_A is the concentration in the retentate tank at any time and c_{A0} is the concentration in the feed. Rearranging for constant k gives

$$t = \frac{V_f}{k A c_{A0}} \int_{c_{A0}}^{c_{Af}} \frac{dc_A}{\ln(c_{Aw}/c_A)} = \frac{V_f \mu R_m}{A c_{A0}} \int_{c_{A0}}^{c_{Af}} \frac{dc_A}{(\Delta P - \pi)}, \tag{8.33}$$

where

$$c_{Af} = \frac{c_{A0} V_0}{V_f}. \tag{8.34}$$

But differentiation of Eq. 8.2 at constant pressure gives

$$dc_A = \frac{c_A}{k} \left(\frac{k}{c_{Aw}} + \frac{d\pi/dc_{Aw}}{\mu R_m} \right) dc_{Aw}, \tag{8.35}$$

which can be written in terms of c_{Aw} only as

$$dc_A = \frac{c_{Aw} e^{-J/k}}{k} \left(\frac{k}{c_{Aw}} + \frac{d\pi/dc_{Aw}}{\mu R_m} \right) dc_{Aw}, \tag{8.36}$$

where J can be written in terms of c_{Aw} using the usual osmotic pressure expression for the flux. Finally, therefore, the time can be expressed as

$$t = \frac{V_f \mu R_m}{A c_{A0}} \int_{c_{Aw0}}^{c_{Awf}} \frac{e^{-J/k}}{\Delta P - \pi} \left(1 + \frac{c_{Aw}}{k} \frac{d\pi/dc_{Aw}}{\mu R_m}\right) dc_{Aw}. \tag{8.37}$$

Use of this equation is illustrated in the next example.

Example 8.5 Time for fed-batch ultrafiltration

A whey solution is to be concentrated from 15 to 90 g/l in a fed-batch operation. The total solution volume is 120 l, of which 20 l is placed in the retentate tank. The system operates at 3 bar, the membrane area is 1.2 m^2, the mass transfer coefficient is 3.0×10^{-6} m/s, the membrane resistance is 4×10^{11} m^{-1}, the trans-membrane pressure is 3 bar and the permeate viscosity is 0.001 Ns/m^2. Osmotic pressure parameters are as given in Table 4.5. Compute the time required for this operation.

Solution. The solution requires numerical integration of the integral in Eq. 8.37. The limits on the integral must be computed first by numerical solution of the governing algebraic equation of the osmotic pressure model. MATLAB code is shown below.

```
function example85
global k DP cA0 cAf a1 a2 a3 u Rm
k=3.0e-6;A=1.2;DP=3e5;u=0.001;Rm=4e11;Vf=0.02;
a1=454;a2=-0.176;a3=0.0082;cA0=15;cAf=90;
guess=250;
a=fzero(@cw0,guess); %find cAW0
b=fzero(@cwf,guess); % fnd CAWf
int=quad(@opint,a,b);
time=Vf*u*Rm/A/cA0*int
function f=cw0(x)
global k DP cA0 a1 a2 a3 u Rm
osm=a1*x+a2*x^2+a3*x^3;
f=log(x/cA0)-(DP-osm)/(u*Rm*k);
function f=cwf(x)
global k DP cAf a1 a2 a3 u Rm
osm=a1*x+a2*x^2+a3*x^3;
f=log(x/cAf)-(DP-osm)/(u*Rm*k);
function f=opint(x)
global k DP a1 a2 a3 u Rm
osm=a1*x+a2*x.^2+a3*x.^3;
dosm=a1+2*a2*x+3*a3*x.^2;
J=(DP-osm)/(u*Rm)
f=(1+x.*dosm/(u*k*Rm))./(DP-osm).*exp(-J/k) % Eq. 8.37
```

Running the code gives an answer of 16 661 s or just over 4.6 hours. In Problem 8.7, you are asked to check this answer by solving the governing ODEs for fed-batch systems.

8.5   Single pass ultrafiltration

In Chapter 5, single pass ultrafiltration (Fig. 5.7) was examined and a key aspect of this problem was the fact that the spatial variation of the solute concentration and tangential flowrate had to be incorporated into the analysis. Since trans-membrane pressure is a key parameter in the osmotic pressure model, it is worth considering the more general situation where this too is spatially dependent.

From basic fluid flow theory, it is well known that when a viscous fluid flows through a channel, it experiences a loss in hydrodynamic pressure. Thus, when a fluid flows though a membrane module, it experiences a loss in pressure that results in a loss in trans-membrane pressure if pressure losses on the permeate side are negligible.

Let us consider UF in a single tubular membrane in which the flow is turbulent. The permeate pressure is essentially zero at all points and thus the fluid pressure in the channel at any point is equal to the trans-membrane pressure at that point. Using a Darcy-type expression [1], the variation in trans-membrane pressure can be expressed as

$$\frac{d\Delta P}{dz} = -\frac{32\rho Q^2 f}{\pi^2 d^5}. \tag{8.38}$$

This assumes, of course, that the porous membrane is behaving like a non-porous tube. In Eq. 8.38, ρ is the fluid density, Q is the tangential flowrate at any position, z, f is the Fanning friction factor and d is the tube diameter. The variation in flowrate can be described by the expression

$$\frac{dQ}{dz} = -\pi d J, \tag{8.39}$$

where J is the permeate flux at that point. The flux can be written as

$$J = \frac{\Delta P - \pi}{\mu R_{\mathrm{m}}}, \tag{8.40}$$

where π is a function of the wall concentration at position, z. A solute balance gives

$$c_A = \frac{c_{A0} Q_0}{Q}, \tag{8.41}$$

where the '0' subscript denotes conditions at the inlet. This leads to

$$\frac{dc_A}{dz} = -\frac{c_{A0} Q_0}{Q^2} \frac{dQ}{dz}. \tag{8.42}$$

Now, differentiating across the osmotic pressure model equation, as done previously, gives

$$\frac{dc_{Aw}}{dz} = \frac{\dfrac{k}{c_A}\dfrac{dc_A}{dz} - \ln\dfrac{c_{Aw}}{c_A}\dfrac{dk}{dz} + \dfrac{1}{\mu R_m}\dfrac{d\Delta P}{dz}}{\dfrac{k}{c_w} + \dfrac{1}{\mu R_m}\dfrac{d\pi}{dc_{Aw}}}, \tag{8.43}$$

where the mass transfer coefficient is assumed to be viscosity independent. Finally, the variation in mass transfer coefficient is given by

$$k = k_0 \left(\frac{Q}{Q_0}\right)^n, \tag{8.44}$$

where n can be taken to be 0.8 and k_0 is the mass transfer coefficient at the inlet. Therefore

$$\frac{dk}{dz} = \frac{nk}{Q}\frac{dQ}{dz}. \tag{8.45}$$

Use of these equations is illustrated in the next example.

Example 8.6 Analysis of a single pass system

A 5 g/l solution of whey undergoes ultrafiltration in a tubular membrane module. The flowrate into each tube is 5 l/min and the inlet pressure is 5 bar. The inlet mass transfer coefficient is 2.5×10^{-5} m/s, the membrane resistance is 5×10^{11} m/s and the permeate viscosity is 0.001 Ns/m^2. The tube diameter is 19 mm and the friction factor can be taken to be constant and equal to 0.008. Osmotic pressure parameters are as given in Example 8.5 and the fluid density can be taken as 1000 kg/m^3. Calculate the exit concentration if each tube is 3.0 m in length.

Solution. This simply requires a numerical solution of the governing system of ODEs (Eqs. 8.38, 8.39 and 8.43) but for that to be done, the inlet wall concentration must be computed in the usual way. MATLAB code is shown below.

```
function example86
format short g
global a1 a2 a3 cA0 DP0 k0 u Rm rho d Q0 fr n
DP0=5e5; k0=2.5e-5;u=0.001;Rm=5e11;L=3.0;a1=454;a2=-
0.176;a3=0.0082;
Q0=5e-3/60; cA0=5;rho=1000;d=0.019;fr=0.008;n=0.8;
guess=250;
cw0=fzero(@cwinit,guess);
[z,y]=ode45(@singlepass,[0 L],[DP0 Q0 cw0])
cA=cA0*Q0./y(:,2)
function f=cwinit(x)
global a1 a2 a3 cA0 DP0 k0 u Rm
osm0=a1*x+a2*x^2+a3*x^3;
```

```
f=log(x/cA0)-(DP0-osm0)/(k0*u*Rm);
function f=singlepass(z,y)
global a1 a2 a3 cA0 k0 u Rm rho d Q0 fr n
%y(1) is DP, y(2) is Q and y(3) is cW
osm=a1*y(3)+a2*y(3)^2+a3*y(3)^3;
dosm=a1+2*a2*y(3)+3*a3*y(3)^2;
k=k0*(y(2)/Q0)^n;
J=(y(1)-osm)/(u*Rm);
cA=cA0*Q0/y(2);
Peqn=-32*rho*y(2)^2*fr/pi^2/d^5;
Qeqn=-pi*d*J;
dkdz=n*k/y(2)*Qeqn
dCAdz=-cA0*Q0/y(2)^2*Qeqn
cweqn=(k/cA*dCAdz-
log(y(3)/cA)*dkdz+Peqn/(u*Rm))/(k/y(3)+dosm/u/Rm)
f=[Peqn;Qeqn;cweqn];
```

Running the code gives an exit concentration of just 6.26 g/l. The output also shows that the variations in pressure drop and wall concentration are negligibly small in this system.

8.6   Constant volume diafiltration

The basic outline of constant volume diafiltration is shown in Fig. 6.5. It was established in Chapter 6 (Eq. 6.23) that when the macrosolute has a rejection coefficient of unity and the microsolute has a rejection coefficient of zero, the time taken for CVD of a volume, V_0, is given by

$$t_{\text{CVD}} = \frac{V_0 \ln(c_{B0}/c_{Bf})}{Ak \ln(c_{Aw}/c_A)}, \tag{8.46}$$

where the 'B' subscript denotes the microsolute. In Chapter 6, the calculation was simple because the wall concentration was given (as c_{lim}). Here, however, it is unknown but can be computed in the usual way with the OP model. This is the subject of Problem 8.8.

8.6.1   Optimisation of UFCVD – constant mass transfer coefficient

Suppose a solution is to be diafiltered to reduce the microsolute concentration from c_{B0} to c_{Bf} and ultrafiltered to increase the macrosolute concentration from c_{A0} to c_{Af}. Then, as shown in Chapter 6, there exists a certain intermediate concentration, c_A, at which the CVD step should be done in order to minimise the process time, where $c_{A0} < c_A < c_{Af}$. The diafiltration time in this case is given by

$$t_{\text{CVD}} = \frac{V \ln(c_{B0}/c_{Bf})}{Ak \ln(c_{Aw}/c_A)}, \tag{8.47}$$

where V is the solution volume when CVD is begun. For a macrosolute rejection coefficient of unity, this becomes

$$t_{CVD} = \frac{V_0 c_{A0} \ln(c_{B0}/c_{Bf})}{A k c_A \ln(c_{Aw}/c_A)}. \tag{8.48}$$

The diafiltration time, and hence the total UFCVD time is minimised when

$$\frac{d}{dc_A}[c_A \ln(c_{Aw}/c_A)] = 0. \tag{8.49}$$

Differentiating gives

$$c_A \left(\frac{1}{c_{Aw}} dc_{Aw} - \frac{1}{c_A} dc_A \right) + \ln \left(\frac{c_{Aw}}{c_A} \right) dc_A = 0. \tag{8.50}$$

Rearranging gives

$$\ln \frac{c_{Aw}}{c_A} + \frac{c_A}{c_{Aw}} \frac{dc_{Aw}}{dc_A} = 1. \tag{8.51}$$

This obviously reduces to the classic result, Eq. 6.30, when $c_{Aw} = c_{\lim} = $ constant. Now as shown earlier, we can use Eq. 8.2 to find

$$\frac{dc_{Aw}}{dc_A} = \frac{1/c_A}{1/c_{Aw} + (d\pi/c_{Aw})/(\mu R_m k)}. \tag{8.52}$$

Combining Eqs. 8.51 and 8.52 gives

$$\ln \frac{c_{Aw}}{c_A} + \frac{1}{1 + c_{Aw}(d\pi/c_{Aw})/(\mu R_m k)} = 1. \tag{8.53}$$

But, using Eq. 8.2, this can be written in terms of c_{Aw} only as

$$\frac{\Delta P - \pi}{\mu R_m k} + \frac{1}{1 + c_{Aw}(d\pi/c_{Aw})/(\mu R_m k)} = 1. \tag{8.54}$$

This equation can be solved numerically to give c_{Aw}^{opt}, while the optimum value of c_A can be obtained from

$$c_A^{opt} = c_{Aw}^{opt} \exp^{-J/k}. \tag{8.55}$$

The next example illustrates this approach.

Example 8.7 Optimum concentration for CVD

A whey solution undergoes UFCVD where the membrane resistance is 5×10^{11} m^{-1}, the trans-membrane pressure is 4 bar, the mass transfer coefficient is 3×10^{-6} m/s and the permeate viscosity is 0.001 Ns/m^2. Osmotic pressure parameters are as given in Table 4.5. Compute the optimum concentration for diafiltration.

Solution. MATLAB code employing **fsolve** to solve Eq. 8.54 is shown below.

```
function example87
global DP u Rm  a1 a2 a3 osm k
```

```
k=3e-6;DP=4e5;u=0.001;Rm=5e11;
a1=454.3;a2=-0.176;a3=0.0082;
cAw=fsolve(@cvdopt,250)
J=(DP-osm)/(u*Rm);
cA=cAw*exp(-J/k)
function f=cvdopt(x)
global DP u Rm  a1 a2 a3 osm k
osm=a1*x+a2*x^2+a3*x^3;
dosm=a1+2*a2*x+3*a3*x^2;
f=(DP-osm)/(u*k*Rm)+1/(1+x*dosm/(u*Rm*k))-1
```

Running the code gives an optimum concentration of 118.23 g/l. Note that the optimum value of the wall concentration is 320.85 g/l, which means that Eq. 6.31 is pretty accurate (320.85/e = 118.03 g/l) as long as one uses the computed value of the wall concentration.

8.6.2 Optimisation of UFCVD – viscosity dependent mass transfer coefficient

If the mass transfer coefficient is not constant, Eq. 8.49 generalises to

$$\frac{d}{dc_A}[c_A k \ln(c_{Aw}/c_A)] = 0. \tag{8.56}$$

For laminar flow and assuming the usual exponential dependence of viscosity on concentration, this becomes

$$\frac{d}{dc_A}\left[c_A e^{\gamma z(c_A - c_{Aw})} \ln(c_{Aw}/c_A)\right] = 0. \tag{8.57}$$

Differentiating gives

$$\frac{dc_{Aw}}{dc_A}\left(\frac{c_A}{c_{Aw}} - \gamma z c_A \ln\frac{c_{Aw}}{c_A}\right) + (1 + \gamma z c_A)\ln\frac{c_{Aw}}{c_A} = 1. \tag{8.58}$$

Now using Eq. 8.28 for dc_{Aw}/dc_A, this becomes a sizeable equation involving both c_A and c_{Aw}. The other equation required is Eq. 8.2, which in this case can be written

$$e^{\gamma z(c_A - c_{Aw})} \ln\frac{c_{Aw}}{c_A} - \frac{\Delta P}{\mu R_m k_0} = 0, \tag{8.59}$$

where J is computed using the osmotic pressure expression. The solution of Eqs. 8.58 and 8.59 is shown in the next example.

Example 8.8 UFCVD optimisation with viscosity dependent mass transfer coefficient

A whey solution undergoes UFCVD where the membrane resistance is 5×10^{11} m^{-1}, the trans-membrane pressure is 4 bar, the uncorrected mass transfer coefficient is 3×10^{-6} m/s and the permeate viscosity is 0.001 Ns/m^2. The flow is laminar, $z = 0.14$ and $\gamma = 0.01$ l/g. Osmotic pressure parameters as as given in Example 8.7. Compute the optimum concentration for diafiltration.

Solution. MATLAB code to solve Eq. 8.58 is shown below.

```
function example88
global DP u Rm  a1 a2 a3 k0 g z
k0=3e-6;DP=4e5;u=0.001;Rm=5e11;g=0.01;z=0.14
a1=454.3;a2=-0.176;a3=0.0082;
x=fsolve(@cvdopt,[100,500])
function f=cvdopt(x)
global DP u Rm  a1 a2 a3 osm k0 g z
%x(1) is cA, x(2) is cAw
osm=a1*x(2)+a2*x(2)^2+a3*x(2)^3;
dosm=a1+2*a2*x(2)+3*a3*x(2)^2;
top=1/x(1)-g*z*log(x(2)/x(1))
bot=1/x(2)-g*z*log(x(2)/x(1))+exp(g*z*(x(2)-
x(1)))/(k0*u*Rm)*dosm
dcwdc=top/bot  % Eq. 8.28
Eqn1=dcwdc*(x(1)/x(2)-
g*z*x(1)*log(x(2)/x(1)))+(1+g*z*x(1))*log(x(2)/x(1))-1
Eqn2=exp(g*z*(x(1)-x(2)))*log(x(2)/x(1))-(DP-
osm)/(u*Rm*k0)
f=[Eqn1;Eqn2]
```

Running the code gives an optimum concentration of 139.09 g/l compared with the 118.23 g/l found when viscosity effects are ignored. The shift to higher concentrations under laminar flow conditions is consistent with findings in Chapter 6. It is once again interesting to note that if the dc_{Aw}/dc_A term is ignored, one gets an optimum concentration of 139.0 g/l. In Problem 8.9, you are asked to carry out a similar optimisation for the turbulent flow situation.

8.7   Ultrafiltration with constant wall concentration

All the analyses so far have focused on constant pressure operation. Let us now consider the case of batch UF with variable pressure. However, while the pressure is varying, it is assumed that the wall concentration is kept constant. This mode of operation has been studied by some researchers and has the advantage of making it easier to quantify and control product losses [2]. Recall that the concentration of product in the permeate is determined by the wall concentration and the intrinsic rejection coefficient. Keeping the wall concentration low minimises product losses but this means that low fluxes are obtained. Ultimately, a compromise must be reached.

Superficially, analysis of constant wall operation of batch UF proceeds in the same manner as limiting flux analysis of that system. However, the limiting flux model made no mention of trans-membrane pressure, assuming that such changes were irrelevant to the analysis. However, with the OP model analysis, further information is gained on changes in the trans-membrane pressure.

In a batch operation, Eq. 8.14 describes the variation of the retentate concentration where, now, the wall concentration is constant. Now, since the flux is given by Eq. 8.15, this means that as the retentate concentration increases, the flux declines. Thus, the trans-membrane pressure declines because

$$\Delta P = \mu R_{\mathrm{m}} J + \pi, \tag{8.60}$$

where π is fixed at a value given by the constant wall concentration. This is illustrated in the next example.

Example 8.9 Batch UF at constant wall concentration

A 60 l volume of a 25 g/l whey solution is to undergo batch UF at constant wall concentration. The initial trans-membrane pressure is set at 2 bar, the mass transfer coefficient is constant and equal to 1×10^{-5} m/s, the membrane area is 1.0 m^2, the membrane resistance is 1×10^{12} m/s, the permeate viscosity is 0.001 Ns/m^2 and the usual virial coefficients apply for whey (Table 4.5). Generate a plot of trans-membrane pressure versus time covering a range of 5400 s.

Solution. The first task is to find the value of the wall concentration. This is done by numerical solution of Eq. 8.2 in the usual way. Once $c_{A\mathrm{w}}$ is computed, Eq. 8.14 can be solved numerically and Eq 8.60 can be used to compute the pressure at each time. MATLAB code which employs **fsolve** and **ode45** is shown below.

```
function example89
global k cAw u Rm cA0 a1 a2 a3 osm DP0 V0 A
Rm=1e12;V0=0.06;A=1.0;k=1e-5;u=0.001;cA0=25;DP0=2e5;
a1=454.3;a2=-0.176;a3=0.0082;
cAw=fsolve(@findcAw,500)
[t,cA]=ode45(@constwall,[0 5400],[cA0]);
J=k*log(cAw./cA);
DP=(u*Rm*J+osm)/1e5;
plot(t,DP,'-k'); xlabel ('Time (s)');ylabel('\DeltaP (bar)');
function f=findcAw(x)
global k u Rm cA0 a1 a2 a3 osm DP0
osm=a1*x+a2*x^2+a3*x^3;
f=log(x/cA0)-(DP0-osm)/(k*u*Rm);
function dcAdt=constwall(t,cA)
global k cAw cA0 a1 a2 a3 V0 A
dcAdt=cA^2/cA0/V0*A*k*log(cAw/cA);
```

The resulting plot is shown in Fig. 8.1, and illustrates the decline in pressure that arises.

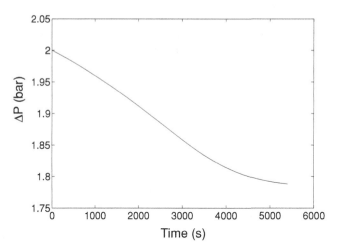

Figure 8.1 Variation of trans-membrane pressure during constant wall batch UF.

The above simulation would, in theory, yield the trajectory of the set-point for pressure-based control of constant wall operation. An alternative approach would be to base the control not on trans-membrane pressure trajectory but on the flux trajectory and this is the subject of Problem 8.10.

8.8   The osmotic pressure model with incomplete rejection

Equation 8.1 is a statement of the osmotic pressure model with incomplete rejection, and it brings us back to the whole question of rejection and rejection coefficients. Recall that there have been three rejection coefficients presented so far. These are:

$$\sigma = 1 - \frac{c_p}{c}, \tag{8.61}$$

$$\sigma_i = 1 - \frac{c_p}{c_w}, \tag{8.62}$$

where σ is the *apparent* rejection coefficient, σ_i is the *intrinsic* rejection coefficient, c_p is the solute concentration in the permeate, c is the solute concentration in the retentate and c_w is the solute concentration at the membrane (wall). The third coefficient is the *osmotic reflection coefficient*, σ_o, defined by the flux expression

$$J = \frac{\Delta P - \sigma_o \Delta \pi}{\mu R_m}. \tag{8.63}$$

This is a somewhat mysterious parameter as it is not at all obvious why it is needed. To understand where it comes from, one must delve into some quite fundamental membrane science, specifically the role of diffusion and convection in the transport of solvent and solute through membranes.

The basic picture of transport in membranes is that the transport of the solute occurs via both hindered diffusion and hindered convection and is coupled with the transport of the solvent. There are a number of ways of formulating the underlying equations but the following is probably the easiest to understand. Let us retain the notation, J, to represent the solvent flux and let J_s represent the solute flux. Then, these fluxes can be expressed as follows [3]:

$$J = L_p(\Delta P - (1 - K_c)\Delta\pi),\tag{8.64}$$

$$J_s = \omega\Delta\pi + K_c c_m J,\tag{8.65}$$

where L_p and ω are constants. In the case of the *solute* flux, Eq. 8.65, the first part of the expression is the diffusion term (diffusion is ultimately driven by chemical potentials which are related to osmotic pressures), while the second is the convection term. The parameter, c_m, represents the average concentration of the solute in the membrane itself. The constant, K_c, is the convective hindrance factor, first mentioned in Eq. 7.4. Perhaps surprisingly, this factor also appears in the expression for the solvent flux and this is an indication of the coupling between solvent and solute flux. If $K_c = 1$, for example, the solute is completely un-hindered and there is no osmotic pressure acting to oppose the solvent flux. If $K_c = 0$, the solvent 'sees' all of the osmotic pressure difference. Clearly, K_c is related to the osmotic reflection coefficient by the expression

$$\sigma_o = 1 - K_c.\tag{8.66}$$

Thus, the osmotic reflection coefficient is a measure of the extent to which a solute is hindered in its convective flow through the membrane pores.

Now, from Eq. 7.4, the membrane Peclet number can be written in terms of the osmotic reflection coefficient by the expression

$$Pe_m = \frac{J\delta_m}{D_{\text{eff}}}(1 - \sigma_o),\tag{8.67}$$

where D_{eff} is the effective diffusion coefficient of the solute in the membrane. It can be shown that the intrinsic rejection coefficient and the osmotic reflection coefficient are related by [3, 4]

$$\sigma_i = \frac{\sigma_o(1 - e^{-Pe_m})}{1 - \sigma_o e^{-Pe_m}}.\tag{8.68}$$

Assuming a constant reflection coefficient, the intrinsic rejection is a complex, and increasing, function of the flux. Thankfully, it is usually reasonable to assume in UF that convection is the dominant transport mechanism in the membrane, i.e. $Pe_m \gg 1$. In that limit, it is clear (and this was mentioned in Chapter 7) that the intrinsic rejection coefficient is constant. It also means, on the basis of Eq. 8.68, that the intrinsic rejection coefficient and the osmotic reflection coefficient become identical, i.e.,

$$\sigma_i = \sigma_o.\tag{8.69}$$

This makes things a bit more manageable and one can begin to think about using the general osmotic pressure model for process analysis and design. In the next three sections, some UF and DF process calculations are tackled using the OP model with incomplete rejection. To keep things from getting excessively complicated, the mass transfer coefficient is assumed constant in all cases.

8.8.1   Single-stage feed-and-bleed UF

To illustrate use of the OP model with incomplete rejection, the starting point is a simple, single-stage continuous feed-and-bleed system. The intrinsic rejection coefficient, σ_i, appears in the osmotic pressure expression for the flux and thus the entire problem is formulated in terms of this parameter, rather than the apparent rejection coefficient.

Using the same notation as in Section 8.2.1, the mass balances for the system work out as (Problem 8.11)

$$Q_0(1 - x_{A1}) - JA\left(1 - \frac{x_{A1}}{y_{A1}}(1 - \sigma_{Ai})\right) = 0. \tag{8.70}$$

Now, assuming a constant mass transfer coefficient, the flux can be written as (Problem 8.12)

$$J = k \ln\left(\frac{c_{Aw1}\sigma_{Ai}}{c_{A1} - c_{Aw1}(1 - \sigma_{Ai})}\right) = k \ln\left(\frac{\sigma_{Ai}}{y_{A1}/x_{A1} - (1 - \sigma_{Ai})}\right). \tag{8.71}$$

Thus, Eq. 8.70 becomes

$$\frac{Q_0}{kA}(1 - x_{A1}) - \left(1 - \frac{x_{A1}}{y_{A1}}(1 - \sigma_{Ai})\right)\ln\left(\frac{\sigma_{Ai}}{y_{A1}/x_{A1} - (1 - \sigma_{Ai})}\right) = 0. \tag{8.72}$$

At the same time, the OP model can be written

$$\ln\left(\frac{\sigma_{Ai}}{y_{A1}/x_{A1} - (1 - \sigma_{Ai})}\right) - \frac{\Delta P - \sigma_{Ai}\Delta\pi}{k\mu R_m}, \tag{8.73}$$

where

$$\Delta\pi = \pi(y_{A1}) - \pi(y_{Ap1}), \tag{8.74}$$

with

$$\pi(y_{A1}) = \left(a_1 c_{A0} y_{A1}^{-1} + a_2 c_{A0}^2 y_{A1}^{-2} + a_3 c_{A0}^3 y_{A1}^{-3}\right), \tag{8.75}$$

and

$$\pi(y_{Ap1}) = a_1 c_{A0} y_{A1}^{-1}(1 - \sigma_{Ai}) + a_2 c_{A0}^2 y_{A1}^{-2}(1 - \sigma_{Ai})^2 + a_3 c_{A0}^3 y_{A1}^{-3}(1 - \sigma_{Ai})^3. \tag{8.76}$$

Use of this model is illustrated in the next example.

Example 8.10   Single-stage analysis with incomplete rejection

This example is an extension of Example 8.1. A 1 l/min feed of a whey solution enters a single-stage feed-and-bleed UF system. The feed enters at 10 g/l, the mass transfer coefficient is 3.5×10^{-6} m/s and the membrane area is 2.7 m². Calculate the exit concentration. Use osmotic pressure data for whey from Table 4.5. The membrane resistance is 1×10^{12} m^{-1}, the permeate viscosity is 0.001 Ns/m^2 and the trans-membrane pressure is 2 bar. The *intrinsic* rejection coefficient is 0.95.

Solution. This problem requires use of the MATLAB function **fsolve** as illustrated below. The initial guesses are the results from Example 8.1.

```
function example810
x=fsolve(@singlestageop, [0.18,0.04])
function f=singlestageop(x)
A=2.7;Q0=1e-3/60;k=3.5e-6;u=0.001;Rm=1e12;DP=2e5
cA0=10; a1=454.3;a2=-0.176;a3=0.0082; si=0.95;
%x(1) is xA1, x(2) is yA1
osm1=a1*cA0/x(2)+a2*cA0^2/x(2)^2+a3*cA0^3/x(2)^3
osmp= a1*cA0/x(2)*(1-si)+a2*cA0^2/x(2)^2*(1-
si)^2+a3*cA0^3/x(2)^3*(1-si)^3
xA1eqn=Q0/k/A*(1-x(1))-(1-x(1)/x(2)*(1-
si))*log(si/(x(2)/x(1)-(1-si)));
yA1eqn=log(si/(x(2)/x(1)-(1-si)))-(DP-si*(osm1-
osmp))/(k*u*Rm);
f=[xA1eqn;yA1eqn]
```

Running the code gives $x_{A1} = 0.212$ and $y_{A1} = 0.042$. Thus, $c_{A1} = 47.17$ g/l and $c_{Aw1} = 238.1$ g/l. Thus the concentration of A leaving the stage is slightly less than one would have obtained in the case of complete rejection where $x_{A1} = 0.184$, as obtained in Example 8.1. In Problem 8.13, you are required to work out the apparent rejection coefficient from the results of this example.

8.8.2 Dynamic modelling of batch UF

In this case, Eqs. 8.12 and 8.14 apply once more. The flux is given in terms of the intrinsic rejection coefficient as follows:

$$J = k \ln \left(\frac{c_{Aw}\sigma_{Ai}}{c_A - c_{Aw}(1 - \sigma_{Ai})} \right). \tag{8.77}$$

Now, defining c_{Ap} as the permeate concentration, the analogous expression to Eq. 8.17 becomes after some considerable algebra

$$\frac{k dc_{Aw}}{c_{Aw}} - \frac{k(dc_A - (1 - \sigma_{Ai})dc_{Aw})}{c_A - c_{Aw}(1 - \sigma_{Ai})}$$
$$+ \frac{\sigma_{Ai}}{\mu R_m} \left(\frac{d\pi}{dc_{Aw}} dc_{Aw} - (1 - \sigma_{Ai}) \frac{d\pi}{dc_{Ap}} dc_{Aw} \right) = 0. \tag{8.78}$$

Rearranging gives

$$\frac{dc_{Aw}}{dt} = \frac{k/(c_A - c_{Aw}(1 - \sigma_{Ai}))}{\left(\dfrac{k}{c_{Aw}} + \dfrac{k(1 - \sigma_{Ai})}{c_A - c_{Aw}(1 - \sigma_{Ai})} + \dfrac{\sigma_{Ai}}{\mu R_m}\left(\dfrac{d\pi}{dc_{Aw}} - (1 - \sigma_{Ai})\dfrac{d\pi}{dc_{Ap}}\right)\right)} \frac{dc_A}{dt},$$

(8.79)

where

$$\frac{d\pi}{dc_{Aw}} = a_1 + 2a_2 c_{Aw} + 3a_3 c_{Aw}^2,$$

(8.80)

and

$$\frac{d\pi}{dc_{Ap}} = a_1 + 2a_2 (1 - \sigma_{Ai}) c_{Aw} + 3a_3 (1 - \sigma_{Ai})^2 c_{Aw}^2.$$

(8.81)

Use of this model is illustrated in the next example. Note that the initial wall concentration must be computed by solution of the appropriate NLAE.

Example 8.11 Dynamic modelling of batch UF with incomplete rejection

A whey solution is to be concentrated from 14 to 56 g/l. The membrane resistance is 5×10^{11} m^{-1}, the trans-membrane pressure is 4 bar, the membrane area is 1.3 m^2, the mass transfer coefficient is 3×10^{-6} m/s and the initial solution volume is 60 l. The intrinsic rejection coefficient is 0.98. Calculate the time required and plot the *apparent* rejection coefficient versus time.

Solution. MATLAB code which employs **fsolve** (to evaluate c_{Aw0}) and **ode45** is shown below. The reader should derive the c_A balance (ceqn) as an exercise.

```
function example811
global cA0 cAw0 k DP u Rm V0 A a1 a2 a3 si
cA0=14;k=3e-6;DP=4e5;u=0.001;Rm=5e11;V0=0.06;A=1.3;
a1=454.3;a2=-0.176;a3=0.0082; si=0.98
cAw0=fsolve(@initialop,28)
options=odeset('events', @eventfcn);
[t,y]=ode45(@batchufop,[0 10000],[cA0,cAw0 V0],options);
s=1-y(:,2)*(1-si)./y(:,1) %apparent rejection coefficient
plot(t,s,'-k')
xlabel('Time (s)');ylabel('\sigma_A')
function f=initialop(x)
global cA0 k DP u Rm  a1 a2 a3 si
osmw0=a1*x+a2*x^2+a3*x^3;
cp=x*(1-si);
osmp0=a1*cp+a2*cp^2+a3*cp^3;
J=(DP-si*(osmw0-osmp0))/(u*Rm);
f=log(si*x/(cA0-x*(1-si)))-J/k;
function dydt=batchufop(t,y)
```

```
global cA0 k u Rm V0 A a1 a2 a3 si
%y(1) is cA; y(2) is cAw; y(3) is V
J=k*log(si*y(2)/(y(1)-y(2)*(1-si)));
cp=(1-si)*y(2);
dosmw=a1+2*a2*y(2)+3*a3*y(2)^2;
dosmp=a1+2*a2*cp+3*a3*cp^2;
ceqn=J*A*(y(1)-y(2)*(1-si))/y(3)
top=k/(y(1)-(1-si)*y(2));
bot=k/y(2)+k*(1-si)/(y(1)-(1-
si)*y(2))+si/u/Rm*(dosmw-(1-si)*dosmp);
cweqn=top/bot*ceqn;
Veqn=-J*A
dydt=[ceqn;cweqn;Veqn]
function [value,isterminal,direction]=eventfcn(t,y)
value=y(1)-56;
isterminal=1;
direction=0;
```

Running the code gives a batch time of 4370 s. The variation of the apparent rejection coefficient with time is shown in Fig. 8.2. Note that if the program is run with $dc_{Aw}/dt = 0$, one gets a time of 4373 s, showing that the variation of wall concentration with time is small and insignificant.

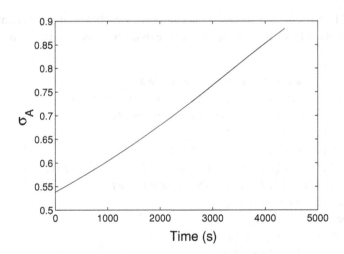

Figure 8.2 Apparent rejection coefficient versus time during batch UF.

8.8.3 Constant volume diafiltration

Let us consider a volume, V_0, of solution. It is assumed that the microsolute (B) has an apparent rejection coefficient of *zero*. The macrosolute has an intrinsic rejection coefficient of σ_{Ai}. The balances for the system are:

$$\frac{dV_w}{dt} = JA,$$ (8.82)

$$\frac{dc_A}{dt} = -\frac{JA}{V_0}c_{Aw}(1 - \sigma_{Ai}).$$ (8.83)

The variation of the wall concentration is given by Eq. 8.79, and CVD stops when (Eq. 6.21)

$$V_w = V_0 \ln \left(\frac{c_{B0}}{c_{Bf}}\right).$$ (8.84)

Computation of the diafiltration time is best done by solving this system of differential equations as shown below.

Example 8.12 Calculation of diafiltration time

A 25 g/l whey solution is to undergo CVD to reduce the concentration of a salt impurity by a factor of 5. The membrane resistance is 5×10^{11} m^{-1}, the trans-membrane pressure is 4 bar, the membrane area is 1.4 m^2, the mass transfer coefficient is 3×10^{-6} m/s and the solution volume is 60 l. The intrinsic rejection coefficient of the macrosolute is 0.99. Calculate the time required and plot the *macrosolute* concentration versus time. Use the same osmotic pressure parameters as in the previous example.

Solution. MATLAB code to implement the model is shown below. It is very similar to the code in Example 8.11 with small changes to the dydt function and the event function.

```
function example812
global cA0 cAw0 k DP u Rm V0 A a1 a2 a3 si
cA0=14;k=3e-6;DP=4e5;u=0.001;Rm=5e11;V0=0.06;A=1.4;
a1=454.3;a2=-0.176;a3=0.0082; si=0.99;Vw0=0;
cAw0=fsolve(@initialop,100)
options=odeset('events', @eventfcn);
[t,y]=ode45(@cvdop,[0 10000],[cA0,cAw0,Vw0],options)
s=1-y(:,2)*(1-si)./y(:,1); %apparent rejection
coefficient, see Prob. 8.17
plot(t,y(:,1),'-k');
xlabel('Time (s)');ylabel('c_A (g/L)')
function f=initialop(x)
global cA0 k DP u Rm  a1 a2 a3 si
osmw0=a1*x+a2*x^2+a3*x^3;
cp=x*(1-si);
osmp0=a1*cp+a2*cp^2+a3*cp^3;
J=(DP-si*(osmw0-osmp0))/(u*Rm);
f=log(si*x/(cA0-x*(1-si)))-J/k;
function dydt=cvdop(t,y)
global cA0 k u Rm V0 A a1 a2 a3 si
```

```
%y(1) is cA; y(2) is cAw; y(3) is Vw
J=k*log(si*y(2)/(y(1)-y(2)*(1-si)));
cp=(1-si)*y(2);
dosmw=a1+2*a2*y(2)+3*a3*y(2)^2;
dosmp=a1+2*a2*cp+3*a3*cp^2;
ceqn=-J*A*y(2)/V0*(1-si);
top=k./(y(1)-(1-si)*y(2));
bot=k./y(2)+k*(1-si)./(y(1)-(1-
si)*y(2))+si/u/Rm.*(dosmw-(1-si)*dosmp);
cweqn=top./bot*ceqn;
Vweqn=J*A;
dydt=[ceqn;cweqn;Vweqn];
function [value,isterminal,direction]=eventfcn(t,y)
global V0;
Vwtot=V0*log(5);
value=y(3)-Vwtot;
isterminal=1;
direction=0;
```

The time required is 6271 s and the plot of the macrosolute concentration versus time is shown in Fig. 8.3, illustrating the loss of product as a result of incomplete rejection.

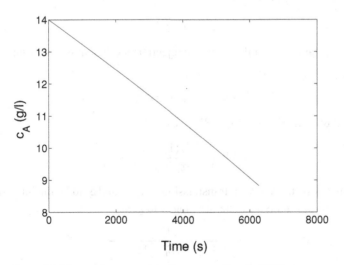

Figure 8.3 Macrosolute concentration versus time in CVD.

8.9 Product transmission in CFMF

At this stage, we are almost finished with the coverage of microfiltration and ultrafiltration processes, at least for situations where membrane fouling is negligible. One process that requires knowledge of both of these separation techniques is soluble product recovery in CFMF. In this section, a very simple model of product recovery is presented. It draws on knowledge gained in the book so far.

Since this is a problem involving cake formation of particles, the following equations, outlined in Chapters 2 and 3, are used to describe flux behaviour in this process:

$$J = \frac{\Delta P}{\mu(R_m + \alpha M)}, \tag{8.85}$$

$$\frac{dM}{dt} = cJ - k_r \tau_w M, \tag{8.86}$$

where c is the solids concentration, α is the specific cake resistance and k_r is the cake removal constant. This clearly assumes that cake formation proceeds in a fashion that is independent of the solute and that the flux is determined entirely by the microfiltration aspect of the problem. Solute transmission then follows from cake formation. Now let c_s be the solute concentration in the retentate and let c_{sc} be the solute concentration at the cake surface. Concentration polarisation theory predicts

$$J = k \ln \frac{c_{sc} - c_{sp}}{c_s - c_{sp}}, \tag{8.87}$$

where c_{sp} is the solute concentration in the permeate. Now, let c_{sw} be the solute concentration at the membrane. Then, assuming a depth filtration model for solute capture in the cake (see Chapter 2), one gets

$$c_{sw} = c_{sc} e^{-\lambda M}, \tag{8.88}$$

where λ is a constant. Now, if the instrinsic rejection coefficient for the solute is σ_{si}, one gets

$$c_{sp} = c_{sw}(1 - \sigma_{si}). \tag{8.89}$$

Putting Eqs. 8.88 and 8.89 into Eq. 8.87 gives

$$J = k \ln \left(\frac{e^{\lambda M}/(1 - \sigma_{si}) - 1}{c_s/c_{sp} - 1} \right). \tag{8.90}$$

Now defining S to be the apparent transmission, c_{sp}/c_s, and S_i to be the intrinsic sieving coefficient, c_{sp}/c_{sw}, Eq. 8.90 can be rearranged to give

$$S = \frac{1}{1 + (e^{\lambda M}/S_i - 1)e^{-J/k}}. \tag{8.91}$$

The dynamics of product transmission is examined in the next example.

Example 8.13 Product transmission in CFMF

A 5 g/l biological suspension undergoes CFMF to recover a soluble product. The specific resistance of cake formed by the suspension is 1×10^{14} m/kg, the membrane resistance is 4×10^{10} m^{-1}, the trans-membrane pressure is 1.5 bar, the permeate viscosity is 0.001 Ns/m^2, λ is 100 m^2/kg and the product $k_r \tau_w$ is 0.005 s^{-1}. The mass transfer coefficient is 1×10^{-5} m/s and the intrinsic sieving coefficient for the solute is 0.9. Generate a plot of apparent transmission, S, versus time for the range 0–500 s.

Solution. MATLAB code using ode45 is shown below.

```
function example813
global DP u Rm alpha
Si=0.9;lam=100;k=1e-5;alpha=1e14;
[t,M]=ode45(@trans,[0 500],[0]);
J=DP/u./(Rm+alpha*M);
S=1./(1+(exp(lam*M)/Si-1).*exp(-J/k));
plot(t,S); xlabel('Time (s)');ylabel('S (-)');
function dydt=trans(t,M)
global DP u Rm alpha
c=5;DP=1.5e5;krtauw=0.005;u=0.001;Rm=4e10;
J=DP/u/(Rm+alpha*M);
dydt=c*J-krtauw*M;
```

The plot of transmission versus time is shown in Fig. 8.4.

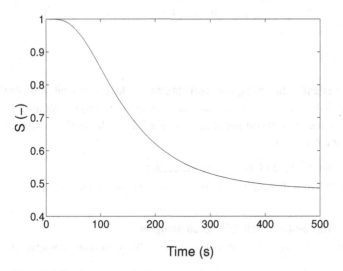

Figure 8.4 Product transmission versus time.

8.10   Conclusions

This chapter has focused on application of the OP model to UF and DF process design and analysis calculations. There is very little precedent for these calculations, so this chapter can be seen, to some extent, as a brief journey into UF and DF research. Although the calculations are more awkward than those that arise when the limiting flux is used, there are no major computational challenges as long as one formulates the problems in a way that is suitable for routine solution with MATLAB.

References

1. Wilkes, O.J. (2005). *Fluid Mechanics for Chemical Engineers with Microfluidics and CFD*, 2nd Edition. Prentice Hall, New Jersey, USA.
2. Van Reis, R., Goodrich, E.M., Yson, C.L., *et al.* (1997). Constant Cw ultrafiltration process control. *Journal of Membrane Science*, **130**, 123–140.
3. Spiegler, K. and Kedem, O. (1966). Thermodynamics of hyperfiltration (reverse osmosis). *Desalination*, **1**, 311–326.
4. Zeman, L.J. and Zydney, A.L. (1996). *Microfiltration and Ultrafiltration. Principles and Applications*. Marcel Dekker, New York, USA.

Additional reading

Foley, G. and Garcia, J. (2000). Ultrafiltration flux theory based on viscosity and osmotic effects: application to diafiltration optimisation. *Journal of Membrane Science*, **176**, 55–61.

Problems

Problem 8.1   Effect of assuming constant wall concentration in feed-and-bleed analysis
Repeat Example 8.1 but use limiting flux methods as employed in Example 5.1. Assume that the wall concentration computed in Example 5.1 is effectively the limiting concentration in your calculations.

Problem 8.2   Single-stage feed-and-bleed design
Using the data from Example 8.1, solve the equivalent design problem assuming $x_{A1} = 0.4$.

Problem 8.3   Three-stage feed-and-bleed analysis
Using the data of Example 8.2, find the exit concentration if *three* stages of 0.9 m^2 are employed.

Problem 8.4   Effect of assuming constant c_w in batch time calculation
Repeat Example 8.3 but use the limiting flux approach of Chapter 5 and take $c_{\lim} = 320$ g/l.

Problem 8.5   Changes in the mass transfer coefficient due to concentration changes
Make a rough estimate of the *average* mass transfer coefficient in Example 8.4. By comparing this value with the infinite dilution value (the value used in Example 8.3), check whether the answer to Example 8.4 is consistent with the answer to Example 8.3.

Problem 8.6   Calculation of batch time in turbulent flow

Repeat Example 8.1 but assume turbulent flow. Assume $k_0 = 2 \times 10^{-5}$ and all other parameters as before.

Problem 8.7   Fed-batch time from the governing ODEs

Check the answer to Example 8.5 by solving the governing ordinary differential equations.

Problem 8.8   Calculation of CVD time

A 20 l volume of a 25 g/l whey solution is to be diafiltered at constant volume to reduce the impurity concentration from 1 to 0.1 g/l. The system is operated at 3 bar with a constant mass transfer coefficient of 1×10^{-5} m/s, a permeate viscosity of 0.001 Ns/m^2 and a membrane resistance of 1×10^{12} m/s. The membrane area is 0.9 m^2. Calculate the time required for this operation. Use osmotic pressure data for whey from Table 4.5.

Problem 8.9   Calculation of CVD time with viscosity dependent mass transfer coefficient

Repeat Example 8.8 with viscosity-dependent mass transfer coefficient and *turbulent* flow.

Problem 8.10   Flux changes during constant wall operation

Using Example 8.9 as a basis, generate the flux trajectory for constant wall operation.

Problem 8.11   Solute balance for single-stage feed-and-bleed

Derive Eq. 8.70.

Problem 8.12   Flux expression written in terms of the intrinsic rejection coefficient

Write the flux as written in Eq. 8.71.

Problem 8.13   Computation of apparent rejection coefficient in single-stage feed-and-bleed

Determine the apparent rejection coefficient in Example 8.10.

Further problems

Problem 8.14   CVD solute balance written in terms of the intrinsic rejection coefficient

Equation 7.61 represents the change in solute concentration during a CVD process in terms of the apparent rejection coefficient. Investigate whether a similarly simple expression can be derived in terms of the *intrinsic* rejection coefficient.

Problem 8.15   Approximate calculation of CVD time

Run the code in Example 8.12 and choose an appropriate value for the effectively constant wall concentration and an average apparent rejection coefficient. Now use the method of Section 7.9.1 to recalculate the CVD time. Compare your answer with that of Example 8.12.

Problem 8.16   Optimum concentration for CVD under limiting flux conditions

Consider Eq. 8.58. Now assume that the wall concentration is equal to the limiting concentration as given by Eq. 4.40. Deduce that Eq. 4.40 *cannot* predict a finite optimum concentration for CVD.

Problem 8.17   A more advanced model of enzyme transmission

This problem could possibly be done as a mini research project In Section 8.9, a simple model of solute transmission was developed in which the flux was determined solely by cake formation and independent of solute concentration. In that scenario, the solute concentration at the cake surface followed from the dynamics of cake formation. Explore how you might modify that model to allow for the fact that the flux is determined not only by the cake but by the solute.

9 Reverse osmosis and nanofiltration

9.1 Introduction to reverse osmosis

Reverse osmosis (RO) is the membrane filtration of solutions of low molecular weight solutes, especially salts [1]. At this small scale, it is appropriate to discard the idea of the membrane having distinct pores but instead to view it as a matrix through which both the solvent and solute diffuse. Diffusion occurs through the space between the chains of the polymer from which the membrane is constructed. However, the membrane may contain a small number of defects resulting in the presence of some pores.

The ability of a species to diffuse through an RO membrane is typically controlled by charge effects rather than size and, consequently, RO is used extensively for desalination processes. It is also used for water softening, i.e., removal of cations, especially calcium, and in production of high purity water such as that required in electronics industries. Smaller scale RO units are often used to produce high purity water for laboratory purposes.

Just like ultrafiltration, the flux of pure water through RO membranes is partly limited by the osmotic pressure of the solution. Osmotic pressure is a *colligative* property, meaning that at a fixed mass concentration, it increases as the molecular weight of the solute decreases. In other words, the osmotic pressure depends on the *number* of solute molecules per unit volume. Thus, the osmotic pressures of salt solutions are much greater than those of macromolecular solutions at the same molar concentration. This contributes to the need for the very high pressures, up to 100 bar, that are used in RO.

Ideally, RO membranes should be mechanically strong while providing high fluxes and high rejection of solutes. However, achieving both high flux and high rejection is difficult as increasing the flux is often accompanied by a reduction in the rejection coefficient. The most commonly used membrane in modern RO processes is the composite polyamide membrane, although some membranes based on cellulose acetate are used as well. This type of composite membrane, which is often referred to as a thin film composite (TFC), is made up of two polymers cast upon a fabric support. The top layer is where all the separation takes place and is composed of polyamide. Below this is a polymer support made typically of polysulphone. The bottom layer is made of fabric. The overall thickness of the membrane is generally in the range 1.5 to 2.0 mm [1]. The basic rationale, therefore, is not unlike that behind the asymmetric membranes employed in ultrafiltration, namely to combine good separation with high fluxes and mechanical strength. The most widely used membranes are probably the FILMTEC membranes produced by Dow and the CPA membranes produced by the Hydranautics Corporation. Most RO membranes boast very high rejections (99% or more) for common salts such as sodium chloride and calcium chloride but have considerably lower rejections for many larger molecules such as organics that one might need to remove in water treatment processes. Nominal values of rejection coefficients for a wide range of solutes are supplied by the manufacturer, but given that rejection is a function of operating conditions (as shown below), testing by the end user is usually required to establish actual rejection coefficients under process conditions.

9.2   Reverse osmosis theory

In Chapter 8, a general description of solvent and solute fluxes in membranes was presented, the main goal being to explain the origin of the osmotic reflection coefficient. In the approach used there, which is known as the *irreversible thermodynamics* approach, transport was assumed to have convective and diffusive components. Crucially, the membrane was viewed as having a porous structure, i.e., having distinct pores. However, the most successful approach to modelling fluxes in reverse osmosis assumes that the membrane is a non-porous (i.e., does not contain distinct pores) structure through which both the solvent and solute can diffuse. Solute and solvent molecules are assumed to 'meander' through the three-dimensional mesh formed by the network of polymer chains.

The most commonly used diffusion model in RO is the *solution-diffusion* model [2]. Solvent and solute are assumed to *dissolve* in the membrane, which is treated as a homogeneous 'phase', and then diffuse through the membrane. Although one normally associates diffusion with concentration gradients, diffusion is, in fact, driven by gradients in *chemical potential*. At a fixed temperature, the change in chemical potential of any component is determined by the change in its concentration *and* the change in pressure. Carrying out a detailed analysis of the dissolution and diffusion process gives the following expression for the flux of any component, i, through the membrane [2]:

$$J_i = \frac{D_i K_i}{\delta} \left(c_{iw} - c_{ip} \exp^{-\frac{v_i \Delta P}{RT}} \right), \tag{9.1}$$

where D_i is the diffusivity of component, i, in the membrane, K_i is the partition coefficient, δ is the membrane thickness, c_{iw} is the wall concentration of i in the *liquid* phase, c_{ip} is the concentration in the permeate, v_i is the molar volume of the component, R is the universal gas constant and T is the absolute temperature.

The partition coefficient, also known as a distribution coefficient, is a parameter that is well known in chemical engineering as it plays a key role in important operations like liquid–liquid extraction. It represents the ratio of the concentration of a solute in one phase relative to its concentration in another phase. In the RO context, it can be defined by

$$K_i = \frac{c_{im}}{c_{iw}}, \tag{9.2}$$

where c_{im} is the solute concentration *in the membrane* at the point that is in contact with the liquid (retentate) phase. In Eq. 9.1, the pressure term is typically negligible and that equation reduces to (Problem 9.1)

$$J_s = \frac{D_s K_s}{\delta}(c_w - c_p) = \kappa_s(c_w - c_p), \tag{9.3}$$

where κ_s is the solute *permeability* and it is understood that all concentrations refer to the single solute present. A similar expression can be derived for the water flux but it is much more useful, from a practical perspective, to express this flux in terms of the applied pressure. Using the fact that $J = 0$ when $\Delta P = \Delta \pi$ gives (Problem 9.2)

$$J = \kappa_w(\Delta P - \Delta \pi) \tag{9.4}$$

where κ_w is the water permeability. Note that the reader needs to be watchful here as when the 'w' subscript is used with concentrations, it refers to the *wall* concentration of the solute. When it is used with a permeability it refers to *water*.

One of the consequences of these equations is that the intrinsic rejection coefficient is a complex function of operating conditions. First, it is noted that the solute flux can be related to the solvent flux by

$$J_s = c_p J. \tag{9.5}$$

Combining this expression with Eqs. 9.3 and 9.4 gives the following expression for the intrinsic rejection coefficient (Problem 9.3):

$$\frac{1}{\sigma_i} = \left(1 - \frac{c_p}{c_w}\right)^{-1} = 1 + \frac{\kappa_s/\kappa_w}{\Delta P - \Delta \pi}. \tag{9.6}$$

The explicit dependence of the rejection coefficient on the permeability of solute and solvent is not surprising but the explicit dependence on pressure is interesting and, perhaps, not so obvious. Indeed Eq. 9.6 is actually a lot more complicated than it looks because the permeate and wall concentrations are contained in the $\Delta \pi$ term. As was shown earlier, use of the osmotic pressure model (see Eq. 9.7 below) in conjunction with Eq. 9.6 implies that solute concentration in the permeate is ultimately a rather complicated function of the operating conditions.

In the next example it is shown how the rejection coefficient at a given pressure can be predicted from a rejection coefficient value obtained at another pressure, such as that employed in a manufacturer's test.

Example 9.1 Calculation of the rejection coefficient under process conditions

A 0.2 M solution of ethanol undergoes reverse osmosis in a module which is operated at a trans-membrane pressure of 55 bar. Concentration polarisation is negligible under the conditions employed. The water permeability of the membrane is 7.0 l/m²h/bar and the manufacturer quotes an intrinsic rejection coefficient of 0.6, obtained at 15 bar with a 0.1 M solution under conditions of negligible concentration polarisation. The osmotic pressure (in Pa) of the solution at the temperature employed can be calculated from the equation

$$\pi = k_{osm}c,$$

where the concentration, c, is expressed as a molarity (mol/l) and

$$k_{osm} = 2.5 \times 10^6 \text{ Pa l/mol.}$$

Calculate the solute permeability and, assuming it is constant, calculate the rejection coefficient at 55 bar.

Solution. First we examine the test conditions. Combining Eqs. 9.3 to 9.5 gives

$$\kappa_s = \frac{c_p J}{c_w - c_p} = \frac{c_p/c_w}{1 - c_p/c_w} = \frac{1 - \sigma_i}{\sigma_i} J,$$

where $c_w = c$ because there is no concentration polarisation. Now, for the test conditions, where $c_p = c_w(1 - 0.6) = 0.1(1 - 0.6) = 0.04$ M,

$$J = \kappa_w(\Delta P - \Delta\pi) = \kappa_w(\Delta P - \pi(0.1) + \pi(0.04))$$
$$= \frac{7.0 \times 1 \times 10^{-3}/3600}{1 \times 10^5}(15 \times 10^5 - 2.5 \times 10^6(0.1 - 0.04)) = 2.63 \times 10^{-5} \text{ m/s,}$$

where consistent SI units are used throughout. Therefore

$$\kappa_s = \frac{1 - 0.6}{0.6} \times 2.63 \times 10^{-5} = 1.75 \times 10^{-5} \text{ m/s.}$$

Now, moving on to the process itself, the goal is to calculate the permeate concentration. Combining Eqs. 9.3 to 9.5 and noting the expression given for the osmotic pressure, we get the following quadratic equation for c_p:

$$c_p^2 + \left(\frac{\Delta P}{k_{osm}} + \frac{\kappa_s}{k_{osm}\kappa_w} - c_w\right)c_p - \frac{\kappa_s}{k_{osm}\kappa_w}c_w = 0,$$

where, again, c_w represents the concentration of the process solution (0.2 M) since concentration polarisation is negligible. Putting in the values of the parameters gives

$$c_p^2 + 2.361c_p - 0.072 = 0.$$

Solving gives $c_p = 0.03$ M. Thus the intrinsic (and apparent) rejection coefficient at this pressure is given by

$$\sigma = 1 - \frac{0.03}{0.2} = 0.85.$$

This increase in intrinsic rejection coefficient with pressure is typical of a membrane process, as shown in an ultrafiltration context by Eq. 8.68. In Problem 9.4, you are asked to investigate this phenomenon in an RO context in more detail. It is worth noting that RO systems are normally operated at low levels of concentration polarisation with the result that the intrinsic and apparent rejection coefficients behave in a very similar manner. UF is generally operated at high levels of concentration polarisation. In that case, the apparent rejection coefficient generally decreases with increasing pressure as demonstrated in Fig. 7.1.

9.2.1   Combining RO theory with concentration polarisation theory

The classic osmotic pressure-type model for the flux that was employed earlier in an ultrafiltration context applies equally well to RO, albeit with some small differences. The basic model that combines concentration polarisation and osmotic pressure approaches to modelling the flux can be written

$$k \ln \frac{c_w - c_p}{c - c_p} - \kappa_w(\Delta P - \Delta \pi) = 0, \tag{9.7}$$

where k is the mass transfer coefficient, c is the bulk solute concentration and it is understood that all concentrations refer to the single solute in the solution. For a given value of c there are two unknowns in this equation, namely c_w and c_p. The second equation that is needed is obtained by expanding out the terms in Eq. 9.5, giving

$$\kappa_s(c_w - c_p) - \kappa_w(\Delta P - \Delta \pi)c_p = 0. \tag{9.8}$$

Use of these two equations is illustrated in the example below. Before proceeding to that example, however, it is just worth mentioning some terminology that tends to be used in RO practice. RO practitioners sometimes talk about a parameter that they call *beta*, i.e., β. This is simply the ratio of wall concentration to bulk concentration, i.e.,

$$\beta = \frac{c_w}{c}. \tag{9.9}$$

This parameter is perhaps more well-known, especially in ultrafiltration, as the *polarisation modulus* and is an effective measure of the degree of concentration polarisation in a system. It should not be confused with the β parameter used in previous chapters to define the reduction in microsolute concentration during diafiltration. RO systems tend to be run with values of beta that are kept below about 1.2. This is generally done to avoid fouling processes such as scale formation, as well as to keep the permeate concentration of the solute as low as possible.

Example 9.2 Calculation of rejection coefficients with concentration polarisation

A 1.0 M solution of a certain solute undergoes reverse osmosis in a module which is operated at a trans-membrane pressure of 40 bar. The water permeability is 3×10^{-11} m/s/Pa, the solute permeability is 2.0×10^{-5} m/s and the osmotic pressure of the solution is given by

$$\pi = k_{osm}c,$$

where π is in Pa, c is in mol/l and k_{osm} is 1.4×10^6 Pa l/mol. If the mass transfer coefficient is estimated at 3.5×10^{-5} m/s, calculate the value of beta and both the intrinsic and apparent rejection coefficients.

Solution. The most logical way to solve this problem is to rearrange Eq. 9.8 and express c_w as a function of c_p as follows:

$$c_w = \frac{\kappa_w k_{osm} c_p^2 + (\kappa_w \Delta P + \kappa_s) c_p}{\kappa_s + \kappa_w k_{osm} c_p}.$$

This expression can be combined with Eq. 9.7 to yield a single equation in one unknown, c_p. MATLAB code, employing **fsolve**, is shown below.

```
function example92
guess=0.1;
cp=fsolve(@ROCP,guess)
function f=ROCP(cp)
kw=3e-11;ks=2e-5;k=3.5e-5;DP=40e5;c=1.0;kosm=1.4e6;
cw=(kw*kosm*cp^2+(kw*DP+ks)*cp)/(ks+kw*kosm*cp);
osm=kosm*(cw-cp);
beta=cw/c
si=1-cp/cw
s=1-cp/c
f=log((cw-cp)/(c-cp))-kw/k*(DP-osm);
```

Running the code gives, $c_p = 0.629$ M, $\beta = 2.255$, $\sigma_i = 0.721$ and $\sigma = 0.371$. In Problem 9.5, you are asked to use this example as a basis for examining the dependence of the water flux on pressure.

9.3   Predicting osmotic pressures

So far, a simple linear expression has been used for osmotic pressure but nothing has been said about where this came from. A simplifying feature of reverse osmosis is that one is generally dealing with dilute solutions and operating at beta values close to 1.0. This means that it is usually sufficiently accurate to approximate the osmotic pressure as a linear function of concentration. In other words, the higher order terms in the virial expansion are insignificant. One equation that has a theoretical basis is the van't Hoff or Morse equation [3] which, for a single solute, can be written

$$\pi = iRTc, \tag{9.10}$$

where i is the van't Hoff factor which is usually taken as the number of ions into which a solute dissociates. For example, $i = 2$ for NaCl. Other parameters in the van't Hoff equation are c, the *molar concentration* (molarity), R the ideal gas constant (which must be expressed as 8314 J/kmol K) and T is the absolute temperature. The RT term in the above equation is an indicator of the theoretical origins of the osmotic pressure, namely the *chemical potential*.

The van't Hoff equation is usually sufficiently accurate for dilute solutions of a single solute. For multi-component systems containing N solutes, the usual approach is to adopt an expression of the form

$$\pi = \zeta RT \sum_{n=1}^{N} M_n,$$ (9.11)

where ζ is a temperature-dependent empirical constant and the M_n are the solute molarities.

9.4 Process configurations in RO

RO installations tend to use multi-stage configurations employing spiral wound cartridges and tend to be operated in single pass mode (Fig. 5.7), although the feed-and-bleed mode (Fig. 5.1) is sometimes used.

As discussed previously, the feed-and-bleed configuration has the disadvantage that it operates at the highest possible solute concentration and, more importantly for an energy intensive operation like RO, it requires two pumps for each membrane element.

As reverse osmosis is typically used for large scale operations such as desalination of water, single pass operation is usually conducted with a large number of modules that are configured in a combined parallel-series arrangement. Commonly the 'Christmas tree' configuration is used, and a 3–2–1 arrangement is shown in Fig. 9.1.

In the *double-pass* arrangement, permeate from one RO unit becomes the feed to a second unit. This is used if rejection in the first unit is not sufficiently high and further 'polishing' of the permeate is needed to meet the final product specification.

In the next two sections, simple analyses are performed on a single-stage feed-and-bleed system and a single pass system containing a single membrane element. As can be seen, the very nature of the underlying flux models makes analysis of RO systems inherently 'messy', even if all of the model parameters are known. Thus, application of this classic approach to more advanced configurations is unrealistic, and the general

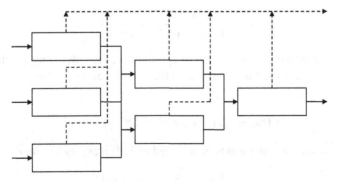

Figure 9.1 'Christmas tree' arrangement for single pass RO. Dashed lines denote permeate streams.

approach employed in industry is to use specialised software as described later. However, the calculations shown below will hopefully help the reader to acquire a deeper insight into RO theory.

9.5   Analysis of a continuous feed-and-bleed system

This configuration (Fig. 5.1) is used less frequently than the single pass mode of operation but it serves as a useful starting point to illustrate practical use of the RO equations. It was seen in Chapter 8 that solution of the feed-and-bleed model for a system described by the osmotic pressure model of ultrafiltration requires simultaneous solution of two non-linear algebraic equations. However, in RO, where there is a linear relation between osmotic pressure and solute concentration, only one non-linear equation must be solved.

Let Q_0 be the inlet flowrate, Q_1 the exit flowrate, A the membrane area, J the water flux through the membrane and J_s the solute flux. The overall balance for the system can be written

$$Q_0 - Q_1 - JA = 0. \tag{9.12}$$

Noting Eq. 9.4, this becomes

$$Q_1 = Q_0 - \kappa_w(\Delta P - \Delta \pi)A = 0. \tag{9.13}$$

Now, letting c_0 be the solute concentration in the feed and c_1 the solute concentration in the exit stream, the solute balance becomes

$$Q_0 c_0 - Q_1 c_1 - J_s A = 0. \tag{9.14}$$

Combining with Eq. 9.12 and noting Eqs. 9.3 and 9.5 gives

$$Q_0(c_0 - c_1) + \kappa_w(\Delta P - \Delta \pi)A(c_1 - c_{p1}) = 0, \tag{9.15}$$

where c_{p1} is the solute concentration in the permeate. This can be rearranged to give

$$c_1 = \frac{Q_0 c_0 - \kappa_w(\Delta P - \Delta \pi)A c_{p1}}{Q_0 - \kappa_w(\Delta P - \Delta \pi)A}. \tag{9.16}$$

Thus, c_1 can be expressed as a function of c_{p1} and c_{w1}, the latter being implicit in the $\Delta \pi$ term. Now, one can use Eq. 9.8 which, in the notation of the feed-and-bleed system, yields

$$\kappa_s(c_{w1} - c_{p1}) - \kappa_w(\Delta P - \Delta \pi)c_{p1} = 0. \tag{9.17}$$

As shown in Example 9.2, this can be rearranged to find an expression for c_{w1}, namely

$$c_{w1} = \frac{\kappa_w k_{osm} c_{p1}^2 + (\kappa_w \Delta P + \kappa_s)c_{p1}}{\kappa_s + \kappa_w k_{osm} c_{p1}}. \tag{9.18}$$

Finally, we can write

$$k \ln \frac{c_{w1} - c_{p1}}{c_1 - c_{p1}} - \kappa_w(\Delta P - \Delta \pi) = 0. \tag{9.19}$$

Using Eqs. 9.16 and 9.18, c_1 and c_{w1} can be expressed as functions of c_{p1}, thus leaving us with one non-linear algebraic equation, Eq. 9.19, to be solved.

Example 9.3 Single-stage feed-and-bleed RO

A 2.6×10^{-4} m^3/s flowrate of a 0.1 M solution of a certain solute undergoes feed-and-bleed reverse osmosis in a test unit with an area of 5.6 m^2. The operating pressure is 20 bar. The water permeability is 2×10^{-11} m/s/Pa, the solute permeability is 3.0×10^{-5} m/s and the osmotic pressure of the solution is given by the van't Hoff equation with $i = 1$. If the mass transfer coefficient is estimated at 5.0×10^{-5} m/s, calculate the retentate and permeate concentrations and the intrinsic and apparent rejection coefficients.

Solution. MATLAB code employing **fsolve** to solve Eq. 9.19 in conjunction with Eqs. 9.16 and 9.18 is shown below.

```
function example93
x=fsolve(@ROFB, [0.05])
function f=ROFB(x)
kw=2e-11;ks=3e-5;k=5.0e-5;DP=20e5;c0=0.1;Q0=2.6e-4;A=5.6;
kosm=1*8314*298;
cw1=(kw*kosm*x^2+(kw*DP+ks)*x)/(kw*kosm*x+ks)  %Eq.9.18
osm=kosm*(cw1-x);
c1=(Q0*c0-kw*(DP-osm)*A*x)/(Q0-kw*(DP-osm)*A)  %Eq.9.16
si=1-x/cw1
s=1-x/c1
f=log((cw1-x)/(c1-x))-kw/k*(DP-osm);
```

Running the code gives $c_1 = 0.192$ M, $c_{w1} = 0.304$ M, $c_{p1} = 0.031$ M, $\sigma_i = 0.898$ and $\sigma = 0.839$. In Problem 9.6, the effect of mass transfer coefficient (i.e. crossflow velocity) on the rejection coefficients is examined.

9.6 Analysis of a single pass system

It was shown in Chapter 8 that solution of the single pass analysis problem requires formulating the underlying equations as a system of ordinary differential equations. The same approach is followed here. However, following from Example 8.6, one simplifying assumption is made, namely, that the pressure drop down the channel is negligible. Thus, ΔP is assumed to be constant throughout the analysis.

To start, the balances are set up as follows:

$$\frac{dQ}{dA} = -J, \tag{9.20}$$

$$\frac{d(Qc)}{dA} = -J_s, \tag{9.21}$$

where dQ and dc are the changes in flowrate and bulk concentration over a differential area, dA. Expanding out Eq. 9.21 and combining with Eq. 9.20 gives

$$\frac{dc}{dA} = \frac{1}{Q}(cJ - J_s) = \frac{J}{Q}(c - c_p). \tag{9.22}$$

As before, the following three equations describe the two fluxes and the relationship between them:

$$J = \kappa_w(\Delta P - \Delta \pi), \tag{9.23}$$

$$J_s = \kappa_s(c_w - c_p), \tag{9.24}$$

$$J_s = c_p J, \tag{9.25}$$

where c_w is the wall concentration and c_p is the permeate concentration. Differentiating these at constant ΔP gives

$$dJ = -\kappa_w \frac{d\pi_w}{dc_w} dc_w + \kappa_w \frac{d\pi_p}{dc_p} dc_p, \tag{9.26}$$

$$dJ_s = \kappa_s dc_w - \kappa_s dc_p, \tag{9.27}$$

$$dJ_s = c_p dJ + J dc_p, \tag{9.28}$$

where π_w and π_p are the osmotic pressure expressed as functions of c_w and c_p respectively. Combining these three equations and dividing by dA gives, after some algebra,

$$\frac{dc_p}{dA} = \theta \frac{dc_w}{dA}, \tag{9.29}$$

where

$$\theta = \frac{\kappa_s + c_p \kappa_w \dfrac{d\pi_w}{dc_w}}{\kappa_w(\Delta P - \Delta \pi) + \kappa_s + c_p \kappa_w \dfrac{d\pi_p}{dc_p}}. \tag{9.30}$$

When the van't Hoff equation applies, this is simplified somewhat by the fact that

$$\frac{d\pi_w}{dc_w} = \frac{d\pi_p}{dc_p} = iRT. \tag{9.31}$$

Now, the basic flux model can be written

$$k \ln \frac{c_w - c_p}{c - c_p} - \kappa_w(\Delta P - \Delta \pi) = 0. \tag{9.32}$$

Differentiating and combining with Eq. 9.30 ultimately gives

$$\frac{dc_w}{dA} = \frac{\dfrac{k}{c - c_p}\dfrac{dc}{dA} - \ln\dfrac{c_w - c_p}{c - c_p}\dfrac{dk}{dA}}{\dfrac{k\theta}{c - c_p} + \dfrac{k(1 - \theta)}{c_w - c_p} + \kappa_w\dfrac{d\pi_w}{dc_w} - \kappa_w\theta\dfrac{d\pi_p}{dc_p}}. \tag{9.33}$$

Now, using the usual expression for the variation in mass transfer coefficient, i.e.,

$$k = k_0\left(\frac{Q}{Q_0}\right)^n, \tag{9.34}$$

where the '0' notation denotes inlet conditions, one gets

$$\frac{dk}{dA} = \frac{nk}{Q}\frac{dQ}{dA}. \tag{9.35}$$

Substituting into Eq. 9.33 gives

$$\frac{dc_w}{dA} = \frac{\dfrac{k}{c - c_p}\dfrac{dc}{dA} - \dfrac{nk}{Q}\left(\ln\dfrac{c_w - c_p}{c - c_p}\right)\dfrac{dQ}{dA}}{\dfrac{k\theta}{c - c_p} + \dfrac{k(1 - \theta)}{c_w - c_p} + \kappa_w\dfrac{d\pi_w}{dc_w} - \kappa_w\theta\dfrac{d\pi_p}{dc_p}}. \tag{9.36}$$

Thus, the complete model involves Eqs. 9.20, 9.22, 9.29, 9.30, 9.34 and 9.36. Solving these equations numerically is no great challenge. The only aspect of the problem that requires some thought is specification of the initial conditions for c_w and c_p. These can be computed quite easily. For a known c_0, c_{w0} can be expressed as a function of c_{p0} precisely as shown in Example 9.2, giving

$$c_{w0} = \frac{\kappa_w k_{osm} c_{p0}^2 + (\kappa_w \Delta P + \kappa_s)c_{p0}}{\kappa_s + \kappa_w k_{osm} c_{p0}}. \tag{9.37}$$

This can be substituted into Eq. 9.32 which, for known c_0, becomes a single equation in one unknown, c_{p0}.

Example 9.4 Analysis of a single pass system

A 2.6×10^{-4} m^3/s feed of a model seawater solution containing 0.5 M NaCl undergoes single pass reverse osmosis in a test unit with an area of 5 m^2. The operating pressure is 60 bar. The water permeability is 3×10^{-11} m/s/Pa, the solute permeability is 1.0×10^{-6} m/s and the osmotic pressure of the solution is given by the van't Hoff equation with $i = 2$. If the mass transfer coefficient is estimated to be 6.5×10^{-5} m/s at the inlet flowrate, calculate the *average* concentration of the *permeate*. Assume that the variation in mass transfer coefficient is given by

$$k = k_0\left(\frac{Q}{Q_0}\right)^{0.8},$$

where k_0 is the value of the mass transfer coefficient at the inlet flowrate Q_0.

Solution. MATLAB code employing **ode45** to implement the model is shown below

```
function example94
global kw ks k0 DP c0 kosm Q0 n cw0
kw=3e-11;ks=1.0e-6;k0=6.5e-5;DP=60e5;c0=0.5;
R=8314;T=298;i=2; Q0=2.6e-4;n=0.8;A=5;
kosm=i*R*T;
guess=0.05;
cp0=fsolve(@ROSP0,guess)
[A,y]=ode45(@ROSP, [0 A], [Q0 c0 cw0 cp0])
cpav=(Q0*c0-y(end,2)*y(end,1))/(Q0-y(end,1))%average
permeate conc.
function f=ROSP0(x)
global kw ks k0 DP c0 kosm cw0
cw0=(kw*kosm*x^2+(kw*DP+ks)*x)/(ks+kw*kosm*x);
osm=kosm*(cw0-x);
f=log((cw0-x)/(c0-x))-kw/k0*(DP-osm);
function dydA=ROSP(t,y)
global kw ks k0 DP c0 kosm Q0 n
%y(1) is Q, y(2) is c, y(3) is cw and y(4) is cp
k=k0*(y(1)/Q0)^n;
osm=kosm*(y(3)-y(4));
J=kw*(DP-osm);
theta=(ks+y(4)*kw*kosm)/(J+ks+y(4)*kw*kosm);
Qeqn=-J;
ceqn=J/y(1)*(y(2)-y(4));
Num=k/(y(2)-y(4))*ceqn-J*n/y(1)*Qeqn;
Denom=k*theta/(y(2)-y(4))+k*(1-theta)/(y(3)-
y(4))+kw*kosm*(1-theta);
cweqn=Num/Denom;
cpeqn=theta*cweqn;
dydA=[Qeqn;ceqn;cweqn;cpeqn];
```

Running the code gives an exit concentration of 0.915 M and an exit flowrate of 1.36×10^{-4} m^3/s. Therefore, the average permeate concentration is given by the following simple solute balance:

$$c_{\text{pav}} = \frac{Q_0 c_0 - Q_1 c_1}{Q_0 - Q_1},$$

which the MATLAB code gives as 0.043 M.

The key source of complexity in the above analysis was determination of the differential equations for c_{w1} and, consequently, c_{p1}. In Problem 9.7, you are asked to present a simplified analysis in which the variations in c_w and c_p are neglected.

9.7   Reverse osmosis software

The brief introduction to reverse osmosis in the previous sections has provided just a glimpse into the complexities of this important membrane filtration process. Reverse osmosis is ultimately concerned with separating multi-component mixtures of ions by taking advantage of the relative permeabilities of solutes through a polymeric membrane. In that sense, it is inherently more complicated than ultrafiltration, and understanding it requires a deep knowledge of the chemistry, including chemical thermodynamics, of highly complex ionic solutions. For that reason, real industrial design of RO systems involves a mix of theoretical and empirical equations, the latter often being of a proprietary nature. Such equations are typically embedded in design software along with databases of physical and chemical property data [1]. Probably the best known of these is the ROSA (Reverse Osmosis System Analysis) package produced by the Dow Corporation. Routine use of software like ROSA makes RO design quite close in spirit to the more traditional chemical engineering separation techniques, such as distillation and liquid–liquid extraction. These separations involve multi-component systems and modern day calculations are done with the aid of one or more of the many design and flowsheeting packages with which chemical engineers are familiar. Just like reverse osmosis, doing distillation and liquid–liquid extraction calculations requires a deep knowledge of the physical chemistry of multi-component systems, along with the availability of extensive and accurate databases of physical and chemical property data.

Ultimately, therefore, the purpose of the calculations shown in the preceding section is to give the reader a feel for the sort of issues that are involved in RO design. In truth, reverse osmosis is a subject that requires deep study, incorporating as it does, many aspects of membrane science, chemistry and chemical engineering.

9.8   Introduction to nanofiltration

Nanofiltration (NF), so named because the membrane pores are in the nanometre range, is one of those awkward processes that are sometimes encountered in chemical engineering. It exists in something of a transition zone between UF and RO, involving solutes in the molecular weight range 100–1000 Da and operating at pressures up to about 40 bar. If you have studied fluid mechanics, you will be aware of *transitional flow*, i.e., that flow that lies between laminar and fully turbulent. Here, the flow contains both laminar and turbulent 'zones', making numerical predictions fraught with difficulty. The transitional nature of NF makes it difficult to define precisely and sometimes causes people to question whether it is really a membrane filtration technique worthy of its own name. It could be argued, quite reasonably, that it is an especially 'leaky' form of RO or a particularly 'tight' form of UF.

Nonetheless, membrane filtration in the 'nano' range is gaining huge importance in a variety of applications. These include water treatment processes of all kinds, including removal of low levels of contaminants such as endocrine disruptors; food processes such as demineralisation of cheese whey; textile processes requiring removal of dyes from effluents; chemical processes such as de-acidification of crude oil and bioprocesses such as antibiotic recovery. This trend is reflected in increasing worldwide NF capacity and in the burgeoning number of publications produced by researchers in industry and academia. For further information on all aspects of nanofiltration, the best available source at present is the book edited by Shafer and co-editors [4]. Much of the factual information in this chapter comes from that source.

So, is NF just a region that membrane engineers have decided to 'create' somewhat arbitrarily? Of course, all definitions are, to some extent, arbitrary but some of the defining characteristics of NF are listed in Table 9.1.

Table 9.1 Characteristics of NF

- Rejection of NaCl in the range 0–70% (RO normally has complete rejection)
- High rejection of ions with more than one negative charge, e.g. SO_4^{2-}
- UF-like dependence of rejection on molecular size and shape for uncharged and positively charged solutes

The transitional nature of NF means that it is an inherently complicated process and, in this book, we can only scratch the surface of the underlying theory. It is a classic example of a chemical engineering process that requires integration of ideas from a number of different fields of chemistry and physics. A further complication is the wide range of nanofiltration membranes that are currently being developed. These membranes, which can be made from ceramic as well as polymer materials, have very individual rejection and fouling characteristics. In keeping with the chemical engineering philosophy of this book, we do not worry too much about the membrane structure, but the interested reader can find further information, including a nice history of NF, in [4].

9.9   Nanofiltration theory

One way of approaching the prediction of solvent and solute fluxes in NF is to use an irreversible thermodynamics approach as outlined in Section 8.8. There it was shown that transport of solute and solvent through the membrane can have diffusive and convective components. Using the RO notation used in this chapter and assuming the van't Hoff equation applies with $i = 1$, the irreversible thermodynamics approach applied to NF becomes

$$J = \kappa_{w}(\Delta P - \sigma_{o}RT(c_{w} - c_{p}))$$ (9.38)

and

$$J_{s} = \kappa_{s}(c_{w} - c_{p}) + (1 - \sigma_{o})Jc_{av},$$ (9.39)

where σ_{o} is the osmotic reflection coefficient (which arises because of the convective aspect of NF) and c_{av} is the average solute concentration in the membrane. This is usually computed as the log-mean average, i.e.,

$$c_{av} = \frac{c_w - c_p}{\ln(c_w/c_p)}.$$ (9.40)

Now using Eq. 9.5, one can combine Eqs 9.38 and 9.39 to get a single equation for c_p, namely

$$\kappa_s(c_w - c_p) + \kappa_w(\Delta P - \sigma_o RT(c_w - c_p))\left((1 - \sigma_o)\frac{c_w - c_p}{\ln(c_w/c_p)} - c_p\right) = 0.$$ (9.41)

Thus, if all the parameters in the irreversible thermodynamics model are known, the permeate concentration can be computed if the wall concentration is known. This is done in the example below.

Example 9.5 Computing the intrinsic rejection coefficient in NF

A 0.5 M solution of non-ionising solute undergoes nanofiltration at 20 bar and 298 K. The water permeability is 7.5×10^{-11} m/s/Pa, the solute permeability is 2×10^{-5} m/s and the osmotic reflection coefficient is 0.9. Calculate the permeate concentration and the rejection coefficient. Assume that concentration polarisation is negligible.

Solution. This requires numerical solution of Eq. 9.41 and MATLAB code is shown below. Note that Eq. 9.41 is written as

$$1 + \frac{\kappa_w(\Delta P - \sigma_o RT(c_w - c_p))((1 - \sigma_o)(c_w - c_p)/\ln(c_w/c_p) - c_p)}{\kappa_s(c_w - c_p)} = 0.$$

```
function example95
guess=0.01
cp=fsolve(@NFREJ,guess)
function f=NFREJ(x)
kw=7e-11;ks=1e-5;so=0.9;DP=20e5;c=0.5;T=298;R=8314;
f=1+kw*(DP-so*R*T*(c-x))*((1-so)*(c-x)/log(c/x)-x)/(ks*(c-x))
```

The result given is $c_p = 0.0794$ M. Thus, the rejection coefficient is given by

$$\sigma = 1 - \frac{0.0794}{0.5} = 0.841.$$

In Problem 9.8, you are asked to use this example as the basis for an investigation into the pressure dependence of the intrinsic rejection coefficient in NF.

9.10 Fed-batch nanofiltration

Here, the irreversible thermodynamics approach is used to model fed-batch nanofiltration. Recall that this involves operating a batch membrane filtration process in such a way that the feed flowrate is exactly balanced by the permeate flowrate (Fig. 5.6). In this way, a large volume of feed can be concentrated, even if the retentate tank has a limited capacity. Let us consider a volume, V_0, containing a solute of concentration c_0. The governing balance for this solute is

$$V_0 \frac{dc}{dt} = -J_s A + J A c_0, \tag{9.42}$$

where A is the membrane area and c_0 is both the initial concentration in the retentate tank and the feed concentration. As before

$$J_s - c_p J = 0, \tag{9.43}$$

where J_s and J are given by Eqs. 9.38 and 9.39. The equation for the flux can be written as

$$\ln \left(\frac{c_w - c_p}{c - c_p} \right) - \frac{J}{k} = 0. \tag{9.44}$$

Thus the model contains a single ordinary differential equation, Eq. 9.42 and two non-linear algebraic equations, Eqs. 9.43 and 9.44. For a given value of c_0, the three unknowns at any time are c, c_w and c_p. Throughout this book, the approach taken at a juncture such as this has been to differentiate the NLAE(s) in order to transform the problem into a system of explicit ODEs. This means that the resulting models can be solved with any ODE solver. Such a strategy is also possible in this case but a slightly different approach is used here, basically because the algebra is simpler. Technically, the system of equations can be classified as a differential algebraic equation (DAE) of index 1 and the key step in solving a system such as this is to write it in matrix form. For the current model, the matrix formulation becomes

$$\begin{pmatrix} 1 & 0 & 0 \\ 0 & 0 & 0 \\ 0 & 0 & 0 \end{pmatrix} \begin{pmatrix} dc/dt \\ dc_w/dt \\ dc_p/dt \end{pmatrix} = \begin{pmatrix} -(J_s A - J A c_0)/V_0 \\ J_s - c_p J \\ \ln((c_w - c_p)/(c - c_p)) - J/k \end{pmatrix}, \tag{9.45}$$

where J and J_s are, of course, functions of c_w and c_p. The 3×3 matrix in this equation is known as the 'mass matrix' and this must be supplied to MATLAB. In this case, it is a singular matrix, meaning that it has no inverse. The best way to illustrate this approach is in an example.

Example 9.6 Fed-batch NF

A 0.5 M solution of a non-dissociating solute is to undergo fed-batch nanofiltration at 298 K. The solution volume is 100 l and the membrane area is 2 m^2, while the operating pressure is 20 bar. The osmotic reflection coefficient is 0.9 for this solute and the membrane has a water permeability of 7×10^{-11} m/s/Pa and a solute permeability of 5×10^{-6} m/s. The mass transfer coefficient is 5×10^{-5} m/s. Generate a plot of solute concentration versus time for one hour of operation.

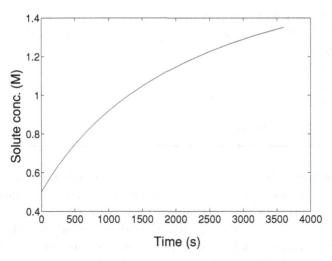

Figure 9.2 Increase in solute concentration during fed-batch NF.

Solution. MATLAB code to solve this problem is given below. Note that **fsolve** must be used to find initial values of c_w and c_p while **ode15s** must be used as the ODE solver in this case.

```
function example96
global kw ks k DP c0 V0 so R T A
kw=7e-11;ks=5.0e-6;k=5e-5;DP=20e5;c0=0.5;
so=0.9;V0=0.1;
R=8314;T=298; A=2.0;
guess=[0.7 0.1];
x0=fsolve(@NFFB0,guess)
M=[1 0 0
   0 0 0
   0 0 0 ];
options=odeset('Mass',M);
y0=[c0 x0]
[t,y]=ode15s(@NFFB,[0 3600],[y0],options)
plot(t,y(:,1));xlabel('Time (s)');ylabel('Solute Conc. (M)');
function f=NFFB0(x)
global kw ks k DP c0 so R T
%x(1) is cw0, x(2) is cp0;
J=kw*(DP-(1-so)*R*T*(x(1)-x(2)));
Js=ks*(x(1)-x(2))+J*(1-so)*(x(1)-x(2))/log(x(1)/x(2));
cweqn=1-x(2)*J/Js;
cpeqn=log(x(1)-x(2))/(c0-x(2))-J/k;
f=[cweqn;cpeqn];
function f=NFFB(t,y)
global kw ks k DP so R T A V0 c0
%y(1) is c, y(2) is cw, y(3) is cp
```

```
J=kw*(DP-(1-so)*R*T*(y(2)-y(3)));
Js=ks*(y(2)-y(3))+J*(1-so)*(y(2)-y(3))/log(y(2)/y(3));
ceq=(-Js*A+J*A*c0)/V0;
cweqn=1-y(3)*J/Js;
cpeqn=log((y(2)-y(3))/(y(1)-y(3)))-J/k;
f=[ceq;cweqn;cpeqn];
```

The program output is shown in Fig. 9.2.

The above examples have shown that even with this simple black box approach, the equations to be solved are getting increasingly complicated and a good deal of empirical information is needed to actually use them. Furthermore, when the solute is charged, the solute flux is determined not only by diffusion and convection but also by electromigration effects, and the complexity of the analysis increases further. It is generally true to say that processes involving solutions of electrolytes involve a step up in complexity from solutions of uncharged solutes.

Modern attempts to model NF have moved beyond the black box, irreversible thermodynamics approach and are based on detailed analysis of the hydrodynamic and electrostatic interactions involved as a solute travels through a membrane pore. The complexity of these models is such that they are very much a work in progress and considerably beyond the scope of this book. Indeed, it will probably be some time before there is any rigorous and widely accepted model of transport in NF membranes.

9.11   Conclusions

This chapter has provided a very brief introduction to the techniques of reverse osmosis and nanofiltration. It should be clear to the reader that these processes are inherently more complicated than ultrafiltration, requiring, as they do, the integration of ideas from chemical engineering and physical chemistry. The intention here has been to give the reader a glimpse into this complexity and to provide some sense of what calculations in this field might entail. In Chapter 10, we return to reverse osmosis when membrane fouling is discussed and analysed in more detail.

References

1. Kucera, J. (2010). *Reverse Osmosis. Industrial Applications and Processes*. Wiley-Scrivener, Massachusetts, USA.
2. Wijmanns, J.G. and Baker, R.W. (1995). The solution-diffusion model: a review. *Journal of Membrane Science*, **107**, 1–21.
3. The van't Hoff / Morse equation is empirical in origin but can be derived from chemical thermodynamics. Actually, the technical difference between the van't Hoff and Morse equations is that in Morse, the concentrations should be expressed in molalities rather than molarities. However, this distinction is often forgotten. More information can be found in the following

source: V.N.T. Dao, P.H. Morris and P.F. Dux (2008). On equations for the total suction and its matric and osmotic components. *Cement and Concrete Research*, **38**, 1302–1305.
4. Shafer, A.I., Fane, A.G. and Waite, T.D. (Eds.) (2006). *Nanofiltration. Principles and Applications*. Elsevier, Oxford, UK.

Additional reading

Prausnitz, J.M., Lichtenhaler, R.N. and Gomes de Azevedo, E. (1998). *Molecular Thermodynamics of Fluid-phase Equilibria*, 3rd Edition. Prentice Hall, NJ, USA.
Rautenbach, R., and Albrecht, R. (1989). *Membrane Processes*. Wiley, New York, USA.
Sablani, S.S., Goosen, M.F.A., Al-Belushi, R. and Wilf, M. (2001). Concentration polarization in ultrafiltration and reverse osmosis: a critical review. *Desalination*, **141**, 269–289.
Soltanieh, M., and Gill, W. (1981). Review of reverse osmosis membranes and transport models. *Chemical Engineering Communications*, **12**, 279–363.

Problems

Problem 9.1 Derivation of expression for solute flux
Using Eq. 9.1 as a starting point, derive Eq. 9.3.

Problem 9.2 Derivation of expression for solvent flux
Using Eq. 9.1 as a starting point, derive Eq. 9.4. Note that for the diffusive processes that occur in RO, it is correct to say that when $\Delta P = \Delta \pi$, $J = 0$.

Problem 9.3 Effect of operating conditions on intrinsic rejection coefficient in RO
Derive Eq. 9.6 from Eqs. 9.3 to 9.5.

Problem 9.4 Effect of pressure on intrinsic rejection coefficient in RO
Using the values given in Example 9.1, generate a plot of intrinsic rejection coefficient versus pressure in the range 15 to 60 bar. Neglect concentration polarisation in all calculations.

Problem 9.5 Effect of pressure on the flux in RO
Using the values given in Example 9.2, generate a plot of flux versus pressure in the range 20 to 80 bar.

Problem 9.6 Effect of mass transfer coefficient on rejection in RO
Using Example 9.3 as a template, investigate the effect of mass transfer coefficient on the apparent rejection in a feed-and-bleed system.

Problem 9.7 Simplified analysis of single pass RO systems
Re-analyse Example 9.4 and compute the exit concentration, assuming that c_w and c_p remain at their inlet values. Compare your answer to that given for that example.

Problem 9.8 Effect of pressure on intrinsic rejection in NF
Using Example 9.5 as a template, investigate the effect of pressure on the intrinsic rejection in NF.

Further problems

Problem 9.9   Effect of pressure on apparent rejection in NF

Modify the analysis of Problem 9.8 to include the effects of concentration polarisation and calculate the apparent rejection coefficient using a mass transfer coefficient of 5×10^{-5}.

Problem 9.10   Modelling of batch NF

A 0.5 M solution of a non-dissociating solute is to undergo batch nanofiltration at 298 K. The solution volume is 300 l and the membrane area is 0.5 m², while the operating pressure is 40 bar. The osmotic reflection coefficient is 0.95 for this solute and the membrane has a water permeability of 2×10^{-11} m/s/Pa and a solute permeability of 5×10^{-6} m/s. The mass transfer coefficient is 5×10^{-5} m/s. Generate a plot of solute concentration versus time for a one-hour operation.

Note that this requires modification of the differential equation for the bulk concentration, c and the addition of an equation for dV/dt. If the latter is the fourth equation in the model, the mass matrix must be written

$$\begin{pmatrix} 1 & 0 & 0 & 0 \\ 0 & 0 & 0 & 0 \\ 0 & 0 & 0 & 0 \\ 0 & 0 & 0 & 1 \end{pmatrix}.$$

Problem 9.11   Modelling of product loss in CVD by NF

Modify Example 9.6 to model product loss in CVD. Make all parameters the same except take the solute permeability to be 5×10^{-7} m/s, take the membrane area to be 0.5 m², the solution volume to be 300 l and let the osmotic reflection coefficient be 0.95.

10 Membrane fouling

10.1 Introduction

Membrane fouling refers to a deterioration in membrane performance due to interactions between the feed and the membrane. This typically involves an increase in membrane resistance and an increase in solute rejection. Cake formation and concentration polarisation are technically not considered to be membrane fouling processes, at least in this book. Cake formation is viewed as forming an *additional* resistance. Concentration polarisation is seen as reducing the *driving force* for filtration.

There is a vast literature on the subject of membrane fouling but despite this, membrane fouling still remains a largely empirical science. This is not surprising given the huge variation in membrane and feed characteristics that is encountered in membrane filtration processes. Furthermore, the underlying physical processes are complicated, involving solute–solute, particle–particle, solute–membrane and particle–membrane interactions. Identifying which interactions are important is usually very difficult indeed, especially when complex mixtures are involved.

Membrane fouling is typically irreversible in that cleaning agents are required to return the membrane to its original state. These can be alkalis, acids or surfactants, the precise one to be used being very much dependent on the membrane and the nature of the foulants. While hydrodynamic cleaning methods such as backflushing and crossflushing promote cleaning of the membrane (see Chapter 3), they are rarely sufficient on their own.

In keeping with the rest of this book, the goal in this chapter is to develop techniques for modelling membrane filtration systems where fouling is significant. Other books provide plenty of information on the actual physical processes involved [1–3].

10.1.1 The 'blocking' view of fouling

The language of membrane fouling tends to be dominated by the term *blocking*. This is the idea that when particles (e.g. cells or protein aggregates) come into contact with the surface of the membrane, they cause the pores of the membrane to become blocked, leading to a reduction in the available area for permeate flow. Another fouling mechanism that is sometimes discussed is the idea of pore *constriction*, a phenomenon whereby solutes penetrate into the membrane and bind to the walls of the membrane pores. In a well-known paper, Hermia [4] recognised that blocking and constriction models of membrane fouling could be unified as special cases of a single equation that also included cake formation as a special case. Hermia's analysis has spawned a plethora of papers applying various fouling models to experimental data. Recently, there has been a trend towards combining various fouling models to create 'super models' that allow for multiple processes occurring simultaneously [5–9]. The work of Ho and Zydney [10] was probably the first of this kind and provides a nice picture of the transition from pore blocking to cake formation. Recently, however, it has been suggested that many of the fouling models in use are physically unrealistic and thus it is likely that they should be taken with something of a 'pinch of salt' and ultimately viewed as purely empirical models [11].

To get a feel for the mathematical basis of blocking models, let us start with the following equation describing the rate of 'destruction' of open membrane pores:

$$\frac{dN}{dt} = -bJc, \tag{10.1}$$

where N is the number of unblocked pores per unit area of the membrane, b is a measure of the blocking efficiency, J is the flux and c is the concentration of the particles causing the blocking. Now, the Hagen–Poiseuille equation [12] states that the volumetric flowrate through a single pore of diameter, d_p, and length, L_p, is given by

$$Q_p = \frac{\pi d_p^4 \Delta P}{128 \mu L_p}. \tag{10.2}$$

Hence the flux, i.e., the total flowrate per unit area of membrane, is given by

$$J = NQ_p = \frac{N \pi d_p^4 \Delta P}{128 \mu L_p}. \tag{10.3}$$

Thus, using the resistance concept, i.e., letting

$$J = \frac{\Delta P}{\mu R_m}, \tag{10.4}$$

we have

$$N = \frac{128 L_p}{R_m \pi d_p^4}. \tag{10.5}$$

Thus, Eq. 10.1 becomes

$$\frac{128 L_p}{\pi d_p^4} \frac{d(1/R_m)}{dt} = -bJc. \tag{10.6}$$

Rearranging, combining with Eq. 10.4 and using the chain rule gives

$$\frac{1}{R_m} \frac{dR_m}{dt} = \frac{bc \Delta P \pi d_p^4}{128 \mu L_p}. \tag{10.7}$$

Integrating gives

$$R_m = R_{m0} \exp \left\{ \frac{bc \Delta P \pi d_p^4}{128 \mu L_p} t \right\}. \tag{10.8}$$

Ironically, this is probably the sort of equation that one would have proposed for modelling membrane fouling on a purely ad hoc basis. Indeed, this raises a somewhat philosophical point. Just because data, especially flux data, agree with a model, this does not mean that this model represents 'truth'. In other words, if one were to find the membrane resistance to increase exponentially with time, this would not necessarily mean that Eq. 10.1 is a uniquely accurate representation of the fouling process. Such a conclusion would probably require some independent measurements, perhaps direct observation of the membrane itself.

The importance of assuming Eq. 10.4 in the above derivation cannot be over-emphasised. In the vast majority of situations, membrane fouling occurs at the same time as cake formation or concentration polarisation and the two processes, fouling and concentration polarisation, for example, are coupled, i.e., they affect each other. In such a case, Eq. 10.4 would not apply. This idea is explored later.

10.1.2 Empirical modelling

The approach that is taken in this chapter is to adopt empirical expressions for the variation in membrane resistance and combine them with cake formation and concentration polarisation models. The simplest approach is to use Eq. 10.8 and write it as

$$R_{\mathrm{m}} = R_{\mathrm{m0}}e^{k_b t}, \tag{10.9}$$

where k_b is a fouling constant. This is mathematically simple but obviously all of the complexity is hidden in the fouling constant. Equation 10.9 also predicts that the membrane resistance will eventually go to infinity but it would perhaps be useful to have an equation that predicted a finite value of R_{m} at large times. This type of model has been used especially in dead-end filtration theory. Using the work of Tiller [13] as a guide, the following model seems reasonable:

$$R_{\mathrm{m}} = R_{\mathrm{m0}} \left[1 + (\delta - 1) \left(1 - e^{-\lambda V_{\mathrm{f}}} \right) \right], \tag{10.10}$$

where δ and λ are constants and V_{f} is the filtrate volume. Clearly, $R_{\mathrm{m}} \rightarrow \delta R_{\mathrm{m0}}$ as $V_{\mathrm{f}} \rightarrow \infty$. This equation predicts a potentially rapid increase in membrane resistance at short times, followed by a slower approach to a finite steady state value. Equations 10.9 and 10.10 are used at various times throughout this chapter to illustrate how fouling impacts on process behaviour.

10.2 Dead-end filtration

In this section, a number of simulations are presented. The two goals of the section are as follows: (i) to show how membrane fouling affects the shape of the t/V_{f} versus V_{f} curve in analysis of dead-end filtration data; and (ii) to show how membrane fouling relates to cake properties via its effect on the cake pressure drop.

10.2.1   Incompressible cakes

Referring to Chapter 2, Section 2.4, the governing equation for dead-end filtration can be written

$$\frac{dV_f}{dt} = AJ = \frac{A\Delta P}{\mu\,(R_m + \alpha c V_f/A)},\qquad(10.11)$$

where α is the specific cake resistance, which is taken to be constant in this case. In the absence of membrane fouling, this equation is easily integrated to give the well-known expression

$$\frac{t}{V_f} = \frac{\alpha\mu c}{2A^2\Delta P}V_f + \frac{\mu R_m}{A\Delta P}.\qquad(10.12)$$

Thus, a plot of t/V_f versus V_f is predicted to be a straight line. As mentioned in Chapter 2, it is generally preferable to analyse the data without prior integration by simply noting that

$$\frac{1}{J} = \frac{\mu R_m}{\Delta P} + \frac{\alpha\mu c}{A\Delta P}V_f.\qquad(10.13)$$

Thus, in the absence of membrane fouling, a plot of $1/J$ versus V_f should yield a straight line. However, this approach often requires numerical differentiation of volume data, a procedure that tends to magnifying experimental errors. A simple simulation of dead-end filtration with membrane fouling is shown below, and the results are plotted according to Eq. 10.13. Note that in this simulation, the fouling constants are assumed to be unaffected by the accumulation of solids on the membrane.

Example 10.1 Dead-end filtration with membrane fouling

A 20 g/l suspension undergoes dead-end filtration in a filtration unit whose area is 0.0017 m^2. The applied pressure is 0.5 bar, the specific cake resistance is 2×10^{11} m/kg, the clean membrane resistance is 5×10^{10} m^{-1} while the filtrate viscosity is 0.001 Ns/m^2. The membrane fouls according to Eq. 10.10 with $\delta = 10$ and $\lambda = 1 \times 10^4$ m^{-3}. Generate a plot of $1/J$ versus V_f for a 3600 s period.

Solution. MATLAB code employing ode45 is shown below.

```
function example101
global u Rm0 A DP alpha c del lam
A=0.0017;Rm0=5e10;DP=0.5e5;c=20;u=0.001;alpha=2e11;
del=10;lam=1e4;
[t,Vf]=ode45(@deffoul,[0 3600],0);
Rm=Rm0*(1+(del-1)*(1-exp(-lam*Vf)));
invJ=u*Rm/DP+alpha*u*c/(A*DP)*Vf;
plot(Vf,invJ); xlabel('V_f (m^3)');ylabel('1/J (s/m)');
function dvfdt=deffoul(t,Vf)
global u Rm0 A DP alpha c del lam
```

```
Rm=Rm0*(1+(del-1)*(1-exp(-lam*Vf)));
dvfdt=A*DP/u/(Rm+alpha*c*Vf/A);
```

The downward curvature obtained (Fig. 10.1) is typical of membrane fouling.

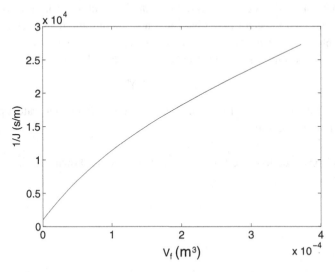

Figure 10.1 $1/J$ versus V_f in DEF with membrane fouling.

10.2.2 Compressible cakes

The analysis in the previous section was straightforward because of the assumption of constant specific cake resistance. However, if a cake is compressible, its specific resistance is a function of cake pressure drop, ΔP_c, where

$$\Delta P_c = \Delta P - \mu R_m J. \tag{10.14}$$

Now, the flux equation can be written as follows:

$$\frac{\Delta P}{\mu J} = R_m + \alpha c V_f / A. \tag{10.15}$$

Differentiating both of these with respect to time and recalling that

$$\frac{dV_f}{dt} = AJ, \tag{10.16}$$

ultimately leads to (Problem 10.1)

$$\frac{dJ}{dt} = -\frac{\dfrac{\alpha \mu c J^3}{\Delta P} + \dfrac{A\mu J^3}{\Delta P}\left(1 - \dfrac{\mu c V_f J}{A}\dfrac{d\alpha}{d\Delta P_c}\right)\dfrac{dR_m}{dV_f}}{1 - \dfrac{c\mu^2 R_m V_f J^2}{A\Delta P}\dfrac{d\alpha}{d\Delta P_c}}. \tag{10.17}$$

This is a generalisation of Eq. 2.49, which applied to compressible cake filtration with no membrane fouling.

Example 10.2 Compressible cake filtration with membrane fouling

A 5 g/l suspension undergoes dead-end filtration in a filtration unit whose area is 0.004 m^2. The applied pressure is 0.5 bar, the clean membrane resistance is 5×10^{10} m^{-1}. The membrane fouls according to Eq. 10.10 with $\delta = 10$ and $\lambda = 1 \times 10^4$ m^{-3}. The specific cake resistance is related to cake pressure drop by the expression

$$\alpha = \alpha_0 \left(1 + k_c \Delta P_c\right)^m ,$$

with $\alpha_0 = 1 \times 10^{11}$ m/kg, $k_c = 1.0 \times 10^{-4}$ Pa^{-1} and $m = 0.55$. Generate a plot of $1/J$ versus V_f for a 3600 s period.

Solution. MATLAB code to solve Eqs. 10.16 and 10.17 is as shown below.

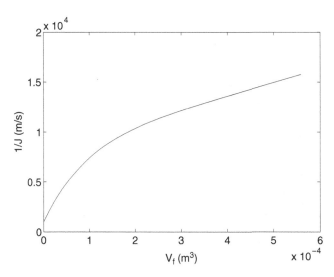

Figure 10.2 $1/J$ versus V_f with membrane fouling and a compressible cake.

```
function example102
global u Rm0 A DP c del lam alpha0 kc m
A=0.0017;Rm0=5e10;DP=0.5e5;c=5;u=0.001;
del=10;lam=1e4;alpha0=1e11;kc=1e-4;m=0.55;
J0=DP/u/Rm0;
[t,y]=ode45(@compfoul,[0 3600],[0 J0]);
Vf=y(:,1); invJ=1./y(:,2);
plot(Vf,invJ); xlabel('V_f (m^3/s)');ylabel('1/J (m/s)');
function dVfdt=compfoul(t,y)
%y(1) is Vf, y(2) is J
```

```
global u Rm0 A DP c del lam alpha0 kc m
Rm=Rm0*(1+(del-1)*(1-exp(-lam*y(1))));
DPc=DP-u*Rm*y(2);
alpha=alpha0*(1+kc*DPc)^m;
dadpc=alpha*kc*m/(1+kc*DPc);
dRmdVf=Rm0*(del-1)*lam*exp(-lam*y(1));
Vfeqn=A*y(2);
top=-alpha*u*c*y(2)^3/DP-A*u*y(2)^3/DP*(1-
u*c*y(2)*y(1)/A*dadpc)*dRmdVf;
bot=1-c*u^2*Rm*y(1)*y(2)^2/A/DP*dadpc;
Jeqn=top/bot;
dVfdt=[Vfeqn;Jeqn];
```

The results of the simulation are shown in Fig. 10.2, and reveal a similar downwards curvature to that seen in Fig. 10.1.

10.3 Crossflow microfiltration

The effect of membrane fouling in CFMF can be simulated in a very similar way to that employed in dead-end filtration. The flux can be related to cake mass through the usual expression

$$J = \frac{\Delta P}{\mu \left(R_m + \alpha M \right)},$$ (10.18)

where M is the cake mass per unit area. Using the kinetic model of cake formation (see Section 3.5 of Chapter 3), one can write

$$\frac{dM}{dt} = cJ - k_r \tau_w M,$$ (10.19)

where k_r is the cake removal constant. To complete the model, one can use any appropriate equation for membrane resistance as a function of time or filtrate volume. Implementing the model is simple and is illustrated in Problem 10.2.

Of more interest, however, is the role of membrane fouling in determining the properties of what one might call the 'apparent specific cake resistance'. This is discussed in the next section.

10.3.1 The apparent specific resistance

The specific cake resistance is a key parameter in the simple filtration theory presented in this book. The factors affecting the specific resistance were discussed in detail in Chapter 2. Of these factors, the applied pressure is, perhaps, the most important. The most common observation is that the specific cake resistance increases with pressure, a phenomenon referred to as cake compressibility.

The specific cake resistance is relatively easy to measure in dead-end filtration. It simply involves measuring the filtrate volume versus time and using Eq. 10.12 or Eq. 10.13. Usually, the data analysis is valid even if membrane fouling occurs, because membrane fouling tends to affect the very early time data only.

In the case of CFMF, however, Eqs. 10.12 and 10.13 do not apply and the specific resistance is determined using direct measurement of M and Eq. 10.18, an approach that is not without its critics among filtration theorists [14]. Incidentally, measuring the cake mass is not easy, as it cannot be inferred from the filtrate volume as in dead-end systems where Eq. 2.24 applies. Use of Eq. 10.18 to evaluate the specific cake resistance requires one to supply a value for the membrane resistance. Rearranging that equation gives

$$\alpha_t = \frac{\Delta P/\mu J - R_m}{M}, \tag{10.20}$$

where R_m is the actual membrane resistance when the cake mass is M and α_t is the *true* specific cake resistance. In practice, the membrane resistance is not known and what is actually measured is an *apparent* specific cake resistance, α_{app}, defined by

$$\alpha_{app} = \frac{\Delta P/\mu J - R_{m0}}{M}, \tag{10.21}$$

where R_{m0} is the clean, or pure water, membrane resistance. Combining these two equations gives

$$\alpha_{app} = \alpha_t + \frac{R_m - R_{m0}}{M}. \tag{10.22}$$

One feature of this equation that stands out immediately is the fact that both α_t and M typically increase with applied pressure. Since one term is a numerator and the other is a denominator, this creates the possibility that α_{app} might show a minimum or maximum with respect to pressure. It also means that α_{app} is time-dependent even if α_t is constant. This is illustrated in the next example.

Example 10.3 Time dependence of the apparent specific resistance

A 20 g/l suspension undergoes CFMF in a filtration unit whose area is 0.0017 m^2. The applied pressure is 0.5 bar, the clean membrane resistance is 5×10^{10} m^{-1}, the specific cake resistance is constant and equal to 2×10^{11} m/kg and the filtrate viscosity is 0.001 Ns/m^2. The membrane fouls according to Eq. 10.10 with $\delta = 10$ and $\lambda = 1 \times 10^4$ m^{-3}. Take the product $k_r \tau_w$ to be 0.005 s^{-1}. Generate a plot of α_{app} versus time over a 1 hour period.

Solution. MATLAB code using **ode45** is shown below. Note that unlike the dead-end case where V_f can be computed directly from M, here it requires solution of the appropriate differential equation, i.e.,

$$\frac{dV_f}{dt} = AJ.$$

```
function example103
global u Rm0 A DP alphat c del lam krtauw
A=0.0017;Rm0=5e10;DP=0.5e5;c=20;u=0.001;alphat=2e11;
del=10;lam=1e4;krtauw=0.005;
[t,y]=ode45(@cfmffoul,[0 3600],[0,0]);
Rm=Rm0*(1+(del-1)*(1-exp(-lam.*y(:,1))));
alphaapp=alphat+(Rm-Rm0)./y(:,2)
plot(t,alphaapp); xlabel('t
(s)');ylabel('\alpha_a_p_p (m/kg)');
function dydt=cfmffoul(t,y)
global u Rm0 A DP alphat c del lam krtauw
%y(1) is Vf, y(2) is M
Rm=Rm0*(1+(del-1)*(1-exp(-lam*y(1))));
J=DP/u/(Rm+alphat*y(2))
vfeqn=A*J;
meqn=c*J-krtauw*y(2);
dydt=[vfeqn;meqn];
```

Results from the simulation are shown in Fig. 10.3, and illustrate the significant increase in apparent specific resistance over time. Of course, whether this is likely to be observed in practice (as it has been [15]), depends on how large the increase in membrane resistance is relative to the cake resistance. There may be situations where membrane fouling occurs but is totally swamped by the effect of cake formation and no unexpected behaviour is observed.

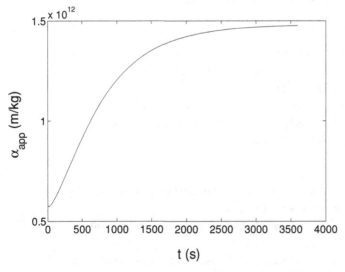

Figure 10.3 Time dependence of the apparent specific cake resistance.

It would be quite reasonable for an experimenter to interpret the time dependence of the specific resistance in terms of cake properties. For example, an increase in specific cake resistance over time is consistent with cake compressibility or preferential deposition of small particles as described in Chapter 3. Ultimately, corroborating evidence would be needed to make a firm conclusion as to the cause of the time dependence of the specific cake resistance.

The above example illustrates the time dependence of the apparent specific resistance at a fixed pressure. But what about the effect of pressure on the 'steady state' apparent specific resistance? This is illustrated in the next example.

Example 10.4 Pressure dependence of apparent specific resistance

A 20 g/l suspension undergoes CFMF in a filtration unit whose area is 0.0017 m². The clean membrane resistance is 5×10^{10} m^{-1} and the filtrate viscosity is 0.001 Ns/m². The membrane fouls according to Eq. 10.10 with $\delta = 10$ and $\lambda = 3 \times 10^3$ m^{-3}. Take the product $k_r \tau_w$ to be 0.005 s^{-1}. The specific cake resistance is related to pressure by the expression

$$\alpha_t = 1.0 \times 10^{11} \left(1 + 1.5 \times 10^{-4} \Delta P\right)^{0.6}.$$

Generate a plot of α_{app} (evaluated at $t = 1$ hour) versus pressure in the range 0.1 to 2.0 bar.

Solution. MATLAB code to solve this problem is shown below.

```
function example104
global u Rm0 A n alphat c del lam krtau
A=0.0017;Rm0=5e10;c=20;u=0.001;
del=10;lam=3e3;krtau=0.005;
for n=1:50;
DP(n)=0.2*1e5+n*0.05*1e5;
[t,y]=ode45(@cfmffoul,[0 3600],[0,0]);
Vf=y(:,1); M=y(:,2);
alphat(n)=1.0e11*(1+1.5e-4*DP(n))^.6;
alphaaap(n)=alphat(n)+Rm0*(del-1)*(1-exp(-
lam*Vf(end)))/M(end);
end
plot(DP, alphaaap);
xlabel('\DeltaP (Pa)');ylabel('\alpha_a_a_p (m/kg)');
function dydt=cfmffoul(t,y)
global u Rm0 A n alphat c del lam krtau DP
%y(1) is Vf, y(2) is M
DP=n*0.1e5; alphat(n)=1.0e11*(1+1.5e-4*DP)^.6;
Rm=Rm0*(1+(del-1)*(1-exp(-lam*y(1))));
```

```
J=DP/u/(Rm+alphat(n)*y(2));
vfeqn=A*J;
meqn=c*J-krtau*y(2);
dydt=[vfeqn;meqn];
```

The effect of pressure, computed using this code, is shown in Fig. 10.4. It confirms that a minimum value occurs, a possibility suggested by Eq. 10.22.

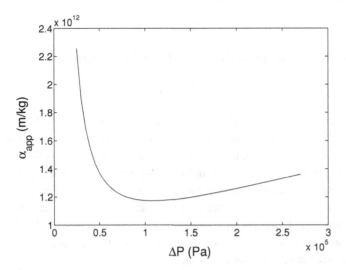

Figure 10.4 Effect of pressure on the apparent specific cake resistance.

As before, it would be reasonable to interpret such a phenomenon, namely the minimum with respect to pressure, as a manifestation of changing cake properties, because the minimum predicted above is also consistent with a combination of preferential deposition and cake compressibility, as explored in Problem 10.3.

10.4   Ultrafiltration

In the sections above, a simple analysis of simultaneous cake formation and membrane fouling was examined. Here, the analogous ultrafiltration problem of simultaneous membrane fouling and concentration polarisation is examined. Since membrane fouling has been defined as an increase in membrane resistance, the analysis must involve this parameter. This means that the osmotic pressure model must be used. Recall that for complete rejection, the osmotic pressure model can be summarised with the following equation

$$k \ln \frac{c_w}{c} - \frac{\Delta P - \pi}{\mu R_m} = 0, \tag{10.23}$$

where k is the mass transfer coefficient, c_w is the wall concentration of solute, c is the bulk concentration, ΔP is the trans-membrane pressure, π is the osmotic pressure (a function of c_w), μ is the permeate viscosity and R_m is the membrane resistance. In all of the previous work presented in this book, the membrane resistance was assumed to be constant. In this section, that assumption is relaxed and some of the resulting theoretical and practical problems are examined.

10.4.1   Membrane fouling and the limiting flux

Recall that the limiting flux is technically the flux for which

$$\frac{dJ}{d\Delta P} = 0.$$ (10.24)

In Chapter 4 (Section 4.6) it was shown that this can occur only if

$$\frac{dJ}{dc_{w}} = 0.$$ (10.25)

This can be true only if one accounts for the viscosity dependence of the mass transfer coefficient. In that case, the existence of the limiting flux is entirely a result of viscosity effects. However, as was shown in Section 4.6.1, this produces predictions that are not entirely consistent with experimental findings. But what if membrane fouling occurs? Anyone who has ever done experimental work with membranes will testify to the inordinate amount of time spent cleaning the membranes after each experiment. It seems that membrane fouling is an almost universal characteristic of ultrafiltration and other membrane filtration operations. So perhaps membrane fouling has some role to play in creating the conditions for a limiting flux.

To explore this possibility, let us differentiate Eq. 10.23 but allow for the fact that the membrane resistance might not be constant. Thus, at constant bulk concentration one gets

$$\frac{k}{c_{w}}dc_{w} - \left[\frac{1}{\mu R_{m}}\left(d\Delta P - \frac{d\pi}{dc_{w}}dc_{w}\right) - \frac{\Delta P - \pi}{\mu R_{m}^{2}}dR_{m}\right] = 0.$$ (10.26)

Rearranging gives

$$\frac{dc_{w}}{d\Delta P} = \frac{1 - \dfrac{\Delta P - \pi}{R_{m}}\dfrac{dR_{m}}{d\Delta P}}{k\mu R_{m}/c_{w} + d\pi/dc_{w}}.$$ (10.27)

Now, the criterion for a limiting flux can be written

$$\frac{dJ}{dc_{w}}\frac{dc_{w}}{d\Delta P} = 0.$$ (10.28)

Perhaps, therefore, a limiting flux arises not because the first term is zero but because the second term is zero, i.e.

$$\frac{dc_{w}}{d\Delta P} = 0.$$ (10.29)

Using Eq. 10.27, and noting Eq. 10.23 gives

$$\ln\left(c_{\lim}/c\right) - \frac{1}{\mu k\left(dR_{m}/d\Delta P\right)_{\Delta P=\Delta P_{\lim}}} = 0.$$ (10.30)

This is a possible new criterion for the existence of a limiting flux. Note that although the problem is couched in terms of changes in membrane resistance, this approach does allow for the fact that fouling may actually involve a 'cake-like' layer, composed, perhaps, of protein aggregates. This is something that is often assumed in resistance-in-series models of ultrafiltration. In that approach, it is assumed that the accumulation of solute at the membrane leads to the formation of a 'gel' layer that behaves like a compressible cake.

In conclusion, it has been shown here that, in principle, a limiting flux could be a consequence of membrane fouling but the precise mechanism(s) of fouling remain unknown in most cases. These ideas are pursued in Problem 10.4.

10.4.2   Flux dynamics at constant composition

Let us consider a situation where the system is operating continuously with $c = $ constant. In the absence of membrane fouling, the flux 'instantaneously' becomes established at a constant value. However, if membrane fouling occurs, the membrane resistance increases. One would obviously expect that the flux would decrease as a consequence. Indeed, that is the usual outcome. However, the relationship between the increase in membrane resistance and the decline in flux is not simple. This is easily seen if Eq. 10.23 is examined. If this equation is to remain true at all times and R_m is changing, that means that c_w is also changing. Indeed, this is how changes in membrane resistance impact on the flux within the context of concentration polarisation theory. An increase in membrane resistance leads to a decrease in wall concentration, resulting in a drop in flux.

It is worth diverting at this point to discuss the interpretation of the gel polarisation model for the limiting flux, mentioned briefly in Section 4.4 of Chapter 4. In this model, the wall concentration is presumed to become so high that a gel-like layer forms on the membrane. This high concentration is referred to as the *gel concentration* and is usually denoted c_g. It is generally assumed to be a property of the solution and not the operating conditions. The limiting flux is then given by

$$J_{lim} = k \ln \frac{c_g}{c}. \tag{10.31}$$

However, if c_g is a fixed concentration, there is no way, at least within the context of concentration polarisation theory, that changes in membrane resistance can affect the flux. Nonetheless, the formation of a gel-like layer can be accommodated within concentration polarisation theory if c_g is not interpreted as the gel concentration per se but the solution concentration *adjacent* to the gel layer. Thus, although the gel may remain at c_g, the concentration within the solution adjacent to it will change as the membrane resistance changes. This idea of a gel layer forming on the membrane is a phenomenon that can be accommodated within the model presented in this section. All it requires is that gel formation be viewed as a special kind of membrane fouling.

Let us now examine how flux decline can be modelled during operation at constant composition and pressure. To keep things simple, the mass transfer coefficient is assumed to be constant. Differentiating Eq. 10.23 gives

$$\frac{k}{c_w}dc_w + \frac{\Delta P - \pi}{\mu R_m^2}dR_m + \frac{1}{\mu R_m}\frac{d\pi}{dc_w}dc_w = 0. \tag{10.32}$$

Rearranging gives

$$\frac{dc_w}{dt} = \frac{-\dfrac{\Delta P - \pi}{\mu R_m^2}\dfrac{dR_m}{dt}}{\dfrac{k}{c_w} + \dfrac{1}{\mu R_m}\dfrac{d\pi}{dc_w}}. \tag{10.33}$$

Thus, as the membrane resistance increases the wall concentration declines. In the context of the osmotic pressure term of Eq. 10.23, the flux declines due to an increase in membrane resistance but this is partly offset by a reduction in osmotic pressure at the wall. From a concentration polarisation viewpoint, i.e. the first term in Eq. 10.23, the decline in flux occurs because of the decline in wall concentration.

Example 10.5 Simulating flux decline during UF at constant composition

A 10 g/l whey solution is ultrafiltered at constant composition at a pressure of 2.0 bar using a membrane with an initial resistance of 1×10^{12} m^{-1} and an area of 1.0 m^2. The mass transfer coefficient is 3×10^{-6} m/s while the permeate viscosity is 0.001 Ns/m^2. Osmotic pressure parameters for whey are given in Table 4.5. The membrane fouls according to the expression

$$R_m = R_{m0}e^{k_b t},$$

where $k_b = 0.001$ s^{-1}. Simulate the flux dynamics for a two-hour time period.

Solution. MATLAB code to solve Eq. 10.33 is shown below. Note that the function **fsolve** is used to compute the initial wall concentration from the osmotic pressure model. The flux is computed using concentration polarisation theory and expressed as a relative value, namely the flux at any time, t, relative to the initial flux.

```
function example105
global k c a1 a2 a3 Rm0 DP kb u
k=3e-6;DP=2e5;u=0.001;c=10;Rm0=1e12;kb=1e-3;
a1=454;a2=-0.176;a3=0.0082;
guess=500;
cw0=fzero(@cwfcn,guess)
[t,cw]=ode45(@uffoul,[0 7200],[cw0])
Jrel=log(cw/c)/log(cw0/c);
plot(t,Jrel); xlabel('Time (s)');ylabel('J_r_e_l (-)');
function f=cwfcn(cw0)
global k c a1 a2 a3 Rm0 DP u
osm=a1*cw0+a2*cw0^2+a3*cw0^3;
f=log(cw0/c)-(DP-osm)/(u*k*Rm0);
function dcwdt=uffoul(t,cw)
```

```
global k a1 a2 a3 Rm0 DP kb u
Rm=Rm0*exp(kb*t);
dRmdt=kb*Rm;
osm=a1*cw+a2*cw^2+a3*cw^3;
dosmdcw=a1+2*a2*cw+3*a3*cw^2;
dcwdt=-(DP-osm)/(u*Rm^2)*dRmdt/(k/cw+dosmdcw/(u*Rm));
```

Results from the simulations are shown in Fig. 10.5.

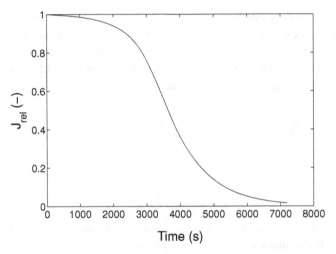

Figure 10.5 Relative flux versus time during UF at constant composition.

The early stages of the process are characterised by a phenomenon that one could describe as 'delayed fouling'.

In Problem 10.5, the behaviour obtained with an alternative fouling model is investigated. It should be pointed out that all of these simulations are based on model parameters that are purely arbitrary, so one should not assume that the predicted behaviour will always be observed. The important point, however, is that the interplay between membrane fouling and concentration polarisation can, in principle, lead to a variety of complex flux behaviours.

In batch systems, the retentate composition also changes. In that case, Eq. 10.32 becomes

$$\frac{k}{c_w}dc_w - \frac{k}{c}dc + \frac{\Delta P - \pi}{\mu R_m^2}dR_m + \frac{1}{\mu R_m}\frac{d\pi}{dc_w}dc_w = 0. \tag{10.34}$$

Rearranging gives

$$\frac{dc_w}{dt} = \frac{\dfrac{k}{c}\dfrac{dc}{dt} - \dfrac{\Delta P - \pi}{\mu R_m^2}\dfrac{dR_m}{dt}}{\dfrac{k}{c_w} + \dfrac{1}{\mu R_m}\dfrac{d\pi}{dc_w}}. \tag{10.35}$$

Simulation of batch UF with membrane fouling is the subject of Problem 10.6.

10.4.3   Dynamics of continuous feed-and-bleed UF

Following from Chapter 8 (Section 8.2.1), the mass balance and osmotic pressure equation for a continuous feed-and-bleed system can be written as

$$\frac{Q_0}{kA}(1 - c_0/c_1) - \ln\left(\frac{c_w}{c_1}\right) = 0, \tag{10.36}$$

$$\ln\frac{c_w}{c_1} - \frac{\Delta P - \pi}{k\mu R_m} = 0, \tag{10.37}$$

where c_1 is the exit concentration and c_w is the exit wall concentration. Now, if the membrane resistance changes, both c_1 and c_w must change with respect to time. Thus, the problem of solving two simultaneous non-linear algebraic equations now becomes one of solving two simultaneous ordinary differential equations. Differentiating these two equations and rearranging gives (Problem 10.7)

$$\frac{dc_w}{dt} = -\frac{\Delta P - \pi}{k\mu R_m^2}\frac{dR_m}{dt} \bigg/ \left[\frac{1}{c_w}\left(1 - \frac{1}{(Q_0/kA)c_0/c_1 + 1}\right) + \frac{d\pi/dR_m}{\mu k R_m}\right], \tag{10.38}$$

$$\frac{dc_1}{dt} = \frac{1}{c_w}\frac{dc_w}{dt} \bigg/ \left(\frac{Q_0}{kA}\frac{c_0}{c_1^2} + \frac{1}{c_1}\right). \tag{10.39}$$

These equations are used in the next example, which is an extension of Example 8.1 to the case where the membrane resistance is not constant.

Example 10.6 Dynamics of a continuous feed-and-bleed system

A 1 l/min feed of a whey solution enters a single-stage feed-and-bleed UF system. The feed enters at 10 g/l, the mass transfer coefficient is 3.5×10^{-6} m/s and the membrane area is 2.7 m^2. Use osmotic pressure data for whey from Table 4.5. The membrane resistance is initially 1×10^{12} m^{-1}, the permeate viscosity is 0.001 Ns/m^2 and the trans-membrane pressure is kept at 3.0 bar. The membrane fouls according to the same exponential model as in Example 10.5 with a fouling constant, k_b, of 5×10^{-6} s^{-1}. Simulate the performance of this system over a ten-day period, illustrating your calculations with a plot of exit concentration versus time.

Solution. MATLAB code for solving this problem is given below. The function **fsolve** is used to determine the intial values of c_1 and c_w while **ode45** is used to compute the time dependence of these terms.

```
function example106
global A Q0 k u Rm0 kb DP a1 a2 a3 c0
A=2.7;Q0=1e-3/60;k=3.5e-
6;u=0.001;Rm0=1e12;DP=3.0e5;c0=10;
a1=454;a2=-0.176;a3=0.0082;kb=5e-6;
x0=fsolve(@initconds, [30,150])
[t,x]=ode45(@contuff,[0,864000],x0);
```

```
plot(t,x(:,1));
xlabel('Time(s)');ylabel('c_e_x_i_t (g/L)');
function f=initconds(x0)
global A Q0 k u Rm0 DP a1 a2 a3 c0
%x0(1) is c10, x0(2) is cw0;
osm=a1*x0(2)+a2*x0(2)^2+a3*x0(2)^3;
x1eqn=Q0/(k*A)*(1-c0/x0(1))-log(x0(2)/x0(1));
x2eqn=log(x0(2)/x0(1))-(DP-osm)/(k*u*Rm0);
f=[x1eqn;x2eqn];
function dxdt=contuff(t,x)
%x(1) is c1, x(2) is cw
global A Q0 k u Rm0 kb DP a1 a2 a3 c0 Rm
Q0rel=Q0/(k*A);
Rm=Rm0*exp(kb*t);
dRmdt=kb*Rm;
osm=a1*x(2)+a2*x(2)^2+a3*x(2)^3;
dosm=a1+2*a2*x(2)+3*a3*x(2)^2;
top=-(DP-osm)/(k*u*Rm^2)*dRmdt;
bot=1/x(2)*(1-1/(1+Q0rel*c0/x(1)))+dosm/(u*k*Rm);
x2eqn=top/bot;
x1eqn=x2eqn/x(2)/(Q0rel*c0/x(1)^2+1/x(1));
dxdt=[x1eqn;x2eqn];
```

The variation in exit concentration is shown in Fig. 10.6. It is worth recalling that the precise shape of the curve is really a function of the fouling model used. The simple exponential model used here creates the effect whereby the decline in performance is small at early times but accelerates rapidly at large times.

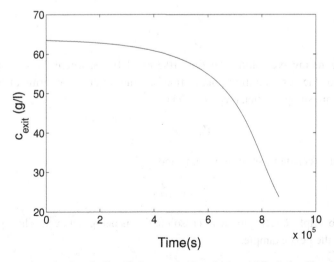

Figure 10.6 Flux decline during continuous feed-and-bleed ultrafiltration.

A deterioration in performance occurs with all membrane filtration systems, and instead of just accepting it as inevitable, it would be normal to try to modify the mode of operation to reduce or even eliminate this deterioration. One way of doing this is to operate at *constant flux*. Suppose c_1 is to be kept at a constant value. Then the required value of the flux is simply given by

$$J = \frac{Q_0}{A}\left(1 - \frac{c_0}{c_1}\right).$$

(10.40)

Therefore, the required value of c_w can be computed using

$$c_w = c_1 e^{J/k}.$$

(10.41)

Now, using the osmotic pressure term in Eq. 10.23, the pressure must follow a trajectory given by

$$\Delta P = \pi\left(c_w\right) + \mu R_m J.$$

(10.42)

The only term in this equation that changes is R_m. Thus, the pressure must be increased to counteract the rise in membrane resistance. This is a strategy that is commonly used in reverse osmosis and is workable as long as fouling is not too severe, leading to a very rapid increase in the required pressure.

10.4.4   Constant volume diafiltration

The problem of calculating the time for a constant volume diafiltration process is really a simple extension of the calculations performed in Section 10.4.2. When the rejection coefficient of the macrosolute is unity, the bulk concentration is constant and thus changes in the wall concentration of that solute are given by Eq. 10.33. At the same time, diluent use is described by the simple ordinary differential equation

$$\frac{dV_w}{dt} = AJ.$$

(10.43)

The calculations involve solution of Eqs. 10.33 and 10.43, where J can be computed using either of the concentration polarisation or osmotic pressure terms of Eq. 10.23. Calculations are stopped when (see Eq. 6.21)

$$V_w = V_0 \ln \beta,$$

(10.44)

where V_0 is the (constant) solution volume and

$$\beta = c_{B0}/c_{Bf},$$

(10.45)

where the subscript, B, refers to the microsolute that is being removed. This approach is illustrated in the next example.

Example 10.7 Time for constant volume diafiltration

A 25 l volume of 10 g/l whey solution undergoes CVD at constant composition at a pressure of 2.0 bar using a membrane with an initial resistance of 1×10^{12} m^{-1} and an area of 1 m^2. The mass transfer coefficient is 3×10^{-6} m/s while the permeate viscosity is 0.001 Ns/m^2. Osmotic parameters for whey are given in Table 4.5. The membrane fouls according to the expression

$$R_{\mathrm{m}} = R_{\mathrm{m0}}e^{k_b t},$$

where $k_{\mathrm{b}} = 5 \times 10^{-4}$ s^{-1}. The concentration of a certain microsolute is to be reduced by a factor of 5. Calculate the time required to perform this operation and show a plot of c_{w} versus time.

Solution. MATLAB code using **ode45** and **fsolve** is shown below.

```
function example107
global k c a1 a2 a3 Rm0 DP kb u beta V0 A
k=3e-6;DP=2e5;u=0.001;c=10;Rm0=1e12;kb=5.e-4;A=1.0;
a1=454;a2=-0.176;a3=0.0082;V0=0.025;beta=5;
guess=100;
cw0=fzero(@cw0fcn,guess)
options=odeset('events', @eventfcn);
[t,x]=ode45(@cvdfoul,[0 10800],[cw0,0],options)
J=k*log(x(:,1)/c)
plot(t,x(:,2))
function f=cw0fcn(cw0)
global k c a1 a2 a3 Rm0 DP u
osm=a1*cw0+a2*cw0^2+a3*cw0^3;
f=log(cw0/c)-(DP-osm)/(u*k*Rm0);
function dxdt=cvdfoul(t,x)
%x(1) is cw, x(2) is Vf
global k a1 a2 a3 Rm0 DP kb u A
Rm=Rm0*exp(kb*t);
dRmdt=kb*Rm;
osm=a1*x(1)+a2*x(1)^2+a3*x(1)^3;
dosmdcw=a1+2*a2*x(1)+3*a3*x(1)^2;
J=(DP-osm)/(u*Rm);
cweqn=-(DP-osm)/(u*Rm^2)*dRmdt/(k/x(1)+dosmdcw/(u*Rm));
Vweqn=A*J;
dxdt=[cweqn;Vweqn];
function [value, isterminal, direction]=eventfcn(t,x)
global beta V0
value = x(2)-V0*log(beta); % stop when desired flux
```

```
is reached
isterminal = 1;
direction = 0;
```

Running the code gives a time of 4396 s. The variation of c_w with time is shown below, illustrating the link between membrane fouling and concentration polarisation.

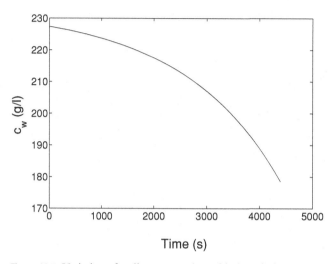

Figure 10.7 Variation of wall concentration with time during CVD.

10.4.5   Ultrafiltration with constant volume diafiltration (UFCVD)

In UFCVD the solution undergoes both concentration steps and constant volume diafiltration steps. In the context of this chapter, the best place to start is to formulate the problem as a generalised variable volume diafiltration process as described in Section 6.4.1, specifically Example 6.11. The equations for the process are:

$$\frac{dV}{dt} = (\alpha - 1)\, JA, \tag{10.46}$$

$$\frac{dc_A}{dt} = \frac{c_A}{V}(1 - \alpha)\, JA, \tag{10.47}$$

$$\frac{dc_B}{dt} = -\frac{c_B}{V}\alpha JA, \tag{10.48}$$

$$\frac{dV_w}{dt} = \alpha JA, \tag{10.49}$$

where V is the solution volume, V_w is the diluent volume and for clarity we use c_A to denote the macrosolute concentration and c_B to denote the microsolute concentration. Here, the macrosolute rejection coefficient is one while that of the microsolute is zero. The quantity, α, is the ratio of the diluent flowrate to the permeate flowrate and should not be confused with specific resistance used earlier in this chapter. UFCVD is most easily defined in this formulation by the pseudocode definition:

```
IF c_A ≥ c_Ai and c_B > c_Bf
THEN α = 1
ELSE α = 0
```

where c_{Ai} is an intermediate concentration somewhere between c_{A0} and c_{Af}. The subscript '0' denotes initial values while 'f' denotes the final, required values.

In the above model, the flux is given by either the concentration polarisation or the osmotic pressure term of Eq. 10.23 while Eq. 10.33 is the required ODE for c_{Aw}. Use of this model is illustrated in the next example.

Example 10.8 Simulation of UFCVD

A 100 l volume of a whey solution is to be concentrated from 12 to 48 g/l and a certain microsolute is to be reduced from 1 to 0.3 g/l at a pressure of 2.0 bar using a membrane with an initial resistance of 1×10^{12} m^{-1} and an area of 1.5 m^2. The mass transfer coefficient is 3×10^{-6} m/s while the permeate viscosity is 0.001 Ns/m^2. The osmotic pressure of the whey solutions is given by the expression in Example 10.7 and the membrane fouls according to Eq. 10.10 with $\lambda = 10$ m^{-3} and $\delta = 5$. Generate plots of the relative solute concentrations versus time if CVD is commenced at $c_A = 24$ g/l.

Solution. MATLAB code, based on that used in Example 6.11 is shown below.

```
function example108
global k cA0 a1 a2 a3 Rm0 DP u cAi A cBf del lam V0
V0=0.1; cA0=12; cB0=1; Vw0=0;cAi=24;cBf=0.3;
k=3e-6;DP=2e5;u=0.001;Rm0=1e12;A=1.5;
a1=454;a2=-0.176;a3=0.0082;beta=10;del=5;lam=10;
cAw0=fsolve(@cAw0fcn,500);
options=odeset('events', @eventfcn,'MaxStep',10);
[t,y]=ode15s(@ufcvd,[0 36000],[V0 cA0 cB0 Vw0 cAw0], options);
time=t(end)/60
cAout=y(:,2)/cA0; cBout=y(:,3)/cB0;
plot(t,cAout,'--k',t,cBout,'-k');
xlabel('Time (s)');ylabel('Conc. (g/L)');
function f=cAw0fcn(cAw0)
global k cA0 a1 a2 a3 Rm0 DP u
osm=a1*cAw0+a2*cAw0^2+a3*cAw0^3;
f=log(cAw0/cA0)-(DP-osm)/(u*k*Rm0);
function dydt=ufcvd(t,y)
global k a1 a2 a3 Rm0 DP u cAi A cBf del lam V0
%y(1) is V, y(2) is cA, y(3) is cB, y(4) is Vw, y(5) is cw
% Note calculation of permeate volume: V0-V+Vw
Vf=V0-y(1)+y(4);
Rm=Rm0*(1+(del-1)*(1-exp(-lam*Vf)));
dRmdVf=Rm0*(del-1)*lam*exp(-lam*Vf);
```

```
osm=a1*y(5)+a2*y(5)^2+a3*y(5)^3;
dosmdcw=a1+2*a2*y(5)+3*a3*y(5)^2;
J=k*log(y(5)/y(2));
if and(y(2)>=cAi,y(3)>cBf);
alpha=1;
else
alpha=0;
end
Veqn=(alpha-1)*J*A;
cAeqn=y(2)/y(1)*(1-alpha)*J*A;
cBeqn=-y(3)/y(1)*alpha*J*A;
Vweqn=alpha*J*A;
dRmdt=dRmdVf*(-Veqn+Vweqn);  %note calculation of Vf
above
top=k/y(2)*cAeqn-(DP-osm)/u/Rm^2*dRmdt;
bot=k/y(5)+dosmdcw/u/Rm;
cweqn=top/bot;
dydt=[Veqn;cAeqn;cBeqn;Vweqn;cweqn];
function [value, isterminal, direction]=eventfcn(t,y)
cAf=48;
value = y(2)-cAf;
isterminal = 1;
direction = 0;
```

Running the code gives a time of just about 332 minutes and the relevant plot is shown in Fig. 10.8. In Problem 10.8, the effect of the intermediate macrosolute concentration, c_{Ai}, is investigated.

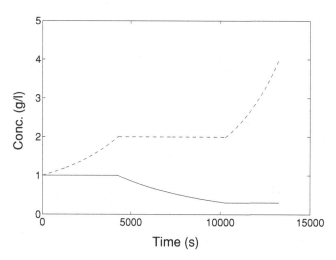

Figure 10.8 Relative concentrations versus time in UFCVD. Dashed curve denotes macrosolute, solid curve denotes microsolute.

10.5   Reverse osmosis and nanofiltration

Many of the fouling mechanisms that apply to microfiltration and ultrafiltration processes apply equally well to reverse osmosis and nanofiltration. The presence of suspended solids can cause fouling by cake formation, microbes can lead to biofilm formation and ionic solutes can lead to *scaling* [16, 17]. Scaling is the formation of a crystal-like deposit on the membrane surface and is a consequence of ionic solutes, such as carbonates and sulphates, attaining wall concentrations that exceed their solubility limit.

Let us consider the simpler case of reverse osmosis and recall that in Chapter 9, the water flux and the solute flux were written in terms of water and solute permeabilities. These permeabilities are ultimately related to the effective diffusion coefficient of each component in the membrane itself. The question is, therefore: how does one incorporate scale formation into the RO or NF flux equations? To begin formulating an answer to this question, let us first write down the RO flux equations as done in Chapter 9:

$$J = \kappa_w (\Delta P - \Delta \pi), \tag{10.50}$$

$$J_s = \kappa_s(c_w - c_p). \tag{10.51}$$

As before, κ_w is the water permeability in the membrane, κ_s is the solute permeability in the membrane, c_w is the wall concentration of the solute and c_p is the permeate concentration of the solute.

The simplest route to modelling fouling in RO (at constant bulk concentration, c) is to assume that the water permeability, κ_w, decreases with time, perhaps in an exponential manner such as

$$\kappa_w = \kappa_{w0} e^{-k_b t}, \tag{10.52}$$

i.e.,

$$\frac{d\kappa_w}{dt} = -k_b \kappa_w. \tag{10.53}$$

So, the model of fouling in RO involves the following equations: first, using Eq. 9.5 we have

$$1 - c_p \frac{J}{J_s} = 0, \tag{10.54}$$

where J is given by Eq. 10.50 and J_s is given by Eq. 10.51. Second, the usual osmotic pressure model applies, i.e.,

$$\ln \frac{c_w - c_p}{c - c_p} - \frac{J}{k} = 0. \tag{10.55}$$

The final equation is the ODE, Eq. 10.53. The entire model can be written as a DAE as

$$\begin{pmatrix} 1 & 0 & 0 \\ 0 & 0 & 0 \\ 0 & 0 & 0 \end{pmatrix} \begin{pmatrix} d\kappa_w/dt \\ dc_w/dt \\ dc_p/dt \end{pmatrix} = \begin{pmatrix} -k_b \kappa_w \\ 1 - c_p \dfrac{\kappa_w (\Delta P - \Delta \pi)}{\kappa_s(c_w - c_p)} \\ \ln((c_w - c_p)/(c - c_p)) - \kappa_w(\Delta P - \Delta \pi)/k \end{pmatrix},$$

$$\tag{10.56}$$

where it is assumed that appropriate osmotic pressure data are available.

10.6   Conclusions

The wide range of membrane fouling mechanisms, and the complexity of fouling processes, means that fouling models are likely to remain largely empirical in the short to medium term. The key point of this chapter has been to show that regardless of the nature of current and future fouling models, methods for incorporating them into engineering equations can be developed relatively easily. Furthermore, the availability of computational software means that there should be no concerns with actually implementing the models.

References

1. Cheryan, M. (1998). *Ultrafiltration and Microfiltration Handbook*, 2nd Edition. CRC Press, Florida, US.
2. Baker, R.W. (2012). *Membrane Technology and Applications*, 3rd Edition. Wiley, West Sussex, UK.
3. Zeman, L.J. and Zydney, A.L. (1996). *Microfiltration and Ultrafiltration. Principles and Applications*. Marcel Dekker, New York, US.
4. Hermia, J. (1982). Constant pressure blocking filtration laws – application to power-law non-Newtonian fluids. *Trans. IChemE*, **60**, 183–187.
5. Palacio, L., Ho, C.C. and Zydney, A.L. (2002). Application of a pore-blockage–cake filtration model to protein fouling during microfiltration. *Biotechnology and Bioengineering*, **79**, 260–270.
6. Bolton, G., LaCasse, D. and Kuriyel, R. (2006). Combined models of membrane fouling: Development and application to microfiltration and ultrafiltration of biological fluids. *Journal of Membrane Science*, **277**, 75–84.
7. Duclos-Orsello, C., Li, W.Y. and Ho, C.C. (2006). A three mechanism model to describe fouling of microfiltration membranes. *Journal of Membrane Science*, **280**, 856–866.
8. Hwang, K.J., Liao, C.Y. and Tung, K.L. (2007). Analysis of particle fouling during microfiltration by use of blocking models. *Journal of Membrane Science*, **287**, 287–293.
9. Sarkar, B. and De, S. (2012). A combined complete pore blocking and cake filtration model for steady-state electric field-assisted ultrafiltration. *AIChE Journal*, **58**, 1435–1446.
10. Ho, C.C. and Zydney, A.L. (2000). A combined pore blockage and cake filtration model for protein fouling during microfiltration. *Journal of Colloid and Interface Science*, **232**, 389–399.
11. Tien, C. and Ramarao, B.V. (2011). Revisiting the laws of filtration: An assessment of their use in identifying particle retention mechanisms in filtration. *Journal of Membrane Science*, **383**, 17–25.
12. Welty, J.R., Wicks, C.E., Wilson, R.E. and Rorrer, G.L. (2008). *Fundamentals of Momentum, Heat and Mass Transfer*, 2nd Edition. Wiley, Hoboken, New Jersey, USA.
13. Tiller, F.M. (1981). Filtering coal liquids: clogging phenomena in the filtration of liquefied coal. *Chemical Engineering Progress*, **77**, 61–68.

14. Tien, C. and Ramarao, B.V. (2008). On the analysis of dead-end filtration of microbial suspensions. *Journal of Membrane Science*, **319**, 10–13.
15. McCarthy, A.A., Walsh, P.K. and Foley, G. (2002). Experimental techniques for quantifying the cake mass, the cake and membrane resistances and the specific cake resistance during crossflow filtration of microbial suspensions. *Journal of Membrane Science*, **201**, 31–45.
16. Kucera, J. (2010). *Reverse Osmosis. Industrial Applications and Processes*. Wiley-Scrivener, Massachusetts, USA.
17. Shafer, A.I., Fane, A.G. and Waite, T.D. (Eds.) (2005). *Nanofiltration. Principles and Applications*. Elsevier, Amsterdam, The Netherlands.

Problems

Problem 10.1   ODE for the flux during DEF with membrane fouling

Derive Eq. 10.17.

Problem 10.2   Simulation of CFMF with membrane fouling

Modify the analysis used for Example 10.2 for the case of CFMF. Take the product $k_r\tau_w$ to be 0.003 s^{-1}. Generate a plot of flux versus time over a 30 minute period.

Problem 10.3   Physical explanation for minimum in α with respect to pressure

Explain why a minimum in α_t with respect to pressure is consistent with cake compressibility combined with preferential deposition. Refer to Chapter 3 if required.

Problem 10.4   Deriving an expression for the limiting flux

Re-read Section 3.5.1 of Chapter 3. Now suppose that ultrafiltration is accompanied by a cake formation process described by the model presented in that section. Develop an expression for the limiting flux in this case.

Problem 10.5   Simultaneous membrane fouling and concentration polarisation

Repeat Example 10.5 but this time use Eq. 10.10 to model the membrane fouling. Take $\delta = 15$ and $\lambda = 20$ m^{-3}, $A = 2.0$ m^2, $k = 1 \times 10^{-5}$ m/s, $\Delta P = 4$ bar and run the simulation for a three-hour period.

Problem 10.6   Simulation of fouling in batch UF

Modify Example 10.5 but assume that UF is done in batch mode. Assume that the initial volume is 60 l, the area of the membrane is 1.3 m^2 and the bulk concentration given in that example is the initial concentration. Generate a plot of flux versus time.

Problem 10.7   Flux decline in continuous feed-and-bleed UF

Derive Eqs. 10.38 and 10.39.

Problem 10.8   Optimum concentration for CVD

For the data in Example 10.8, except that $c_{Af} = 140$ g/l, investigate the effect of c_{Ai} on the total process time. Is there an optimum value?

Further problems

Problem 10.9　Operation of a feed-and-bleed system with constant exit concentration

Using the information given in Example 10.6, determine the pressure trajectory to ensure that the exit concentration does not depart from its initial value.

Problem 10.10　Fouling dynamics in reverse osmosis

Write a MATLAB code to implement the model presented in Section 10.5. Using suitable values for the various parameters, test the code and investigate how the flux changes with time.

Appendix

Mathematical and computational background

A.1   Introduction

A key part of the modelling and analysis of chemical and bioprocess engineering systems is the formulation of mass and/or energy balances. A large part of the skill of an engineer is his or her ability not only to state these balances in words, but more importantly, to express them in mathematical form. Once this has been done, the remainder of the problem is essentially one of applying well-established mathematical and/or computational techniques. In steady state processes, the system balances are in the form of systems of algebraic equations, usually non-linear in nature. An example is multi-stage continuous feed-and-bleed ultrafiltration that is covered in Chapter 5. In a batch process, the balances are usually formulated as systems of ordinary differential equations. An example is batch crossflow microfiltration that is covered in Chapter 3.

This book is aimed at a wide audience, from chemical engineers with a strong background in mathematics and computing, to students studying interdisciplinary subjects like biotechnology or bioprocessing who might be a little less mathematically literate. The aim of this appendix is to provide a brief summary of the main mathematical and computational techniques that are employed throughout the remainder of the text. Coverage of the mathematical topics is brief but provides the student with sufficient information that he or she can review the key topics in more depth using appropriate sections of their preferred mathematics text if desired. Coverage of the computational topics provides just enough background so that those who do not have a strong background in numerical methods will have some appreciation of how a computational tool such as MATLAB actually solves a system of differential or algebraic equations. Again, the student can consult a numerical methods text to obtain a deeper understanding of these topics if needed. However, this should not be necessary for the problems encountered in this book.

A.2   Calculus and symbolic computation

Familiarity with the basic tools of calculus, specifically integration and differentiation, is essential for effective analysis of engineering systems. In a sense, calculus can be viewed as the natural language of these systems. Differentiation is the natural tool for describing the rate of change of parameters such as concentrations or volumes, while integration is the correct technique for computing the total amount of a quantity, such as a filtrate volume, produced in a time period. Throughout this book, both of these techniques are used routinely to formulate and solve process problems.

A.2.1 Differentiation

Quantitative analysis of bioprocess engineering systems inevitably requires competence in the basic tools of differentiation. Most readers should have considerable experience of this technique and should be aware of the basic rules. However, as a reminder, some of the derivatives that occur repeatedly in this book are given in Table A.1.

Table A.1 Some common functions and their derivatives.

Function	Derivative
$f(x) = x^n$	$f'(x) = nx^{n-1}$
$f(x) = \ln \frac{a}{x}$	$f'(x) = -1/x$
$f(x) = e^{ax}$	$f'(x) = ae^{ax}$

In analysing process systems one usually encounters terms in the equations that can be viewed as combinations of functions. As a result there is a need for techniques for differentiating these combinations. For example, the total amount of solute in a tank of fluid is the product of the volume of the solution and the concentration of the solute. It is often important to be able to relate the rate of change of the total amount of solute to the individual rates of change of the volume and the concentration. Fortunately, there are well established rules for doing this type of calculation. In differentiation, the key rules are the product rule, the quotient rule (although this is really just the product rule 'in disguise') and the chain rule. These are defined, with examples of each provided, in Table A.2, where $f(x)$ and $g(x)$ are arbitrary functions of the independent variable, x.

Table A.2 Some basic rules of differentiation.

Rule	Definition	Process example
Chain	$\dfrac{df}{dz} = \dfrac{df}{dx}\dfrac{dx}{dz}$	relating flux decline to cake accumulation in CFMF: $\dfrac{dJ}{dt} = \dfrac{dJ}{dm}\dfrac{dm}{dt}$
Product	$\dfrac{d}{dx}(g(x)f(x)) = g(x)f'(x) + f(x)g'(x)$	rate of change of total solute in batch UF: $\dfrac{d(Vc)}{dt} = V\dfrac{dc}{dt} + c\dfrac{dV}{dt}$
Quotient	$\dfrac{d(g(x)/f(x))}{dx} = \dfrac{f(x)g'(x) - g(x)f'(x)}{(f(x))^2}$	rate of change of constant pressure UF flux in terms of changes in osmotic pressure and membrane resistance: $\dfrac{d}{dt}\left(\dfrac{\Delta P - \pi}{\mu R_m}\right) = \dfrac{-\mu R_m\dfrac{d\pi}{dt} - (\Delta P - \pi)\mu\dfrac{dR_m}{dt}}{(\mu R_m)^2}$

A technique that proves useful in analysis of batch ultrafiltration processes is *implicit differentiation*. This is a differentiation technique that is used when one function cannot be written as an explicit function of the other, i.e., with the dependent variable, y, on the left hand side of the equation and only terms involving the independent variable, x, on the right hand side. The main use of this technique in this book is that it allows one to turn a non-linear algebraic equation (NLAE) into an ordinary differential equation (ODE), a very useful 'trick' when the model of the system contains both types of equation. An example of the use of implicit differentiation is given in Example A.1.

Example A.1 Implicit differentiation

The non-linear algebraic equation shown below describes the relationship between two process parameters, y and x which vary with time, t:

$$a \ln \frac{y}{x} + by^n = 0.$$

Assuming that a, b and n are constants, find an expression for dy/dt in terms of dx/dt.

Solution. The key to solving this problem is to use implicit differentiation as shown below. First we expand out the logarithmic term and write

$$a \ln y - a \ln x + by^n = 0.$$

Note that this equation is *not explicit* because we cannot rearrange it to be in the form $y = f(x)$. Now differentiating implicitly, we get

$$ad(\ln y) - ad(\ln x) + bd(y^n) = 0.$$

Now using the chain rule we can write this as

$$a\frac{d \ln y}{dy}dy - a\frac{d \ln x}{dx}dx + b\frac{dy^n}{dy}dy = 0.$$

Doing the differentiations gives

$$\frac{a}{y}dy - \frac{a}{x}dx + bny^{n-1}dy = 0.$$

Now rearranging and dividing across by dt gives

$$\frac{dy}{dt} = \frac{a/x}{a/y + bny^{n-1}} \frac{dx}{dt}.$$

A key area where differentiation arises is in finding the maximum or minimum of a function. This is an important aspect of process optimisation where one is searching for the best way to design or perform a given process. One can find the maximum or minimum of a function, $f(x)$, by solving the equation

$$\frac{df(x)}{dx} = 0. \tag{A.1}$$

If the solution to this equation is denoted x_{opt}, then there is a maximum at this point if

$$\left.\frac{d^2 f(x)}{dx^2}\right|_{x=x_{opt}} < 0, \tag{A.2}$$

and a minimum if

$$\left.\frac{d^2 f(x)}{dx^2}\right|_{x=x_{opt}} > 0. \tag{A.3}$$

In most circumstances, one will not have to check either of these criteria because if the mathematical model captures the underlying physics of the system, it will reproduce only what is physically meaningful. Generally it will be clear which type of optimum should occur. Optimisation of a function is illustrated in the next example.

Example A.2 Optimising a function
Locate the optimum of the following function:

$$f(x) = x^a \ln(b/x),$$

where a and b are positive constants.

Solution. Differentiating using the product rule

$$\frac{df}{dx} = x^a(-1/x) + ax^{a-1}\ln(b/x).$$

Therefore, the optimum is located by solution of the equation

$$x_{opt}^a(-1/x_{opt}) + ax_{opt}^{a-1}\ln(b/x_{opt}) = 0.$$

Simplifying gives

$$1 - a\ln(b/x_{opt}) = 0$$

and therefore

$$x_{opt} = be^{-1/a}.$$

In most of our calculations, we are dealing with functions of a single variable. For example, the osmotic pressure in ultrafiltration is a function of the wall concentration only (assuming the membrane resistance is constant). However, the mass transfer coefficient in ultrafiltration is a function of both the wall concentration and the bulk concentration and there will be situations where we will have to cope with this complication when performing implicit differentiation.

Suppose then that a function f depends on x and z. Then the implicit differentiation formula is

$$df = \left(\frac{\partial f}{\partial x}\right)_z dx + \left(\frac{\partial f}{\partial z}\right)_x dz.$$

This equation states that the change in f is due to the changes in both x and z. The terms in brackets are *partial derivatives* and represent the rate of change of f with respect to one variable while the other is kept constant. An example of the implementation of this formula is given in Example A.3.

Example A.3 Implicit differentiation of a function of two variables

Find the differential, df, for the function

$$f = x/z.$$

Solution. Carrying out the partial differentials gives

$$\left(\frac{\partial f}{\partial x}\right)_z = \frac{1}{z},$$

$$\left(\frac{\partial f}{\partial z}\right)_x = -\frac{x}{z^2}.$$

Therefore

$$df = \frac{1}{z}dx - \frac{x}{z^2}dz.$$

A.2.2   Series approximations to functions

We often encounter situations in modelling engineering processes where it is useful to approximate a complicated function with a more simple function. As well as making practical calculations more straightforward, this often helps to highlight key trends predicted by a function. One of the most powerful techniques for approximating a function close to a known value of the function is the Taylor Series. This is defined by the following expression:

$$f(x) = f(a) + \frac{f'(a)}{1!}(x - a) + \frac{f''(a)}{2!}(x - a)^2 + \frac{f'''(a)}{3!}(x - a)^3 + \ldots, \quad \text{(A.4)}$$

where a is a value of x at which the value of the function is known and

$$f'(a) = \left.\frac{df}{dx}\right|_a, f''(a) = \left.\frac{d^2f}{dx^2}\right|_a, f'''(a) = \left.\frac{d^3f}{dx^3}\right|_a. \quad \text{(A.5)}$$

The factorial function is defined by

$$n! = n \times (n-1) \times (n-2) \times \cdots \times 1. \quad \text{(A.6)}$$

The series expansion is particularly useful when the variable, x, has values < 1 as in this case the function can sometimes be approximated by a highly truncated form of the series. Table A.3 shows the series expansion for a number of functions

Table A.3 Series approximations to some functions.

Function	Series definition	Truncated form
$f(x) = 1/(1-x)$	$f(x) = 1 + x + x^2 + x^3 \cdots$	$f(x) = 1 + x$ for $|x| \to 0$
$f(x) = \ln x$	$f(x) = (x-1) - \dfrac{1}{2}(x-1)^2 + \dfrac{1}{3}(x-1)^3 + \cdots$	$f(x) = x - 1$ for $|x-1| \to 0$
$f(x) = 1/x$	$f(x) = 1 + (1-x) + (1-x)^2 + (1-x)^3 + \cdots$	$f(x) = 2 - x$ for $|x-1| \to 0$
$f(x) = e^x$	$f(x) = 1 + x + \dfrac{x^2}{2!} + \dfrac{x^3}{3!} + \cdots$	$f(x) = 1 + x$ for $|x| \to 0$

encountered in this book. An example of the use of this technique is given in the following example.

Example A.4 Series approximation to a function

Show that when $|x - 1| \to 0$, the value of the following function approaches zero:

$$f(x) = \ln x + \frac{1}{x} - 1.$$

Verify that this answer is sensible by evaluating the exact value of the function for $x = 0.9$.

Solution. Using the approximations in Table A.3, we get

$$f(x) = (x - 1) + (2 - x) - 1 = 0.$$

The exact answer is given by

$$f(0.9) = \ln(0.9) + \frac{1}{0.9} - 1 = 0.00575.$$

Thus the approximation seems 'quite good' in this case but whether it is actually a good approximation or not really depends on the precision required in any given calculation.

A.2.3   Integration

Integration is the opposite of differentiation and for most students it represents a much greater challenge than differentiation. Solving integrals in symbolic form is essentially a question of remembering the appropriate techniques or 'tricks'. Two simple aspects of integration that are encountered on a number of occasions in this book are worth mentioning here. The first involves changing the order of the limits on an integral and can be described by the following formula

$$\int_a^b f(x)\,dx = -\int_b^a f(x)\,dx. \tag{A.7}$$

This rule is employed in the book on a number of occasions simply to put the final design or analysis equation in a logical form. In general this means eliminating minus signs from the equations because we are dealing with physical quantities for which it makes more sense to have everything expressed in terms of positives.

The other aspect of integration that we mention here is both useful and aesthetic. One often manipulates an integral to put it in a neater form that perhaps makes the solution a little more obvious. Suppose, for example that we want to evaluate the integral

$$y = \int_a^b \ln\left(\frac{z}{x}\right) dx. \tag{A.8}$$

To put this in a tidier form, we define

$$s = \frac{z}{x}, \tag{A.9}$$

which implies

$$x = \frac{z}{s}. \tag{A.10}$$

Therefore, implicit differentiation gives

$$dx = -\frac{z}{s^2} ds. \tag{A.11}$$

Consequently the integral becomes

$$y = -\int_{z/a}^{z/b} \frac{z \ln s}{s^2} ds = z \int_{z/b}^{z/a} \frac{\ln s}{s^2} ds. \tag{A.12}$$

The advantage of this approach is that the integral is in a neat, standard form and it makes coding of numerical integration routines simpler.

In this book we encounter a small number of integrals that the average student may not have met before, or if they have, they may not recall the tricks involved in solving them. This, however, should not pose a problem for the resourceful student. Instead, the Symbolic Math Toolbox within MATLAB can be used. 'Symbolic computation' or 'computer algebra', including symbolic differentiation and integration, is a remarkable development in computing that provides the solution to many types of algebraic and computational problems in algebraic form as shown in the next example.

Example A.5 Symbolic integration

Integrate the following function of x using the MATLAB symbolic integrator:

$$f = \frac{ax}{b + x},$$

where a and b are constants.

Solution. The following instructions are inserted at the Command Window of MATLAB:

```
syms a b x
f=a*x/(b+x);
int(f,x)
```

MATLAB returns

```
ans =
a*(x - b*log(b + x))
```

One of the concepts that arises in doing integrations is that of a *special function*. Most students are familiar with the idea of an *elementary function*. An elementary function can be a combination of nth roots, exponentials, logarithms and trigonometric functions. We normally consider that an analytical solution to an equation exists if it can be written in terms of elementary functions. However, there is a broad field of mathematics that deals with so-called special functions. These functions crop up regularly but less frequently than elementary functions in both pure and applied mathematics. They are often named after their discoverer. The Bessel function is probably the special function most familiar to engineers. Usually the value of these functions can be expressed as an infinite series and, in that sense, they are not inherently different from elementary functions such as exponentials or trigonometric functions. One of the novel aspects of this book is the use we make of a particular special function, the exponential integral, rather than always resorting to numerical integration. Use of the exponential integral within MATLAB is described in detail where it arises in Chapter 5.

A.3   Numerical solution of non-linear algebraic equations

There are many techniques and a vast literature on solution of systems of non-linear algebraic equations (NLAEs). These have been developed and improved over the years to deal with much more challenging problems that we meet in this book. In this section the classic Newton–Raphson method is described and it is then explained how NLAEs can be solved with the MATLAB routines **fzero** and **fsolve**. The main purpose here is to give the student with a less comprehensive mathematics background a sense of how software packages solve equations of this type, albeit with more sophisticated algorithms.

A.3.1   The Newton–Raphson method

The Newton–Raphson algorithm can be derived using the idea of a Taylor Series expansion. Suppose we want solve the following equation:

$$f(x) = 0. \tag{A.13}$$

This means that we want to determine the actual numerical value of x that will make the above statement true. Now suppose we have a current estimate of this value, and let us call it x_n. Then the Newton–Raphson algorithm predicts that a (hopefully) better estimate of the solution, defined as x_{n+1}, is given by

$$x_{n+1} = x_n - \frac{f(x_n)}{f'(x_n)}, \tag{A.14}$$

where

$$f'(x_n) = \left. \frac{df}{dx} \right|_{x=x_n}. \tag{A.15}$$

By repeated application of this algorithm, one converges, ideally, on a solution. A simple example of the implementation of this algorithm is given in Example A.6 below.

Example A.6 The Newton–Raphson algorithm

Develop the Newton–Raphson algorithm for solving the following NLAE:

$$f(x) = 3.0(1 - x) - \ln x - 4.2 = 0.$$

Solution. Differentiating gives

$$\frac{df}{dx} = -3 - \frac{1}{x}.$$

Therefore, the Newton–Raphson algorithm for this problem becomes

$$x_{n+1} = x_n + \frac{3(1 - x_n) - \ln x_n - 4.2}{3 + 1/x_n}.$$

Obviously the Newton–Raphson method requires the analytical derivative of the function, and this is often not available especially if the method is extended to solve systems of NLAEs. Normally, therefore, algorithms used by software packages will evaluate the derivative (or Jacobian in the case of systems of equations) numerically.

It is important to note that, in general, we cannot say with certainty that the Newton–Raphson algorithm, or indeed any numerical method, will *converge*. Mathematicians can, of course, derive precise criteria but here we mention just a couple of useful tips to help in attaining convergence. While most of the problems we encounter are not particularly challenging, difficulties may be encountered. For example, a method may not converge with a particular initial estimate (x_0) but may converge if a different estimate is used. In many problems, however, there will be natural physical bounds on the initial estimates so choosing one that works should not be a major problem. The second, more important point to make is that convergence may be affected by how the equation is written. To understand this, imagine that the algorithm being employed has a built-in tolerance such that iteration will stop when

$$|f(x_n)| \leq \varepsilon, \tag{A.16}$$

where ε is a small number. Now suppose $f(x)$ takes the form

$$f(x) = h(x) - g(x) \tag{A.17}$$

and suppose that

$$|h(x_n)| = O(\varepsilon) \tag{A.18}$$

and

$$|g(x_n)| = O(\varepsilon), \tag{A.19}$$

where the $O(\varepsilon)$ notation means 'of the same order of magnitude as ε'. Then even with the first guess, one might find that Eq. A.16 is satisfied. However, this will not be a good estimate of the solution because the error will be of the same order of magnitude as the individual terms in the equation. This is illustrated in the next example where the MATLAB function **fzero** is used. This function does not, in fact, use the Newton–Raphson method but an algorithm based on the *bisection method*. This is a search-based method in which two adjacent estimates for x are found, one of which gives $f(x) > 0$ and the other which gives $f(x) < 0$. Once two such points are found, it is not hard, although it can be a slow process, to 'zoom in' on the value of x for which $f(x) = 0$.

Example A.7 Solving a single NLAE with fzero

Use **fzero** to first solve the equation

$$5 \times 10^{-5}(1 - x) - 8 \times 10^{-5} \ln x - 20 \times 10^{-5} = 0,$$

where, for physical reasons, x must lie between 0 and 1.

Now, divide across the equation by 5×10^{-5} to get the following equation:

$$(1 - x) - 1.6 \ln x - 4 = 0.$$

Solve this, identical equation, again with **fzero**. In both cases, use an initial estimate of $x = 0.25$.

Solution. MATLAB code for solving the first version of the equation using an initial estimate of 0.5 is shown below.

```
function exampleA7a
x0=0.5;
x=fzero(@fun,x0)
function f = fun(x)
f=5e-5*(1-x)-8e-5*log(x)-20e-5;
```

Running the code and checking the Command Window gives $x = 0.1405$.

The code for the second version can be written

```
function exampleA7b
x0=0.5;
x=fzero(@fun,x0)
function f = fun(x)
f=(1-x)-1.6*log(x)-4;
```

Running the code gives precisely the same answer. Therefore, **fzero** is capable of solving both versions of this equation. However, if you are familiar with the **Goal Seek** tool within Microsoft Excel, and try solving the two versions of this equation, you will find that with Goal Seek, the solution to the first version does not move away from the initial guess, whereas the correct answer is obtained if the second version is used. As a general rule, therefore, it is good practice to write each term of a NLAE so that it is at least $O(1)$.

Finally, it is worth pointing out that **fzero**, like most MATLAB routines, has additional options to give the user more control of the algorithm parameters, thus allowing in some cases for faster computations and/or more accurate answers. The reader can refer to MATLAB documentation for further information if required.

The MATLAB function **fzero** is usually good enough for solving a single NLAE but to solve systems of NLAEs one needs to use **fsolve**. This uses a much more advanced Levenberg–Marquardt algorithm. This is a complex algorithm that uses, amongst other things, a variant of the Newton–Raphson method known as the Gauss–Newton method. Of course, **fsolve** can also be used to solve a single NLAE. Use of **fsolve** to solve a system of two equations is illustrated below.

Example A.8 Solution of a system of two NLAEs using fsolve

Solve the following two equations using **fsolve**:

$$\frac{1-x}{y-0.1} - \ln x - 3 = 0,$$

$$\frac{x-y}{y-0.1} - \ln y - 3 = 0.$$

Use initial guesses of 0.5 for both x and y.

Solution. MATLAB code, employing the function **fsolve** is shown below.

```
function exampleA8
[x]=fsolve(@twoeqns,[0.5 0.5])
function f=twoeqns(x)
%x(1) is x, x(2) is y
eqn1=(1-x(1))/(x(2)-0.1)-log(x(1))-3;
eqn2=(x(1)-x(2))/(x(2)-0.1)-log(x(2))-3;
f=[eqn1;eqn2];
```

Running the code gives $x = 0.5731$ and $y = 0.2747$.

A.4   Numerical solution of ordinary differential equations

An ordinary differential equation (ODE) is an equation of the form

$$\frac{dy}{dt} = f(y, t). \tag{A.20}$$

In the problems encountered in this book, we deal with a particular type of ODE referred to as an initial value problem (IVP). This means that to solve the equation, an initial condition must be specified, i.e.,

$$y = y_0 \ @ \ t = 0. \tag{A.21}$$

The above ODE essentially says that the dependent variable (y) is a function of one independent variable only, time in this case. It may also be a function of itself but ultimately it is time that is the 'cause' and y the 'effect'. Equations in which the dependent variable varies with more than one variable (for example time and position in a diffusion problem) are referred to as partial differential equations (PDEs). These are much more challenging mathematically and they do not arise in this book.

In many instances, a physical system requires more than one ODE to fully describe its behaviour. In that case a system of ODEs is required that can be solved with techniques that are a generalisation of the single equation methods.

As mentioned above, an equation is said to have an analytical solution if it can be solved purely in terms of symbols. Analytical solutions are obviously the preferred result as they give a completely general description of system behaviour, allowing one to make predictions for any process scenario. However, many mathematical models, especially those involving systems of equations, cannot be solved analytically. In that case, they must be solved numerically. The numerical 'solution' of a differential equation is simply a numerical prediction of the dependent variable as a function of the independent variable for the particular parameter values used to the carry out the calculations. Numerical solutions are generally presented in tabular or graphical form, e.g. a plot of y versus t, rather than an equation. The simplest example of a numerical method for solving ODEs is Euler's method, which is outlined below. The aim here is to give the reader with a less advanced background in mathematics a feel for what an ODE solver package is doing when it produces tables of numbers or graphs. It is important to realise, however, that the Euler method would rarely, if ever, be used in practice as much faster and more accurate techniques, many based on Runge-Kutta algorithms, have been developed and are routinely coded into commercial packages.

A.4.1   Euler's method for numerical solution of an ODE

Consider the differential equation

$$\frac{dy}{dt} = f(y), \tag{A.22}$$

where $f(y)$ denotes some function of y. Suppose the value y is known at some time, t, i.e., we know y_t. Then, the value of y at time $t + \Delta t$, i.e., $y_{t+\Delta t}$ is given by

$$y_{t+\Delta t} = y_t + \Delta y, \tag{A.23}$$

where Δy is the change in y over the time interval, Δt. But

$$\Delta y = \frac{\Delta y}{\Delta t} \Delta t. \tag{A.24}$$

Hence

$$y_{t+\Delta t} = y_t(t) + \frac{\Delta y}{\Delta t} \Delta t. \tag{A.25}$$

Now, if Δt is sufficiently small, then

$$\frac{\Delta y}{\Delta t} \approx \frac{dy}{dt}. \tag{A.26}$$

Therefore

$$y_{t+\Delta t} = y_t + \frac{dy}{dt} \Delta t, \tag{A.27}$$

i.e.,

$$y_{t+\Delta t} = y_t + f(y)h, \tag{A.28}$$

where $h = \Delta t$ is referred to as the *step size*. Therefore, if y is known at some initial time, and given that the differential equation is specified (i.e. $f(y)$ is known), values of y at subsequent times can be computed by a simple 'stepping' procedure. The following example illustrates the method.

Example A.9 Accuracy of Euler's method

Consider the ordinary differential equation

$$\frac{ds}{dt} = s^2 \text{ with } s = 1 \text{ @ } t = 0.$$

Solve this equation using the exact (analytical) solution and using ten steps of Euler's method with step sizes, h, of 0.01 and 0.005. Compute the percentage absolute error at each time step for both step sizes.

Solution. The analytical solution for this ODE is

$$s = \frac{1}{1-t}.$$

The Euler approximation to the solution is

$$s_{t+\Delta t} = s_t + s_t^2 h.$$

The relevant calculations are probably best done in a spreadsheet package. Results are shown in Table A.4 below. It is clear from the results that the smaller the step size, the more accurate the calculation. It should also be noted that although reducing the step

Table A.4 Accuracy of Euler method.

	$h = 0.005$				$h = 0.01$		
Time	Exact	Euler	% Error	Time	Exact y	Euler	% Error
0	1.0000	1.0000	0.0000	0.0000	1.0000	1.0000	0.0000
0.005	1.0050	1.0050	0.0025	0.0100	1.0101	1.0100	0.0100
0.01	1.0101	1.0101	0.0050	0.0200	1.0204	1.0202	0.0203
0.015	1.0152	1.0152	0.0076	0.0300	1.0309	1.0306	0.0309
0.02	1.0204	1.0203	0.0102	0.0400	1.0417	1.0412	0.0419
0.025	1.0256	1.0255	0.0129	0.0500	1.0526	1.0521	0.0531
0.03	1.0309	1.0308	0.0156	0.0600	1.0638	1.0631	0.0648
0.035	1.0363	1.0361	0.0183	0.0700	1.0753	1.0744	0.0768
0.04	1.0417	1.0414	0.0211	0.0800	1.0870	1.0860	0.0891
0.045	1.0471	1.0469	0.0239	0.0900	1.0989	1.0978	0.1019
0.05	1.0526	1.0524	**0.0268**	0.1000	1.1111	1.1098	**0.1151**

size increases the accuracy, it also means that more calculations have to be performed to reach a specified time. Thus, if a step size of 0.005 is used, twice as many calculations are needed to reach a given time than if a step size of 0.01 is used. In general, the choice of step size is a trade-off between accuracy and computer time. It should be pointed out, however, that none of the problems we examine are very challenging in a numerical sense and we generally do not have to worry about sacrificing accuracy to ensure manageable computing times.

The Euler method is known as a *first order* method, i.e., the errors in the calculations are proportional to h. Routine calculations are normally done with a fourth order method at least, where the error is proportional to h^4. Remember that h is typically a number much less than 1, so h^4 should be very small. The standard fourth order method for solving systems of ODEs is the 4th order Runge-Kutta method. Other techniques have been developed, including those that automatically adjust the step size, h, depending on how the solution is evolving, and so-called *stiff methods* which are used when there are very rapid changes in a variable.

A.4.2   Solution of systems of ODEs with the MATLAB function ode45

The following is an example of solution of a system of two ordinary differential equations with **ode45**. As in the case of non-linear algebraic equations, using MATLAB is simple and once a user has done one problem, he/she can usually just copy previous code and modify it for the next problem.

Example A.10 Solution of ODEs in MATLAB

Solve the following two equations and calculate the time for c_1 to reach 0.4. Use the initial conditions, $c_1 = 1$ and $c_2 = 1$ at $t = 0$, all parameters being dimensionless. Plot the full concentration-time profiles:

$$\frac{dc_1}{dt} = -0.01c_1^2 \ln(25/c_1),$$

$$\frac{dc_2}{dt} = -0.03c_2 \ln(25/c_1).$$

Solution. MATLAB code to solve this problem is shown below. Note the use of the **event** function to ensure that the calculations stop when $c_1 = 0.4$. This is not required if the goal is simply to simulate a system over a set period of time. Note also that the plot (Fig. A.1)has been edited post simulation using the editing tools within the figure window. This is generally more convenient than coding manually every detail of the formatting. The final dimensionless time works out to be 40.24.

```
function exampleA10
options=odeset('events', @eventfcn);
[t,c]=ode45(@eqns,[0 100],[1,1],options)
plot(t,c(:,1),'-k',t,c(:,2),'--k');
xlabel('Time (-)'); ylabel('Concentration (-)');
function f = eqns(t,c)
c1eqn=-0.01*c(1)^2*log(25/c(1));
c2eqn=-0.03*c(2)*log(25/c(1));
f=[c1eqn;c2eqn];
function [value, isterminal, direction]=eventfcn(t,c)
value = c(1)-0.4;
isterminal = 1;
direction = 0;
```

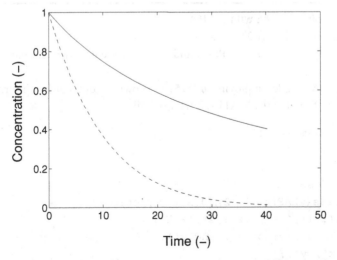

Figure A.1 Solution of a system of ODEs with **ode45**. Solid curve denotes c_1, dashed curve denotes c_2.

A.4.3   Differential algebraic equations

Quite often in this book, process models are encountered that combine an ordinary differential equation (ODE) with a non-linear algebraic equation (NLAE). The usual strategy when this occurs is to differentiate the NLAE to transform it into an ODE. The result is a system of ODEs rather than a mixed ODE–NLAE system. The differentiation is often a little messy when one adopts this approach and, in rare instances, notably in Chapter 9, the complexity of the differentiation is such that it is not worth the effort. In that case, it is simpler to write the model equations as a differential algebraic equation (DAE). This is a matrix equation and it is best illustrated by an example. Suppose we have the following ODE:

$$\frac{dx}{dt} = \frac{F}{H}(x_f - x) - \frac{V}{H}(y - x), \tag{A.29}$$

where F, H, V and x_f are constants. Furthermore, we have

$$y - \frac{\alpha x}{1 + (\alpha - 1)x} = 0, \tag{A.30}$$

where α is another constant. (This is actually a simple model of a flash distillation unit.) These two equations can be written in DAE form as

$$\begin{pmatrix} 1 & 0 \\ 0 & 0 \end{pmatrix} \begin{pmatrix} dx/dt \\ dy/dt \end{pmatrix} = \begin{pmatrix} \frac{F}{H}(x_f - x) - \frac{V}{H}(y - x) \\ y - \frac{\alpha x}{1 + (\alpha - 1)x} \end{pmatrix}, \tag{A.31}$$

Technically, this is a DAE of Index 1. The 2×2 matrix is referred to as the 'Mass' matrix and remarkably, the MATLAB solver **ode15s** can be used to compute x and y as functions of time. MATLAB syntax for doing this is demonstrated in the next example.

Example A.11 Solving a DAE with ode15s
Use **ode15s** to solve Eq. A.31 using initial values of $x = 0.2$ and $y = 0.3$, $H = 4$, $F = 12$, $x_f = 0.5$, $V = 1.5$, $\alpha = 3$. Plot x and y versus time up to $t = 2$. (See Fig. A.2.)

Solution. MATLAB code employing **ode15s** is shown below. Further information on DAEs can be be found in the MATLAB 'Help' facility.

```
function exampleA11
M=[1 0
   0 0];
options=odeset('Mass',M);
[t,z]=ode15s(@DAEfcn,[0 2],[0.2 0.3],options)
plot(t,z(:,1),'-k',t,z(:,2),'--k');
xlabel('Time');ylabel('x and y');
legend('x','y');
function f=DAEfcn(t,z)
%z(1) is x, z(2) is y
```

```
H=4;F=12;xf=0.5;V=1.5;alpha=3;
xeqn=F/H*(xf-z(1))-V/H*(z(2)-z(1))
yeqn=z(2)-alpha*z(1)/(1+(alpha-1)*z(1))
f=[xeqn;yeqn];
```

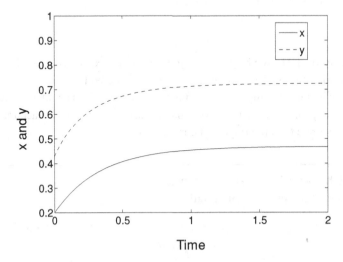

Figure A.2 Variation in x and y by solving equations expressed as a DAE.

As an exercise, you can simply solve this problem by substituting Eq. A.30 into Eq. A.29 and using **ode45** to solve the single ODE that results. You will, of course, get the same answer.

A.5   Numerical integration

Numerical integration, i.e., the computation of a definite integral, can often be a useful tool in the engineering and science toolkit and there are many techniques available for this task. The trapezoidal rule and Simpson's rule are probably the best known, and most often used, in basic engineering calculations. There are a number of processes in both ultrafiltration and microfiltration in which numerical solution of an integral is a valid and sometimes convenient way to solve the problem in question.

The simplest way to compute an integral in MATLAB is to use the **trapz** function. Let us suppose that we need to evaluate the integral

$$y = \int_1^5 x^2 dx. \tag{A.32}$$

The MATLAB commands and the result for this calculation (assuming 10 equal intervals) with **trapz** are as follows:

```
x=1:(5-1)/10:5;
f=x.^2;
y=trapz(x,f)
y=41.44
```

If 20 intervals are used, i.e. if one defines x as

$$x = 1 : (5 - 1)/20 : 5;$$

one gets an answer of 41.36 which is getting closer to the exact answer of 41.333...

The trapezoidal rule is classified as a second order method. For more accurate integrations, one is better off using the **quad** function. This employs an adaptive Simpson's rule method. Simpson's rule with a fixed interval size is fourth order accurate so it is clearly more accurate than the trapezoidal rule.

Example A.12 Numerical integration with quad

Evaluate the following integral with **quad**:

$$y = \int_1^5 x^2 dx.$$

Solution. Simple MATLAB code is shown below.

```
function exampleA12
y=quad(@fun,1,5)
function f=fun(x)
f=x.^2;
```

This yields an answer of 41.333..., i.e., it yields the exact answer. This is actually not surprising as the function being integrated is second order while Simpson's rule is based on approximating the actual function between any two points as being a second order curve. This is in contrast to the trapezoidal rule which approximates each line segment as a straight line.

A.6   Numerical differentiation

A mathematical technique that is often used in data analysis, especially in membrane filtration experiments, is numerical differentiation. Very often, raw data are obtained as a cumulative permeate volume, using a digital balance connected to a computer in most cases. Turning these raw data into permeate fluxes requires numerical differentiation. Suppose that readings of a process parameter, y, are taken at constant time intervals, Δt. Then, the estimate of the rate of change of y with respect to t at any given time, t, is given, in the simplest approximation, by

$$Q_t = \frac{y_{t+\Delta t} - y_t}{\Delta t}.$$ (A.33)

Technically, Q_t is the 'flowrate' at a time, t_m, where

$$t_m = \frac{(t + \Delta t) + t}{2} = t + \frac{\Delta t}{2}.$$ (A.34)

Use of this approach to numerical differentiation is illustrated in the next example where Eq. A.34 is computed with the **diff** function in MATLAB.

Example A.13 Numerical differentiation of data using the *diff* command
The following dimensionless data have been obtained in a membrane experiment:

t	1	2	3	4	5	6	7	8
y	1.0	0.85	0.73	0.63	0.55	0.49	0.45	0.43

Generate a plot of $-dy/dt$ versus t. (see Fig. A.3.)

Solution. MATLAB code and the resulting code are shown below.

```
% exampleA13
t=[1 2 3 4 5 6 7 8];
y=[1.0 0.85 0.73 0.63 0.55 0.49 0.45 0.43];
dydt=-diff(y)./diff(t)    % approximation to derivative
tm=t(2:end)+diff(t)./2    % midpoint of interval
plot(tm,dydt,'ok');
xlabel('t');ylabel('-dy/dt');
```

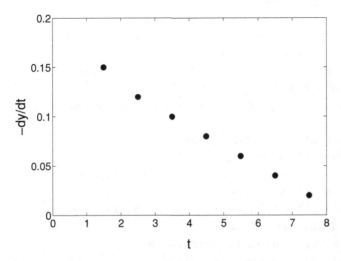

Figure A.3 Numerical differentiation of data.

A.7   Non-linear regression

There are a few places in this book where it is required to fit an equation to a set of data and this equation will not be a simple polynomial. Polynomials can be easily fit to data using MATLAB's **polyfit** function or using the plot editing facility within the figure window. We describe here a simple method for fitting an arbitrary function with an arbitrary number of adjustable parameters to a set of data. The basic approach here is to use MATLAB function **fminsearch** to minimise the sum of the squares of the differences between a predicted value of some parameter and the actual value. The approach is best illustrated by providing an example.

Example A.14 Non-linear regression with *fminsearch*
The following dimensionless data have been obtained during an experiment on enzyme kinetics:

x	2	4	6	8	10	12	14
y	0.6	1.0	1.25	1.37	1.50	1.66	1.73

Using a least squares method, fit the following function to the data:

$$y = \frac{ax}{b+x}.$$

Solution. MATLAB code for solving this problem is shown below.

```
function exampleA14
x=[2 4 6 8 10 12 14];
y=[0.6 1.0 1.25 1.37 1.50 1.66 1.73];
guess=[1 1]; % initial guesses of model parameters
params=fminsearch(@fitfun,guess,[],x,y)
xpred=linspace(0,14); % x values for computing
predicted alphas
ypred=params(1)*xpred./(xpred+params(2));%predicted y
values
plot(x,y,'ok',xpred,ypred,'-k');
xlabel('x');ylabel('y');
function sse=fitfun(params,x,y); % function to
compute sum of square errors
err=y-params(1)*x./(params(2)+x); % difference
between actual and predicted
sqerr=err.^2; %square of difference
sse=sum(sqerr); % sum of squares
```

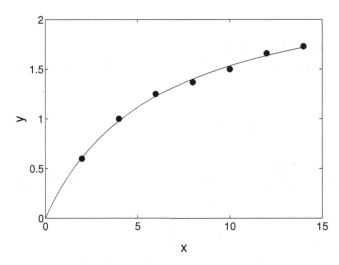

Figure A.4 Non-linear regression on the x–y data.

From the command window, we find that $a = 2.4623$ and $b = 6.0495$ and the graph in Fig. A.4 shows that the fit looks 'good'.

A.8 Conclusion

In this appendix, the key computational and mathematical tools required for the analyses presented in this book have been presented. It is clear that the number of techniques is small and one can achieve a lot by being proficient at solving systems of relatively straightforward ODEs and NLAEs. On the analytical side, the ability to be able to differentiate implicitly across an equation is a key technique that allows us to convert NLAEs into ODEs.

Further reading

Attaway, S. (2009). *MATLAB: A Practical Introduction*. Butterworth-Heinemann, Massachusetts, USA.

Constaninides, A. and Mostufi, N. (1999). *Numerical Methods for Chemical Engineers with MATLAB Applications*. Prentice-Hall, New Jersey, USA.

Cutlib, M.B. and Shacham, M. (2008). *Problem Solving in Chemical and Biochemical Engineering with POLYMATH, Excel and MATLAB*, 2nd Edition. Prentice Hall, New Jersey, USA.

Stroud, K.A. (2007). *Engineering Mathematics*, 6th Edition. Palgrave MacMillan, New York, US.

Practice problems

Problem A.1   Implicit differentiation

Two process parameters y and x are related by the following equation:

$$e^{-\alpha x} \ln \frac{y}{x} - (a - bx) = 0,$$

where α, a and b are constants. Find an expression for dy/dx.

Problem A.2   Optimisation

In a certain filtration process, the productivity (P), i.e., the amount of product produced per unit time, is given by the expression

$$P = \frac{V}{t_w + KV^2},$$

where V is the product volume and K and t_w are constants. Find an expression for the optimum value of V in terms of t_w and K and state whether this minimises or maximises the productivity.

Problem A.3   Differentiating a function of two variables

A certain process parameter, y, is related to two other parameters, x and z, by the following expression

$$y = e^{\alpha(x-z)},$$

where α is a constant. If x and z are functions of time, find an expression for dy/dt in terms of dx/dt and dz/dt.

Problem A.4   Taylor series approximation to a function

Consider the following equation:

$$K_1(1 - x) - \ln x - K_2 = 0,$$

where K_1 and K_2 are constants and $x < 1$ in all cases. In general, this requires a numerical solution. However, assuming x is close to 1, find an analytical expression for x in terms of K_1 and K_2. Check the accuracy of your answer for $K_1 = 3$ and $K_2 = 3$.

Problem A.5   The Newton–Raphson algorithm

The following equation describes the performance of a continuous chemical reactor with non-integer kinetics:

$$1 - x - Dx^n = 0,$$

where x is a certain ratio of concentrations such that $x < 1$ in all cases while D and n are constants. Solve this equation by implementing the Newton–Raphson algorithm with your own MATLAB code (i.e., not using **fzero** or **fsolve**). Use $n = 0.5$ and $D = 1.5$.

Check your answer by comparing it with the analytical solution. Note that to solve this equation analytically, you need to convert it into a quadratic equation.

Problem A.6   Solution of NLAEs with fzero and fsolve

Use **fzero** to solve the following equations *consecutively* for x_1 and x_2:

$$1 - x_1 - 0.9x_1^2 = 0,$$

$$x_1 - x_2 - 0.9x_2^2 = 0.$$

The variables x_1 and x_2 are positive and < 1 in all cases. Do the problem again by solving the equations simultaneously with **fsolve**.

Problem A.7   Solution of a system of non-linear algebraic equations with fsolve

Solve the following system of non-linear algebraic equations using **fsolve**:

$$1 - x_1 - \frac{x_1}{1 + 2x_1x_2} = 0,$$

$$1 - x_2 - \frac{1.5x_2}{1 + 2x_1x_2} = 0.$$

Both x_1 and x_2 are process parameters that must be in the range 0–1.

Problem A.8   Solution of a system of ODEs using ode45

The following equations describe the performance of a certain controlled heating system:

$$\frac{dT}{dt} = 2 \times 10^{-5}Q - 0.002(T - 10),$$

$$\frac{dQ}{dt} = -500\frac{dT}{dt} + 33(20 - T),$$

where $Q = 1000$ at $t = 0$ and $T = 15$ at $t = 0$ and you need not worry about the units of these parameters. Using **ode45**, generate a plot of T versus t in the range $0 < t \le 1000$ time units.

Problem A.9   Numerical integration with quad

Integrate the following using the **quad** function within MATLAB:

$$y = \int\limits_{10}^{75} \frac{dT}{120 - T}.$$

Compare your answer with the analytical solution.

Problem A.10   Numerical differentiation

The data below represent the decline in reactant concentration during an experiment performed in a batch reactor. Use the MATLAB function **diff** to plot dc/dt versus time.

t	1	2	3	4	5	6	7	8	9	10
c	0.61	0.37	0.22	0.14	0.08	0.05	0.03	0.02	0.01	0.007

Index

Printed in the United States
By Bookmasters